Mountain environments:
an examination of the physical geography of
mountains

Mountain environments:

An examination of the physical geography of mountains

John Gerrard

The MIT Press
Cambridge, Massachusetts

First MIT Press edition 1990

© A.J. Gerrard 1990

First published in Great Britain by
Belhaven Press, London

Library of Congress Cataloging-in-Publication Data

Gerrard, John, 1944–
  Mountain environments : an examination of the physical geography of
mountains / by John Gerrard.
     p.   cm.
  ISBN 0–262–07128–2. — ISBN 1–85293–080–2 (Belhaven Press)
  1. Mountains   I. Title.
GB501.2.G47 1990   89–13516
551.4′32–dc20   CIP

Printed and bound in Great Britain

CONTENTS

ACKNOWLEDGEMENTS

We are grateful to the following individuals and organizations who have given permission for the reproduction of copyright material.

Almqvist Periodical Company (Fig. 5.5); Table 4.3 reproduced from *Fluvial Forms and Processes* (D. Knighton) by kind permission of Edward Arnold Ltd; Balkema (Fig. 5.10); D. Barsch (Fig. 5.1); R.G. Barry (Table 4.1, Fig. 4.4); J.C. Bathurst (Fig. 4.8); J.B. Bird (Fig. 1.2); Biuletyn Peryglacialny (Fig. 3.2); Blackie & Son Ltd. (Fig. 6.5); S.F. Burns (Tables 2.3, 2.4, Fig. 2.4); N. Caine (Figs. 5.1, 5.10); Chapman and Hall Ltd. (Fig. 7.7); C.M. Clapperton (Figs. 6.8, 6.9); D.R. Coates (Fig. 7.7); D.M. Cruden (Table 3.2); E. Derbyshire (Fig. 6.5); Elsevier Science Publishers, Physical Sciences and Engineering Division (Tables 1.6, 1.7, 1.8, Fig. 1.5, Figs. 7.4, 7.5, 7.6); R.W. Fairbridge (Table 1.2); C.R. Fenn (Fig. 4.6); R.I. Ferguson (Figs. 4.2, 4.3, 4.4); P.W. Francis (Table 7.1); Franz Sterner Verlag (Figs. 2.7, 2.8, 2.9, Tables 2.7, 2.8, Fig. 8.1); H.M. French (Fig. 5.2); H.F. Garner (Fig. 1.3); Gebruder Borntraeger (Fig. 5.3, Table 5.1, Fig. 5.13); A.M. Gurnell (Fig. 4.9, Table 4.4); International Association of Hydrological Sciences (Fig. 4.10, Tables 4.5, 4.6); Institute of British Geographers (Fig. 4.8); J.D. Ives (Table 4.1, Fig. 4.4, Fig. 8.1); J.D. Ives as editor of Mountain Research and Development (Table 2.1, Figs. 2.3, 5.1, 8.2, 8.5, 8.6); L.C. King (Table 1.1); D. Knighton (Table 4.3); A.K. Lehre (Fig. 4.10, Tables 4.5, 4.6); Longman Group Ltd. (Table 4.2, Fig. 4.7, Fig. 5.2); M.A. Love (Fig. 6.5); B. Messerli (Figs. 2.5, 8.1, 8.2); B.P. Moon (Figs. 5.4, 5.5); K.F. O'Connor (Table 2.1); Oxford University Press, New York (Fig. 1.3); M. Morisawa (Table 4.2, Fig. 4.7); T.L. Pewe (Table 3.1, Fig. 3.1); Reinhold (Table 1.2); The Royal Geographical Society (Figs. 3.2, 4.3, 4.4); M.J. Selby (Table 5.1, Figs. 5.5, 5.6, 5.7); L. Siebert (Figs. 7.4, 7.5, 7.6); O. Slaymaker (Table 4.1, Fig. 4.4); Springer Verlag (Table 7.1);

The Geological Society (Table 8.2); J.J. Tonkin (Table 2.3, Table 2.4, Fig. 2.4); Tables 2.2, 2.5 and 2.6 reproduced by permission of the Regents of the University of Colorado from Arctic and Alpine Research vol. 5 (1973); Figure 2.5 reproduced by permission of the Regents of the University of Colorado from Arctic and Alpine Research vol. 5, (1973); Table 3.1 and fig. 3.1 reproduced by permission of the Regents of the University of Colorado from Arctic and Alpine Research vol. 15, (1983); University of Poone, India (Table 9.1); University of Tokyo Press (Figs. 7.2, 7.2); Tables 2.3, 2.4 and Fig. 2.4 taken from *Space and Time in Geomorphology*, edited by Colin E. Thorn, reproduced by kind permission of Unwin Hyman Ltd; G.L. Wells (Table 7.1); Fig. 5.4 reproduced from B.P. Moon in Earth Surface Processes and Landforms 9, by permission of J. Wiley and Sons; Table 1.5, Fig. 4.6, Fig. 4.9, Table 4.4, Figs. 5.6, 5.7, Tables 5.3, 6.1, 6.3, Figs. 6.8, 6.9 reproduced by kind permission of John Wiley and Sons Limited; J. Wiley and Sons, Canada (Fig. 1.2).

INTRODUCTION

There has recently been a great upsurge in interest, both scientific and general, concerning mountain environments. A summary of this upsurge and reasons for it up to 1985 have been admirably provided by Ives (1985). Ives has argued that much of this interest was stimulated by the Unesco Man and the Biosphere Programme (Project 6: Study of the Impact of Human Activities on Mountain Environments) and the conference on the Use of Mountain Environments, sponsored by the German Agency for International Development (GTZ), held in Munich in 1974. Other notable developments have been the setting up of the International Mountain Society in 1980 and the appearance of its influential journal, *Mountain Research and Development*, in 1981. Price's (1981) book, *Mountains and Man*, and a number of monographs concerned with specific mountain areas — for example, Caine's (1983) study of the mountains of northeastern Tasmania, Hurni's (1982b) treatise on the Simen Mountains, Ethiopia, and the analysis of the Tista Basin by Mukhopadhyay (1982) — have also stimulated interest.

A number of other important developments include the appearance of the journal *Mountain Research*, which first appeared in the People's Republic of China in 1983, published jointly by the Chinese Academy of Science and Chengdu Institute of Geography; and the journal *Himalayan Research and Development*, published by the Central Himalayan Environment Association (Naini Tal, Kumaun Himalaya, Uttar Pradesh). In addition, there have been a large number of edited volumes concerned with mountain environments (e.g. Ives, 1980; Lall and Moddie, 1981; Pierdie and Noble, 1983; Sarkar and Lama, 1983; Brugger *et al.*, 1984; Lauer, 1984b; Messerli and Ives, 1984b; Singh and Kaur, 1985; Joshi, 1986). Specific mention must also be made of the report of the International Karakoram Project (Miller, 1984).

The upsurge of interest reflects the belief that it is important to understand the workings of environmental systems in mountains. As Messerli (1983) points out, approximately 10 per cent of the world's population live in mountain regions and more than 40 per cent are dependent in some way on mountain resources. Mountain ecosystems appear to be more sensitive to natural perturbations, such as climatic change, seismicity, meteorological events, and human activities such as deforestation and land-use changes, than other areas. Mountains possess highly sensitive landscapes, partly the result of steep slopes, high altitudes and high relative relief, and partly the result of current and past processes such as relics from former glacial phases. This sensitivity is easily disrupted and massive landscape changes may be initiated. Thus it is important to understand the workings of environmental systems so that the present situation can be understood and the future predicted. The studies mentioned earlier have provided much information to enable this to be achieved and it is hoped that this book will also contribute.

Messerli (1983) has also stressed that changes in mountain environments frequently occur as the result of slow-acting processes that are often not identifiable within a single generation. So it must not always be assumed that the high-energy and often catastrophic process must necessarily be the most significant. A balanced approach to the workings of mountain environmental systems must be achieved. Ives (1985) identified a number of major research themes — namely, mountain hazards mapping, the quest for integrated mountain research, the problems of research application, and attempts to bridge the gap between human and physical sciences. These themes can only be tackled if adequate knowledge of mountains exists. Muir-Wood (1984), in arguing for a science of mountains, has stressed that if such a science is to be of any significance, it should be able to illuminate the interconnections and provide new conceptual ways of seeing which are the most fundamental problems. Thus:

The beauty of mountains and their intellectual challenge are the consequence of the paradox of order and a high degree of dynamic balance among the diverse elements of mountain landscapes . . . To understand fully and describe the mountains of the world will be a task for many generations yet if we understand what they are and how they are organised many projects can be completed in our own time which will find immediate and very significant use (Thompson, 1964, p.8).

It is hoped that this book provides some of the understanding of the physical geography of mountains.

Chapter 1

THE NATURE AND DISTINCTIVENESS OF MOUNTAINS

DEFINITION

Mountains are extremely important components of the Earth's surface. Fairbridge (1968) has estimated that about 36 per cent of the land area of the world is composed of mountains, highlands and hill country. Numerous definitions of what constitutes a mountain have been proposed but mountains are extremely diverse landforms and it has proved difficult to achieve consistency in description and analysis.

Several criteria have been used, such as elevation, volume, relief, and steepness, as well as spacing and continuity. The *Oxford English Dictionary* defines a mountain as 'a natural elevation of the earth surface rising more or less abruptly from the surrounding level and attaining an altitude which, relatively to adjacent elevation, is impressive or notable'. In *Webster's Dictionary* a mountain is 'any part of a land mass which projects conspicuously above its surroundings'. Both definitions are extremely vague. Even more subjectively, Peattie (1936) argues that mountains should be impressive, possess individuality and should enter into the imagination of the people who live near them.

Some of these inconsistencies have been summarised by Strahler (1946). Descriptions in the writings of early geologists, such as Davis (1909) and Lobeck (1926; 1939), deprived mountains of their usual meaning as topographic features and redefined them as any large mass of deformed rock. Tarr and Martin (1914) followed Davis in restricting mountains to parts of the Earth's crust disturbed by diastrophic movement, but there was increasing recognition that the term should not be restricted to particular geological criteria. Powell (1895) used the term to signify a type of topography and Fenneman (1928) and Cotton (1922) made the distinction between mountainous terrain and the rock structures beneath. Salisbury (1907) also recognised that erosion could produce mountains from horizontal strata such as in the Catskill Mountains of New York state. The major criticism of using geological criteria is that undeserved grandeur can be imparted to structures of small size or limited deformation (Strahler 1946), a good example being Lobeck's (1939) use of the term 'dome mountain' for the Weald uplift in South-East England.

Most early geographers used the term to emphasise rugged terrain. Preston James (1935) argued that a mountain should possess sufficient relief to produce a marked vertical zonation of vegetation and Finch and Trewartha (1936) saw mountains as being characterised by steep slopes, small summit areas and strong relief. However, for consistency of approach, a more rigorous definition is required.

Altitude alone is not sufficient to define mountains. The high plateaux of Bolivia and Peru and the high interior Tibetan Plateau are not mountains or mountainous areas. It is generally agreed that any definition should include relative relief and perhaps also slope steepness and land volume. Thus an isolated inselberg might not be a mountain but a hill. Irrespective of how to estimate relative relief (e.g., Waldbaur, 1952; Rutkis, 1971), the question arises as to what value of relative relief to use. As Price (1981) points out, a relative relief of 900 m would include the European Alps, Pyrenees, Caucasus, Himalayas, Andes, Rockies, Cascades and Sierra Nevada but not the Appalachians or the more degraded mountains of Scandinavia and the British Isles.

Price (1981) has also argued that mountains should be dissected areas which would eliminate plateaux and plains with an occasional area of high relief. Thus the horizontal distances between ridges and valleys are just as fundamental to the delineation of mountains as are the vertical distributions that establish the relief. On the basis of these deliberations Price (1981, p. 5) defined a mountain as 'an elevated landform of high local relief e.g. 300 m (1000') with much of its surface in steep slopes, usually displaying distinct variations in climate and associated biological phenomena from its base to its summit'. Most scientists working in mountain areas are of the opinion that relative relief should be greater than 300 m and a consensus seems to be emerging that a figure of 700 m is more realistic (see, for example, Derrau, 1968; Temple, 1972; Brunsden and Allison, 1986).

A distinction is sometimes made between high mountains (the German *Hochgebirge*) with altitude over 1500 m and relative relief over 1000 m, and lower and intermediate mountains (the German *Mittelgebirge*). The only areas of western Europe with relative reliefs greater than 1500 m are the Alps and the central Pyrenees. Other areas such as the Scandinavian and Carpathian Mountains would be excluded. Hammond (1954; 1964) used a local relief of 3000 ft to separate high mountains from low mountains in North and South America. Low mountains were defined as having a local relief of 1000–3000 ft and included the Appalachian Mountains, the Guyana Highlands and the coastal mountains of Brazil.

Many workers have suggested a geoecological approach for the determination of the lower limit of high mountain environments such that high mountains should rise above the Pleistocene snow line, should extend above the regional timber line, and should display cryonival processes (Troll, 1973b; Hollermann, 1973a). This has often been termed the 'Alpine' zone, although its use has been disputed (for example, by Love, 1970; Ives and Barry, 1974; Barsch and Caine, 1984; Ives, 1985; 1987). The lower limit of the Alpine zone will vary with latitude. Thus the lower limit for cryonival processes is 4000 m in the southern Rocky Mountains and European Alps and 700 m in Alaska, Iceland, northern Scandinavia and Labrador. This implies that some high mountains in the tropics will not possess a high mountain landscape. Also, although the upper timberline and lower limit of solifluction appear to be very similar, the Pleistocene snow line was often much lower. This seems to have been true of most middle-latitude mountain areas, but in areas nearer the Equator the Pleistocene snow line was well above the timber line as in the mountains of North Africa and Mexico.

There are also difficulties in applying these criteria to arid and semi-arid mountains because the ecology of the upper timber line in arid mountains is different to timberline ecology elsewhere. In fact the highest elevation at which trees grow anywhere is at about 30°N or 30°S in the arid zones of the Andes and Himalayas, not, as might be expected, in the humid tropics (Troll, 1973a). The timberline in arid areas appears to be determined largely by slope exposure (Hollermann, 1973a). It is also difficult to locate the lower limit of solifluction in areas where there is a water deficiency and the Pleistocene snow line was usually far above the upper limit of tree growth. The identification of a snow line is also difficult in areas where glaciation occurred on only one side of the mountains. However, Hollermann (1973a) has stressed that there are many similarities between alpine and arid mountains even when the arid mountains do not fulfil the criteria advocated by Troll (1973b).

Debris mantles produced by mechanical weathering are extensive in both alpine and arid mountains and soils in both are thin and poorly developed. Also solifluction in alpine mountains and slope wash in arid mountains may produce similar slope profiles. The role of rock structure and lithology is equally conspicuous in both environments because of discontinuous soil and vegetation cover. It can be concluded that high mountains are mountains which reach altitudes that create land forms, plant cover and soil processes which in the classical region of the European Alps are perceived as high-alpine (Troll, 1972a). The physical geography of high mountains is different to that of any other major landform types. The major mountain systems of the world are shown in Figure 1.1.

Mountains can be distinguished not only by altitude considerations but also on the basis of areal extent. A mountain can be a single isolated feature or a feature outstanding within a belt of mountains. A mountain

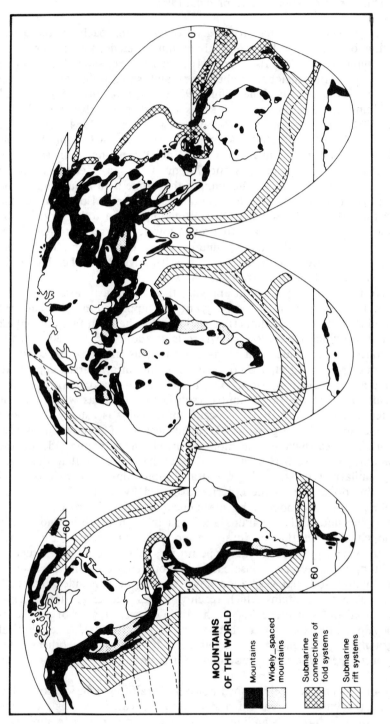

Figure 1.1. Distribution and types of mountains.

range is a single ridge but a mountain chain, although also a linear feature, implies major features that persist for hundreds or even thousands of miles. A mountain mass is usually a group of mountains of irregular shape not characterised by simple linear trends. They may be block-faulted terrain, volcanic areas, or uplifted areas of complex igneous rocks such as batholiths. The term 'mountain system' is usually reserved for the greatest continent-spanning features. Mountain systems usually have complex histories and comprise sub-divisions of ages and types such as combinations of chains, ranges and masses.

ORIGIN

Mountains are created essentially by tectonic forces but their individual shapes are determined by weathering, mass movement, glacial and fluvial action — the combined action of denudational agencies. The major mountain types are volcanic and combinations of folded or faulted rock mass. Volcanic action can produce a specific type of mountain. The origin of mountains has been a major concern of geologists and geophysicists for many years. Mountain belts are composed mainly of marine sediments, though often altered by metamorphism and affected by volcanic activity. The nature of the sediments also indicates that generally they were deposited in quite shallow seas. These depositional areas, adjacent to continents, were called 'geosynclines' and it was assumed that as sedimentation increased, the underlying rock was gradually depressed to accommodate more material being eroded off the land. Such sediments can achieve thicknesses of 12 000 m.

Studies of ancient mountain systems indicate that geosynclines are usually composed of two parts, an inner part of gently folded rocks termed the 'miogeosyncline' and an outer more intensely deformed part, termed the 'eugeosyncline'. Sediments currently found off the eastern coast of North America on the continental shelf and continental rise may represent modern equivalents of the miogeosyncline and eugeosyncline, respectively.

Early theories suggested that these sediments were squeezed and uplifted to form mountains but, until recently, the mechanisms involved were largely unknown. Recent discoveries also indicate that the simple geosynclinal model is inappropriate and a simple folding mechanism is inadequate to form mountains. Mountain-building is now seen to fit into the general, all embracing, theory of sea-floor spreading and plate tectonics. Plate tectonics is the concept that the Earth is composed of a number of rigid plates moving in various directions. Where plates collide, rock material is squeezed and deformed into mountains. One plate may descend beneath the other with the rock being absorbed back inside the Earth in what is known as a subduction zone. Where plates are moving apart, new volcanic material is extruded to fill the gap.

The Earth is composed of six major and several minor plates. The major plates are Eurasia and the adjacent ocean floors; Africa with the eastern Atlantic floor; the Pacific floor; the Americas and the western Atlantic floor; India with Australia and New Zealand; and Antarctica and the surrounding ocean floor. Minor plates have been recognised in the Mediterranean and Arabian areas; the Philippines; in the Caribbean; and off the Pacific coast of the Americas. Faults, young fold mountains, trenches and ridges are all related to the nature of the plate boundary. Two types are related to the situation where plates are converging. Collision boundaries are created when two plates converge but neither plate is consumed. Such a boundary occurs along the line of the Himalayas where the Indian and Eurasian plates meet. A destructive boundary occurs where, on collision, one plate sinks below another and is consumed or subducted. Constructive or diverging boundaries are created when two plates are moving apart and new material is formed. This is typical of mid-ocean ridges. A conservative boundary occurs where two plates are slipping sideways past each other. Relative plate movement is thus parallel to the boundary. The San Andreas Fault, in California, is thought to be such a boundary, with the Pacific side moving north relative to the south-moving Americas plate.

Mountain-building or *orogenesis* is far from the simple process it was once thought to be. It is now known to include 'compressional folding, severe overthrusting, intrusion of magma on a massive scale, large-volume sliding of masses of sediment under the influence of gravity, broad uparching and very large-scale faulting and uplift of crustal blocks' (Selby, 1985, p. 69). There are two types of orogeny: a cordilleran type, involving subduction in which geosynclinal sediments are deformed and intruded by magma; and continental collision orogeny, resulting from the trapping of oceanic crust and sediments between two masses of continental crust.

The stages in a cordilleran orogeny have been outlined by Dewey and Bird (1970). Initially the situation is stable, with a passive continental margin and the development of miogeosynclines and eugeosynclines. This gradually becomes active as the plates move towards each other and a subduction zone is created by the ocean crust being dragged beneath the continent. In general, the oceanic crust breaks seaward of the contact between the oceanic and continental lithospheres and the eugeosynclinal sediment remains attached to the continental plate. This attached slice of oceanic crust is then uparched and broken into a number of overthrust slices. Changes also take place at depth. As the oceanic plate is subducted, melting of its upper surface occurs, creating basaltic magmas that rise behind the uparched oceanic crust as submarine volcanoes or island arcs. The volcanic arc continues to grow and widen, and the nature of the rising magma and the surface volcanoes changes. As the basaltic material of the oceanic crust is further subducted, heat and pressure build up and the material is changed to eclogite, rich in pyroxenes and garnet. New magma, of andesitic composition, is then created by partial melting of the eclogite, fractional crystallisation of basaltic magma and contamination of the magma with trench sediments and water.

These andesitic magmas create new volcanic action on the continental side of the orogen. As subduction continues, the continental plate edge is transformed and metamorphosed developing a mobile core of mafic (gabbro) and intermediate (diorite) composition. This core expands, causing thermal uparching and the arc gradually rises above sea level. Mild heating and compression of the sediment above the core initiates metamorphism and the production of schists and gneisses. During the final stages intense thrusting and overthrusting of the miogeosynclinal sediments occurs with gravity sliding and folding of younger sediments above the thrust blocks. Granite batholiths or plutons may also form and reach the surface in andesitic and rhyolitic volcanoes. This sequence of events is applicable, in part or wholly, to a number of mountain chains but especially to the form and development of the Andes.

The classic example of a continental collision orogeny was the result of the collision between the Indian–Australian plate and the Eurasian plate. This collision created the Himalayas, the Zagros Mountains of Iran, the European Alps and the Atlas Mountains of North Africa. The mechanisms involved are somewhat simpler than those involved in cordilleran orogenies. As the ocean basin between the converging continental blocks narrows the sediments are compressed and deformed and the basaltic oceanic crust is broken into thrust-faulted blocks and pushed up into the sediments forming imbricate structures. As compression continues the upper parts of each thrust sheet are bent over to form elongated flat-lying folds or 'nappes'. Erosion of the rising thrust sheets produces fine-grained sediments which may accumulate in narrow ocean troughs to produce a material known as 'flysch'. This may become incorporated into younger nappes. In the later stages the high mountains create coarser sands and gravels which are spread across the continental forelands or deposited in lakes and seas as molasse.

Mountains can also be formed by faulting and volcanic activity. Faulted mountains are created by the elevation of one block relative to another. The higher blocks are often called 'horsts' and a central downfaulted trough is known as a 'graben'. The Rhine Valley of Germany is a graben and the Vosges and Black Forest ranges are horsts. The Basin and Range Province and the Sierra Nevada of North America are good examples of block faulting. The Basin and Range Province started to develop in late Oligocene times by the formation of major north–south faults. Individual fault-bounded blocks, tens of kilometres long and many kilometres wide were raised or dropped as much as 3 km. The Sierra Nevada Range consists of many individual portions which have been tilted to create a block 650 km long by 80 km wide. The blocks dip gently west, with east-facing slips rising abruptly along fault-scarps producing the highest mountain front in the continental United States.

Volcanoes may often form impressive mountains, especially where they occur in groups, such as in the Cascades of North America, but they do not form major ranges such as are created by orogenies. Volcanic mountains are examined in greater detail in Chapter 7 and so only a brief summary is provided here. The two most important types of volcanic mountain are composite cones and shield volcanoes. Composite cones develop along convergent plate boundaries where subduction is occurring, whereas shield volcanoes are usually associated with boundaries where plates are moving apart. Consequently, shield volcanoes are formed of basaltic lavas, such as Mauna Loa on Hawaii which rises to an altitude of 4169 m on a base with a diameter in excess of 100 km. Composite volcanoes are composed of alternate layers of lava and aerially ejected pyroclastic material. This combination of material provides greater strength and explains the steepness of the angles of most composite volcanoes. Vesuvius, Kilimanjaro and Fujiyama are classic examples.

CLASSIFICATION

As has been seen, mountains exhibit great variability because of differences in tectonic history, primary geological structure and lithology. Mountain landscapes also differ because of climate and vegetation history and the nature of their erosional development. It is not surprising that tectonic history and geological structure have been used as the basic criteria for many classifications of mountains. Such schemes, including those of J. Geikie (1898; 1914), A. Geikie (1903) and Supan (1930), have been summarised by Fairbridge (1968).

One of the most comprehensive classifications of orogenic mountain types has been produced by King (1967), largely based on the work of Argand (1922) and Huang (1946) (see Table 1.1). This classification only considers orogenic mountain types and does not examine the manner in which denudational agencies may modify mountain form. A more comprehensive classification has been produced by Fairbridge (1968) (Table 1.2). Volcanic mountains usually display a radial symmetry. They can be isolated or sometimes occur in clusters. Fold and nappe mountains are generally linear, often with bilateral symmetry. Belts of folds and nappes may be arched into anticlinoria or synclinoria. Fold mountains are formed by lateral compression and uplift. They are underlain by a thin covering of sedimentary rocks resting on a basement of granite or gneiss. Simple fold mountains are formed when rocks of the sedimentary cover have been folded by sliding laterally over the basement rocks. They exhibit straight, parallel valleys and ridges, of which the Jura of France and Switzerland are classic examples. Fold-fault mountains are more complex and form where compressive forces act upon both the basement

Table 1.1. *Classification of orogenic mountains (after King, 1967).*

TYPE A. Folds without ascertainable basement (Alpinotype)
1. Geosynclinal folds (nappe mts and folded mts — Himalaya)
2. Offshore folds (Indonesia)

TYPE B. Folds on ascertainable basement (Germanotype)
1. In ancient geosynclinal areas
    (a) Fold-block mountains (Tien Shan; New Zealand)
    (b) Fault-block mountains (Kirghiz)
    (c) Parageosynclinal (Hunan)
2. In continental masses
    (a) With thin sedimentary cover
        (i) Thin mantle-folds, often overthrust (Tatsing Shan)
        (ii) Fault-block mountains (Shantung)
        (iii) Synorogenic (epeirogenic movements, synchronous with orogenesis)
    (b) With thick sedimentary cover
        (i) Faulted and folded mountains (Saharan Atlas)
        (ii) Folded mountains (Jura)

Table 1.2. *Classification of mountain types (after Fairbridge, 1968)*

TYPE I Structural, tectonic or constructional forms
(a) Volcanic mountains
(b) Fold or nappe mountains
(c) Block mountains
(d) Dome mountains
(e) Erosional uplift or outlier mountains
(f) Structural outlier or klippe mountains
(g) Polycyclic tectonic mountains (Alpinotype)
(h) Epigene mountains

TYPE II. Denudational, subsequent, destructional or sequential forms
(a) Differential erosion or relict mountains
(b) Exhumed mountains
(c) Igneous (plutonic) and metamorphic complexes
(d) Polycyclic denudational forms

and cover rocks. The rigid basement breaks into fault wedges and the relatively mobile cover is folded into narrow systems of parallel folds such as in the Cordillera Oriental of Colombia.

Block mountains, as noted earlier, are raised along linear fracture zones and are often asymmetric in profile (e.g., Sierra Nevada). They possess rectilinear borders and rise abruptly from adjoining lowlands. A variation of block mountains are mountains bordering large rifts. Such rifts may be up to 350 km wide and several thousands of kilometres long. Mountain ranges often border such rifts and possess steep dissected fault scarps such as the Middle Rhine, Red Sea and East African rifts. Mts Ruwenzori, Uluguru and Livingstone are examples in East Africa. Mt Ruwenzori, on the Uganda–Zaire border, rises 5110 m above sea level. It is a deeply dissected block mountain elevated between major faults at the junction of two branches of the western rift system.

Dome mountains can be simple intrusions such as stripped laccoliths (for example, the Henry Mountains, Utah), plutons, or exhumed domal uplifts (for example, the Black Hills, South Dakota). Erosional uplift mountains or outliers are usually simple horizontal structures uplifted and regionally dissected. They are often the erosional margins of little dissected plateaus. Structural outlier or klippe mountains are rather rare tectonic phenomena where a mountain of older rocks has been thrust or has slipped in front of a tectonic belt. Parts of the Helvetic Limestone Alps of central Switzerland are of this type.

Polycyclic tectonic mountains (Alpinotype) are developed from the core or axial regions of the major orogenic belts and have suffered intense metamorphism. They are complex, asymmetric belts of folded and faulted rocks which often possess near-vertical dips producing the sharp-pointed 'aiguilles' and 'dents' characteristic of Alpine landscapes. Such mountains rise abruptly from an adjacent foreland and can be sub-divided into an external zone adjoining the foreland and an internal zone away from the foreland. The external zone is usually composed of metamorphic rocks such as schists and gneisses that have been deformed with complex overthrusts and great overturned folds. Generally, the external zone of alpine belts is composed of miogeosynclinal rocks and the internal zone is composed of eugeosynclinal rocks. The physiography of such belts is conditioned by recent tectonic activity and structure. The main valleys generally follow late orogenic warps. Complex nappe structures give rise to complex drainage patterns.

'Epigine mountains' was the name given by James Geike to accumulations of glacial or aeolian sediments. It is unlikely that they ever achieve the dimensions to be called true mountains. Differential erosion or relict mountains owe their development and isolation to uplift and differential weathering. Structures of ancient origin are exposed in exhumed mountains such as the Harz of West Germany, the Central Massif of France and the Precambrian mountain ranges of the Colorado and Wyoming Rockies. In the southern Appalachians, the Ridge and Valley landscape follows an old fold structure and in the St Francois Mountains of the Missouri Ozarks, Precambrian hills have been stripped of their sedimentary cover.

Igneous (plutonic) or metamorphic complex mountains are exposed in a relatively pristine state in the younger orogenic belts such as the Idaho Batholith or the Coast Ranges of British Columbia. Polycyclic denudational mountains are mountains that have been revived along old or new structural lines. Any mountain type that is of Mesozoic age or older is liable to be called polycyclic.

The classification scheme outlined above separates primary from secondary mountains. In structural mountains, the tectonic processes have played the primary role. In the denudational categories, tectonic activity still plays the primary role but the various classes of denudational mountains are so characterised by secondary denudational events that primary tectonic history can only be inferred. Also a large number of minor third-order events will change the character of a mountain. These events may reflect climatic history, lithology or denudational activity. The larger the mountain system the more numerous the structure or denudational types are likely to be.

MOUNTAIN SYSTEMS

Many of the points summarised earlier can be illustrated by examining the nature and evolution of the major mountain systems. This review relies heavily on the excellent summaries provided by Fairbridge (1968) and Selby (1985). There have been three major mountain-building episodes, two in the Palaeozoic Period (Caledonian and Hercynian/Appalachian) and one overlapping the Cainozoic–Mesozoic Periods (Alpine). The Caledonian orogeny occurred in the early Palaeozoic Period and has been most clearly recognised in Scotland and Scandinavia, although similar rock units have been found in the northern Appalachians and east Greenland. The rock structures indicate a continent-to-continent collision. The orogeny spanned a considerable period of time starting in the late Cambrian Period, reaching a peak in Ordovician–Silurian times and extending into the Devonian Period. The Hercynian (Variscan) orogeny of late Palaeozoic times, involving several colliding continental plates, was even more extensive. North Africa collided with Europe, producing a series of NW–SE trending systems which are now isolated blocks such as South-West England, Brittany, the Central Massif, the Vosges and the Black Forest. The Appalachian mountain system is thought to be the result of continent-to-continent collision involving large-scale low-angle thrusting. The last major phase of mountain building occurred in Mesozoic and Cainozoic times and created two major orogenic belts; the circum-Pacific belt and the Alpo–Himalayan–Indonesian belt.

All the Palaeozoic mountain systems have been destroyed by denudation but the remnants of such ancient mountains are sometimes re-elevated by renewed tectonic activity into new highlands. The summits of such highlands often show summit planes either broadly warped or sharply dislocated such as the Rocky and Bighorn Mountains of Wyoming. Most of the mountain ranges of late Mesozoic and early Cainozoic time appear to have been reduced to plains or undulating areas seldom exceeding 300 m. Uplift occurred again in Pliocene–Pleistocene times resulting in youthful mountains, dissected by deep narrow valleys. There is little structural control on the landforms of the young mountains but as erosion progresses the more resistant elements can be expected to dominate the landscape. Eventually in older mountain systems these differences become less obvious. The effects of the various mountain-building episodes can be assessed by summarising the basic characteristics of the main mountain systems.

North America

The *Appalachian–Ouachita system* is composed of a series of generally NE–SW trending arcs stretching from New England and the eastern USA to Texas. The main orogenies occurred in the middle and late Palaeozoic Periods with block faulting and revival in the Triassic Period and some Cainozoic upwarping. The *North American Cordilleran system* on the western side of the continent is generally 600–900 km wide except where it narrows to about 300 km in southern California and Mexico. Its form is dominated by two deformed Alpine belts of late Mesozoic and Cainozoic age, separated by a less severely deformed intermontane zone of plateaux, basins and lower ranges. The eastern belt, often known as the Rocky Mountain system, consists of several arcuate mountain chains facing the North American platform that have been active in a number of mountain-building episodes. Activity was initiated in the mid-Precambrian with block faulting in the late Mesozoic and revival in the Cainozoic Periods. There is a general NE–SW trend which includes the Brooks Range, Mackenzie and Franklin Mountains and the Rocky Mountains of Alberta and Montana. The Sierra Madre Orientale is an arcuate cordillera forming the eastern border of the Mexican Plateau. The main orogeny occurred in late Mesozoic to early Cainozoic times.

The Pacific Coast Mountain or Western Cordilleran system stretches from Alaska to California and includes the Coast Mountains and the Selkirk Mountains, continuing as the western border of the Mexican Plateau in the Sierra Madre Occidentale. The main orogenies occurred in the mid-Mesozoic to Cainozoic Period and involved massive intrusion of granite batholiths as well as thrusting and faulting. Volcanic activity has been considerable, especially in the states of Washington and Oregon. The general trend is NW to SE.

In the north the Alaskan Rockies diverge in a fan westward to the Bering Sea with open ranges and block-faulted basins in the west. Tectonic activity has ranged from the Palaeozoic to Mesozoic Periods. The Aleutian Arc–Alaskan Range is a continuation of the Pacific Coast system with extensive youthful revival and vulcanism.

Many of the drainage patterns in western North America did not survive the tectonic activity but there appear to be a number of antecedent gorges such as the Shoshone through the Big Horn Mountains and the North Plate in Colorado. Remnants of two early planations have been recognised and differential warping often exceeds 1600 m.

Central America

The mountains of Central America and the Caribbean Arc connect the cordilleras of North and South America. The *Caribbean Arc* includes the larger Caribbean islands such as Cuba and Puerto Rico and extends through the Virgin and Leeward islands to Trinidad. The structural lineaments trend E–W and were created in mid-Mesozoic to Cainozoic times. The *Central American Cordillera* is basically a chain of Cainozoic volcanic mountains connecting Guatemala with Panama and the Andes.

South America

The *Venezuelan Cordillera* and *Caribbean Ranges*, formed in Mesozoic and Cainozoic times, continue the Caribbean Arc to join the main Andes in Colombia and Peru. The *High Cordillera* of the Andes stretches from Colombia to Tierra del Fuego and is also of Mesozoic to Cainozoic age. It is essentially one mountain chain in Chile but widens in Peru and Bolivia to split into two, the East and West Cordillera (known as the Cordillera Oriental and Cordillera Occidental) separated by the high plateau of the altiplano. In Colombia and Venezuela the system branches to form the Western, Central and Eastern Cordillera. Only the Western Cordillera of Colombia, part of the Eastern Cordillera of Venezuela and the southernmost High Cordillera of Chile can be described as Alpine-type belts. The remainder form fold-fault mountains.

The Alpine belts consist of thick shales, greywackes, cherts and volcanic rocks of Triassic, Jurassic and early Cretaceous age (130–225 million years ago) which were strongly folded in the Late Cretaceous and again in the Tertiary Period. The remainder of the systems consists of basement complexes of Palaeozoic and Precambrian rocks covered by thick andesites and interbedded sedimentary rocks. Intense folding was limited to parts of the West and East Cordilleras. Fault basins were filled with thick conglomerates and sandstones derived from the rising mountain chains. Huge granite masses were intruded. Numerous active volcanoes crown many of the summits and recent uplift can be established in many areas (see, for example, Myers, 1976). Some of the larger river valleys such as the Magdalenas and Cauca Valleys in Colombia and the Huallaga Valley in Peru are structurally controlled. Much of the drainage is thought to be antecedent although the evolution of the relief elements is still open to different interpretations (Cotton, 1960). The *Buenos Aires Ranges* of Argentina are very subdued late Palaeozoic mountains.

Europe

The *Caledonian system* of mid-Palaeozoic age consists of subdued mountains trending in a NE–SW direction through Northern Ireland, Scotland, western Scandinavia, Spitzbergen and north-east Greenland. There has been some local volcanic revival in the Tertiary Period. The *Hercynian (Variscan) system*, as noted earlier, stretches approximately east–west through western and central Europe including Iberia, the Massif Central, the Harz Mountains and the Ardennes. These mountains have been worn down considerably by erosional agencies.

The *Alpine system* is composed of two main subsystems; a northern belt which includes the Alps, Carpathians and Balkan Mountains extending into the Caucasus and Elburz Ranges, and a southern section which includes the Apennines in Italy and the Dinaric chains of Yugoslavia, Albania and Greece extending into the Taurus and Zagros Mountains and the mountains of Oman and Baluchistan in Asia. Several belts of fold-fault mountains, such as the Pyrenees, Cantabrian, Iberian and Catalonian Mountains in Spain and the Jura of France and Switzerland, are set in the Alpine foreland zone.

The evolution of the Alpine belt involves the movement of many plates over at least 200 million years. Development began in the Jurassic Period with the creation of the Tethys Sea between the Eurasian and African plates. Eventually compression in late Cretaceous times cut off slices of continental crust underneath the Tethys Sea. These slices were pushed northwards in early Tertiary times causing more thrust structures. The Jura was compressed into folds and the Aar Massif uplifted. The elevation of the Alps results from uplift during the Pliocene and Quaternary Periods averaging $1-2$ mm yr^{-1} (Selby, 1985). The alpine belt in Miocene times suffered severe erosion which might explain the partial planation surfaces which are conspicuous in many parts of the Alps. The courses of the main Alpine valleys are generally independent of the structure of the folds and nappes and it is from these valleys that the disintegration of the whole mountain system has occurred (Penck, 1953).

Africa

The *Atlas system* of Morocco, Algeria and Tunisia is an extension of the Betic chains of Spain and the Apennines of Italy and Sicily. The *Cape system* of the Cape of Good Hope is the remnant of the east–west trending late Palaeozoic (Hercynian) orogeny. In addition, there are the largely volcanic mountains of East Africa.

Asia

The main mountain system in Asia is the *Himalayan system*. It is an extension of the Alpine system of Europe and North Africa with double loops in Anatolia, the Crimea and the Caucasus and the Elburz-Zagros chains of Iran. There are multiple chains in Baluchistan and Afghanistan with the northern belt merging into the Pamirs and Karakoram Mountains and the southern belt becoming the Himalayas. The Himalayas consist of large thrust sheets formed from the basement of the advancing edge of the Indian continent. Topographically and structurally the Himalayas consist of three units east–west aligned. The southern, frontal range, including the Siwaliks, consists of folded molasse-like sedimentary rocks. Folding occurred in Pliocene–Pleistocene times. The frontal ranges are separated from the overlying rocks of the Lower Himalaya by the Main Boundary Thrust. The Lower or Lesser Himalaya are thrust sheets composed of Precambrian schists and gneisses derived from the Indian shield. The Higher Himalaya, formed of Precambrian gneisses intruded with Tertiary granites, has been thrust over the rocks of the Lower Himalaya along the Main Central Thrust. The main drainage pattern is essentially antecedent with rivers such as the Sutlej, Indus and Brahmaputra commencing on the northern flanks of the Himalayan system and flowing to the south through deep gorges. This shows the recent nature of the uplift and how different the topography must have been prior to the post-Pliocene tectonic movements. The *Hindu-Kush– Pamir–Karakoram system* is the outer loop of the Himalayas extending from Afghanistan to western Tibet.

The *Baikal system* is an arcuate belt of ancient mountains, mostly of Cambrian age, to the south and west of the Angara Shield. The mid- to late Palaeozoic *Ural system*, trending north to south, includes Novaya Zemlya and the Taimyr Belt. The *Tien Shan–Altai–Nan Shan systems* include many of the mountain chains extending into western China and Mongolia. The original tectonic episodes took place in the Palaeozoic Period but with revivals during Mesozoic and Cainozoic times. There are a number of South-East Asian systems, of age varying from late Palaeozoic to Triassic times, that occur in a belt from China to Malaysia, Borneo and the Philippines.

The *Indonesian Mountain system* is simply a further extension of the Alpo-Himalayan System. The *Verkoyansk–East Siberian system* is a late Mesozoic to early Cainozoic loop encircling the ancient shield of east Siberia. It can be sub-divided into a continental belt, including the mountains of Verkhoyansk and a Pacific belt which ranges from Anadyr to Kamchatka. The *Sikhote–Alin–East Mongolian Arcs* occur east of the Sea of Okhotsk and are late Palaeozoic to Mesozoic in age. The *Koryat– Kamchatka system* of easternmost Siberia is an extension of the Aleutian–Alaska range and continues through the Kuriles to Japan. The *West Pacific Arcs* are largely submarine but form mountain islands in

Japan, Taiwan and the Philippines. The age of the orogenies varied from middle Palaeozoic to Cainozoic times.

Australia and the South-West Pacific

The Adelaide–Flinders–MacDonnel *system* extends through central Australia to south Australia and is essentially of Cambrian age. The *East Australia system* is a heterogeneous belt of mountains, mid- to late Palaeozoic in age, stretching from east of the Cape York Peninsula to Tasmania. The *Papua–New Caledonia–New Zealand Arc* is mostly drowned and is late Palaeozoic to Mesozoic in age. The *New Guinea–Solomons–Fiji–Tonga–New Zealand Arcs* are extremely youthful mountain systems that are still active in places. In New Zealand there are traces of an advanced summit planation of Plio-Pleistocene age. Since uplift, river erosion has changed a rolling lowland into high mountains.

Antarctica

The *Victoria Land Mountain system* cuts diagonally across the continent and includes the Queen Maud Mountains. The system is of Cambrian age with some Mesozoic and Cainozoic faulting. The *Marie Byrd Land system* is a chain of discontinuous mountains largely covered by ice stretching from the Ross Sea to the Antarctic Peninsula. The main orogeny took place in late Mesozoic to early Cainozoic times. The *Scotia Arc* is a series of mountain islands connecting the Antarctic Peninsula with the Marie Byrd system and the Andes. The system ranges in age from Mesozoic to Quaternary.

MOUNTAIN LANDSCAPES

It has been suggested that mountains are the most complex landforms on Earth, especially cordilleran and collision-type mountain systems (Garner, 1974). This complexity and distinctiveness is due to the interplay between tectonic and structural influences and the activity of surface processes. Tilted or folded sequences of sedimentary rocks produce distinctive patterns of homoclinal ridges, hogback ridges and cuestas (Gerrard, 1988a). Drainage patterns tend to follow the dip and strike of the rocks with occasional rivers cutting across the structural grain. These transverse systems may be antecedent or superimposed systems. The Ridge and Valley Province of the Appalachians is the classic area in which to examine the influence of folding on topography. The landscape is characterised by parallel ridges and valleys with some major transverse valleys producing a trellised drainage pattern. The 'Appalachian' type of structure is usually thought to be alternating open anticlines and synclines but many of the folds are plunging, closed, overturned and thrust-faulted with the development of secondary folds. The area between the Susquehanna and James Rivers is dominated by a regular repetition of similar folds producing a landscape dominated by parallel, even-crested mountain ridges. The Nittany Valley is an anticlinal valley enclosed by homoclinal ridges and the pattern of the Zig Zag Mountains in eastern Pennsylvania has been produced by strongly compressed plunging folds. Six possible topographic expressions are commonly found: anticlinal valleys; anticlinal ridges; synclinal valleys; synclinal ridges; homoclinal valleys; and homoclinal ridges. Landform–structure relationships will change as erosion uncovers varying structural units.

The form of individual mountains is often governed by rock structure and lithology. Heim (1913) has described a number of different mountain types in Switzerland based on underlying structure. Thus there are numerous examples of anticlinal, synclinal and monoclinal mountains. Djebel Bou Daoud is an anticlinal mountain and Djebel Rhoundjaia a synclinal mountain only 5 km apart in the Ksour Mountains of Algeria. In the Western Cordillera of North America Bird (1980) has also noted a variety of mountain types (Figure 1.2). Mount Assiniboine has been cut into from all sides by cirque glaciers to form a horn peak. A repetition of peaks occurs where there is one major resistant bed in steeply dipping rocks as in the Sawback Range. Isolated pinnacle peaks may occur in vertical rock structures as along the Amiskwi Valley in Yoho Park. Mountains in complex structure tend to be less distinctive. Other mountain shapes are found where rocks are flat-lying (Mount Eisenhower) or dip less steeply (Mount Rundle). Synclinal mountains, such as Mount Keskeslin in Jasper National Park, are also distinctive.

One of the most comprehensive syntheses of the relationship between rock structure, elevation and mountain type has been provided by Ashley (1935) for the Appalachians. He established that:

(a) a low dipping monocline was higher than a steeply dipping monocline;
(b) an anticline was higher than a monocline;
(c) a broad anticline was higher than a narrow anticline;
(d) a syncline was higher than an anticline;
(e) two monoclinal ridges close together were higher than the same ridges separated;
(f) the point of junction of two monoclinal ridges of a syncline or of a breached anticline is higher than either ridge elsewhere.

There are two conspicuous landscape elements of large mountain masses that have captured the attention of geomorphologists for many years. These are conspicuous level surfaces at high altitudes and major systems of transverse drainage. Flattened spurs, crest-lines and accordant summits in many mountains appear to be part of a continuous erosional

Horn Mountain : Mt. Assiniboine

"Sawtooth" mountains–extended vertical beds:
Sawback Range

Mountains of complex structure

"Dogtooth" mountains–vertical structure:
Amiski area. Yoho

Horizontal beds: Mt. Eisenhower

Synclinical mountains:
Mt. Keskeslin, Jasper National Park

Dipping sediments: Mt.Rundle

Figure 1.2 Mountain types in the Western Cordillerra of Canada.
Source: Bird (1980)

envelope (Hewitt, 1972). Extensive portions of the European Alps, the Himalayas and the cordilleras of North and South America form peak plains (gipfelfuren). Thompson (1962) considered that the levels were controlled by the regional timber line with slopes above the timber line being degraded by avalanching, solifluction and freeze-thaw activity.

Also some of the platforms may be the result of high ice action. Many workers have argued that the accordant levels are remnants of planation surfaces (see, for example, Davis, 1923; King, 1967). But in areas where erosion rates exceed 1 mm yr^{-1} and tectonic activity is common it is difficult to imagine the survival of accordant summits for millions of years. Other workers have stressed vertical development of drainage networks rather than horizontally developed surfaces. A long time ago Daly (1905) regarded high-level accordance as simply a side-effect of regularly spaced drainage channels with valley slopes intersecting to form approximately uniform watersheds. Also, contemporary rivers in orogenic regions of recent massive uplift may link areas whose recent erosional development was separate. The uncertainty merely indicates the lack of knowledge concerning the landscape evolution of young mountains.

Similar uncertainty surrounds the evolution of drainage systems transverse to major mountain systems such as occurs in the Zagros Mountains of Iran and the Appalachians. The most common explanations invoke either the process of superimposition or the process of antecedence. A superimposed river is one whose course was determined on a higher rock surface and has been imposed on a different rock and structure by downcutting. An antecedent river is one which has been able to maintain its course across active tectonic zones. The Arun, Tista and Brahmaputra Rivers of the Himalayas are good examples of antecedent rivers, evidence being provided by upwarped river terraces. But it is often difficult to find evidence to substantiate these hypotheses and the origin of much of the transverse drainage systems remains a mystery.

A number of early geomorphologists proposed developmental sequences for the evolution of mountain topography. Penck (1953) proposed a scheme where sporadic uplift was reflected in structural scarps and erosional flats, and Davis (1889) envisaged a staged sequence to explain inversion of relief in folded mountains such as occurs in parts of the Appalachians. Major problems in intepreting the broad elements of mountain relief are the long periods of time involved and the fact that denudation will commence as soon as the uplifted mountain mass achieves appreciable elevation and relative relief. Climates will change much faster than portions of the Earth's crust move and the mountain mass itself will affect the climatic environment.

Garner (1974) has attempted to consider most of these points in presenting a series of syntheses for mountain topography. He makes the fundamental distinction between mountains in humid areas and mountains in arid climates. He equates humid mountains with mountains covered with selva-type vegetation such as Sumatra, Borneo, New Guinea, the north coast of Ecuador and much of the Sierra del Norte of Venezuela. In these environments outcrops are sparse, zones of decomposition are locally thick and all slopes are contiguous with existing drainage lines so there is little undrained land. Garner envisages a topography with narrow V-shaped valleys and knife-edge ridges usually mantled by chemically decomposed regolith and drained by rivers carrying mainly fine-grained sediments and solutes.

This general landscape type was confirmed by Loffler (1977) in Papua New Guinea, where he was able to distinguish landscape types based on rock type. Landscapes on ultrabasic rocks are composed of massive ridges with long, straight or slightly convex slopes. The dissection pattern is coarse, gullies are not common on the upper slopes and incision is shallow. V-shaped valleys form only at lower altitudes. Landscapes on rocks such as granodiorite, diorite, gabbro and basic volcanics also tend to be massive with straight slopes of 35–38°, but the dissection pattern is finer than on ultrabasic rocks. If the rocks are highly weathered, ridges have broad rounded crests and convex side slopes. Slopes of metamorphic rocks are more irregular, with a dense network of small streams following foliation planes. Landforms on sedimentary rocks exhibit great variability, governed by differences in composition, degree of induration, bedding and homogeneity within the layers. Landforms on soft, fine-grained rocks, such as marl, mudstone and siltstone, possess a dense dissection pattern and highly irregular slopes. Landforms on coarser-grained sedimentary rocks such as greywacke, sandstone and conglomerate, show a dense pattern of ridges and valley.

A number of mountain ranges experience predominantly arid regimes, although many show signs of previous climatic variation. Detailed information is available about a number of them such as the Macdonnell Ranges of central Australia (Mabbutt, 1966), the High Atlas Mountains of Algeria and Morocco (Dresch, 1941; 1952; Joly, 1952), the Zagros Mountains of Iran (Oberlander, 1965) and the western Andean Front Ranges in southern Peru and northern Chile (Jenks, 1948; 1956; Garner, 1959). Removal of material in arid mountains is intermittent and tends to collect in alluvial fans, playas and intermontane basins. Thus arid mountains tend to bury their own footslopes. The landscape is dominated by mountain fronts, pediments, pediplains and infilled basins. Few drainage lines are deeply incised and weathering appears to be dominantly mechanical in effect.

As stressed earlier, many mountain ranges have experienced several changes of climate. Garner (1974) considers a mountain mass influenced either by complete changes of climate from arid to humid or influenced by the respective climates on different parts of the mountain. Such a situation would be typified by ridge-ravine topography modified by alluvial fills in the valleys. If the climate changes with altitude, drainage lines in the low-elevation, dominantly arid zone, would be gravel strewn and the valleys would be broader and more open than valleys in the selva. The valleys would narrow and deepen upslope where humidity was dominant.

A diagrammatic impression of the sort of landscapes and changes discussed above is shown in Figure 1.3. Sequence 1 shows the theoretica

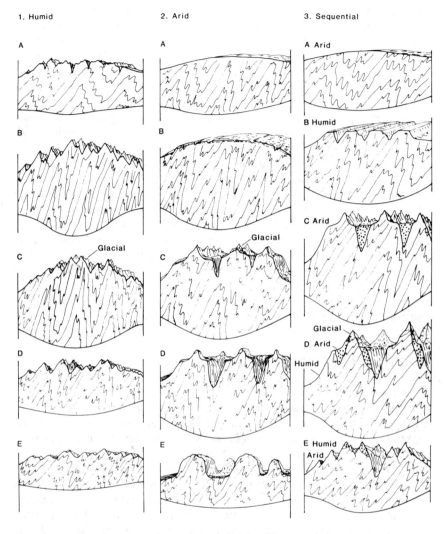

Figure 1.3 Development of mountain relief in different environments.
Source: Garner (1974)

effects of mountain uplift and denudation under a continuously humid regime at relatively low latitudes. Sequence 2 provides a synthesis of mountain uplift and degradation in an arid, possibly mid-continent, situation. Sequence 3 provides the events where an uplifted mountain experiences low-level aridity, intermediate-level humidity, high-level aridity and alpine glaciation during uplift, and the converse during downwearing. In all the sequences glaciation may occur when uplift has raised the mountain mass to sufficient heights. Garner (1974) has emphasised that as uplift continues the mountain mass may enter the zone of moisture-depleted air and then the zone of frost. Mountain degradation under a single subaerial environment must be extremely rare.

A developmental model for the evolution of Japanese mountains involving sequential changes in mean height caused by the interaction of uplift and denudation has been presented by Yoshikawa *et al.* (1981). The first stage is when mean height, dispersion of altitude and sediment delivery rate all increase with time; the second stage occurs when mean height attains a certain critical level and remains constant, together with dispersion of altitude and sediment delivery rate as both rates of uplift and denudation become equal; and the third stage is when mean height, dispersion of altitude and sediment delivery rate gradually decrease. The Japanese Alps are in the earliest second stage, possessing an accordance of summit height nearly at their critical levels. The mountains in the Outer Zone of south-west Japan are in the middle first stage with rapidly increasing mean heights and low denudation rates. The Teshio and Taihei Mountains are in the early first stage with gradually increasing mean heights and the Soya Hills are in the earliest first stage possessing the characteristics of a primary peneplain. All these examples demonstrate the need to understand the interaction of physical processes that contribute to the distinctiveness of mountain landscapes.

MOUNTAIN GEOMORPHOLOGICAL SYSTEMS

Mountains are high-energy environments characterised by instability and variability. Casual observations, especially in high mountain environments, indicate numerous examples of active landslides and mudflows, rockfall events and erosion in general. Mountain streams also appear to carry great quantities of material (Figure 1.4). The literature on mountain streams emphasises the importance of severe, short-lived floods due to heavy rains, snow-melt or natural dam burst. The River Indus transports approximately 90 per cent of its annual sediment load in about two months. As stressed by Hewitt (1972), in his review of the distinctiveness of mountain environments, the vigour of erosion and the frequency of larger erosional events is never so productive of high erosive rates in lowlands as in mountains.

There seems to be no doubt that denudation operates at its greatest rate in high relief areas and that mountains are essentially ephemeral features. A much quoted statement by the American geologist J.W. Powell (1876, p. 193), working in the western United States, emphasises this point: 'We may now conclude that the higher the mountain, the more rapid its degradation; that the high mountains cannot live much longer than low mountains, and that mountains cannot remain long as mountains: they are ephemeral topographic forms.' The high correlation between erosion rate and relief has been established by a number of

Figure 1.4 High turbulence and large bed roughness typical of mountain torrents. (Photo: T.R. Slater.)

researchers (see, for example, Schumm, 1963; 1965; Ohmori, 1983). Gibbs (1967), in an analysis of the Amazon and sixteen major tributaries, found that relief was the most important factor in determining variability in salinity and suspended sediment concentration. Ahnert (1970), using data from twenty river basins, found the following linear relationship between denudation rate D (measured in metres per thousand years) and mean relief h:

$$D = 0.0001535h - 0.01088$$

while Schumm (1963) found that sediment yield S and denudation rate D increased exponentially with relief/length ratio R and relief H, respectively, according to the equalities

$$\log S = aR - b$$
$$\log D = cH - d$$

where a, b, c and d are constants.

It is difficult to isolate one factor from the many that control denudation rate but it has also been established that the difference in erosion rate between plains and mountain rivers is greater than that between climatic regions (Corbel, 1959; 1964; Strakhov, 1967). Erosion rates for mountain rivers appear to be highest in cold climates and lowest in temperate areas. Some indication of denudation rates in mountain areas is provided in Table 1.3.

The Himalayas and the Karakorams possess some of the highest rates of regional denudation (Table 1.4). Values exceed 1 mm yr^{-1} in the Upper Indus (180 000 km^2) and Kosi (60 000 km^2) basins (Hewitt, 1967). There is some uncertainty concerning the accuracy of the figures and it is difficult to compare one figure with another because of the different ways of estimating denudation rate. Some of the measurements are based solely on suspended sediment and no allowance is made for catastrophic events outside the period of investigation. A good example of such an effect is Starkel's (1972) estimation that denudation rates during catastrophic storms in the Darjeeling area are 10–20 mm yr^{-1} compared to the more usual rates of 0.5–5.00 mm yr^{-1}. Rates of more than 2 mm yr^{-1} have been reported in the central mountains of Japan (Ohmori, 1983), and rates of 0.6 mm yr^{-1} have been reported in Alaska, the Canadian Rockies and European Alps (McPherson, 1971). Another way of considering these figures is the suggestion that the European Alps have suffered 30 km of denudation in the last 30 million years (Clark and Jager, 1969).

Detailed analysis of individual mountain areas demonstrate considerable variations in denudation rates that must be related to factors other than elevation. The rivers draining the western slopes of the Southern Alps of New Zealand have specific annual yields ten times higher than world average rates for mountain areas (Griffiths, 1979). These rivers drain short steep catchments rising to over 3000 m supporting dense

Table 1.3. *Estimated rates of denudation for some mountain areas.*

Source	Location	Rate of denudation (μm yr^{-1})
Khosla (1953)	Himalayas	980
Hewitt (1967)	Karakorams	1000
McPherson (1971)	Canadian Rockies	600
Menard (1961)	Appalachians	8
Moberly (1963)	Hawaii	13
Yoshikawa (1974)	Japan	2000

Table 1.4. *Selected denudation rates for the Himalayan region.*

Drainage basin	Denudation rate (mm yr^{-1})	Source
Ganges/Brahmaputra	0.7	Curray and Moore (1971)
River Hunza	1.8	Ferguson (1984)
River Tamur	5.1	Seshadri (1960)
River Tamur	4.7	Ahuja and Rao (1958)
River Tamur	2.6	Williams (1977)
River Arun	1.9	Pal and Bagchi (1975)
River Arun	0.5	Williams 1977 (after Das 1968)
River Sun Kosi	2.5	Pal and Bagchi (1975)
River Sun Kosi	1.4	Williams 1977 (after Das 1968)
River Sapta Kosi	1.0	Williams 1977 (after Das 1968)
River Karnali	1.5	UNDP (1966)

unmodified pedocarp hardwood and beech forests below 1000 m. The world average figure for mountains at 1300 t km^{-2} yr^{-1} (Young, 1974) is more typical of the rivers draining the eastern slopes. In this case precipitation values explained over 95 per cent of the variation in denudation rates. The relationships obtained by Griffiths (1979) in New Zealand may be typical of other very high rainfall areas such as the Himalayas (Starkel, 1970; Curray and Moore, 1971), New Guinea (Wentworth, 1943; Pain and Bowler, 1973), Japan (Tanaka, 1976), Java (Verstappen, 1955) and Taiwan (Li, 1976). In the hardwood forest covered, steep Central Range of Taiwan, average suspended sediment yields of 13 000 t km^{-2} yr^{-1} have been recorded. Possibly the highest recorded specific yield in the world, 31 700 t km^{-2} yr^{-1}, was recorded on a small mountain river in Taiwan.

The important questions that need to be asked are why and how the denudation systems in mountains are different. Mountains possess greater 'relief energy', that is the mass of material above sea level. Tada (1934) has demonstrated the denudational stages of Japanese mountains

Table 1.5. *Altitudinal range between ridges and valleys in the Hengduan Mountains.*

Relative relief (m)	% Total area
0–499	3.3
500–999	15.2
1000–1499	19.3
1500–1999	21.8
2000–2499	17.9
2500–2999	13.7
3000–3499	6.9
3500–3999	2.0
4000–4500	0.6
> 4500	0.2

Source: Liu Shuzheng and Zhong Xianghao (1987)

by diagrams where the maximum height and relief energy in a unit area of about 16 km^2 were plotted. The diagrams give a good visual representation of the denudational features of mountains. A variation on this technique was introduced by Ohmori (1978; 1983) who used the standard deviation (dispersion) of height frequency distribution in a unit area. Using a relationship established between sediment delivery rate and dispersion of altitude, he calculated that the mean denudation rate of twenty-six Japanese mountain ranges was 329 mm kyr^{-1}.

The relief energy available in mountains can be seen well in the Hengduan Mountains, a series of mountain ranges oriented north–south in western Sichuan and Yunnan Provinces of south-west China (Table 1.5). In over 73 per cent of the area, the altitudinal range is 1000–3000 m. In a small part the altitudinal range is 4000–6000 m. The eastern slopes of the Gongga Mountain fall 6400 m in 29 km and the eastern slopes of the Meli Xue Mountain fall 4760 m in 12 km.

Water has greater potential energy at high levels. But the crucial factor appears to be the degree of dissection of the available relief. For a given range of altitude it is the better-dissected mountains which have the higher erosion rates. Thus, comparing the Pamir Mountains with the Karakorams, the former possess the greater positive mass but the latter are more dissected and have the greater denudation rates (Hewitt, 1972). This has also been demonstrated in the study by Ruxton and McDougall (1967) of rates of erosion on a volcanic mountain in Papua New Guinea. They reconstructed the former volcanic form using generalised contours and estimated the age of the rocks by the potassium-argon dating method. Estimated denudation rates for the last 650 000 years ranged from 0.08 mm yr^{-1} at 61 m above sea level to 0.52 mm yr^{-1} at 533 m above sea level. The relationship between denudation rate and altitude was almost perfectly linear. But perhaps more interestingly, denudation rates were approximately 0.1 mm yr^{-1} greater for more dissected areas than for less dissected areas at the same altitude.

There seems little doubt that mountains experience extremely high erosion rates and that these rates are due to a combination of high absolute and relative relief. But the fundamental question is whether it is possible to define a characteristic set of forms and processes in mountains. Are there distinctive variants of processes in mountainous areas? It must also be recognised that mountains may be as different from one another as from lowland areas. Landscapes at similar altitudes may be sheltered or exposed, rocky or soil-covered, shady or sunny, snow-free or snow-covered in winter depending on aspect and topography. However, there are certain aspects of mountains that differentiate them from other types of landscape. As Thompson (1960; 1961; 1962; 1964) has demonstrated, mountain landscapes are not without order as long as climate remains constant. Mountains possess· specific weathering relationships, much bare rock, thus good structural and lithological control, and are often tectonically active with all the implications for denudation rates and river activity. High relief is associated with steep slopes on which rapid and large mass movements occur and leads to higher annual precipitation and runoff with the higher slopes sufficiently cold for snowfall and glaciers which cause additional erosion and improve the coupling between valley-side and valley-floor sediment transport systems.

It may also be the case that high-intensity uplift is the requisite for high mountain denudational processes to operate. It is well known that earthquakes may trigger catastrophic landslides (see, for example, Hadley, 1964; Simonett, 1967; Morton, 1971; Solenko, 1972). Even if a slope failure is not induced directly, long-term instability may be associated with fault zones (Nossin, 1967; Khaire, 1975). Yoshikawa (1974) has shown that rates of degradation in Japan increase with increasing amounts of Quaternary uplift. In most of the catchments studied, rates of uplift were higher than rates of erosion, except from the highest mountain regions where rates of sediment loss were higher than rates of uplift. Seismic activity may also be responsible for the sudden increase in the sediment loads of mountain streams. Pain and Bowler (1973) have estimated that as much as 70 per cent of the total denudation in the highly erodible mountain areas of Papua New Guinea may be attributed to seismic events.

A good example of the long-term effect of earthquake-induced landslides occurred near Mount Ontake in Japan on 14 September 1984 (Okunishi *et al.*, 1987). An earthquake of Richter magnitude 6.8 caused a great number of landslides, including a rockslide at 2500 m on the southern flank of Mount Ontake. A volume of 34 × 10^6 m^3 of debris surged down the Denjo River for 13.5 kilometres, stopping in the main stream of the Otaki River. This amount of debris will affect the development of the Denjo River for many centuries. Also the increased amount of coarse debris will produce elevated channel beds in downstream

Table 1.6. *Some characteristics of the Alpine-Himalayan mountain chain.*

Mountain system	Seismicity	Vulcanicity	Erosion
Europe/Mediterranean			
Greece-Yugoslavia (inc. Dinaris Alps, Pindus and Balkan Mts)	2-3	1	3
North central Europe (inc. Carpathians, Transylvanian Alps)	2	0	2-3
Alps, Jura, Vosges	1-2	0	3
Apennines, Calabria, Sicily	3	1	3
Pyrenees	1	0	2
Western Europe and North Africa (inc. Cantabrian, Catalonian, Atlas Mts)	1-2	0	2
Asia, Middle and Far East			
Karakoram and Greater Himalayas	3-4	0	5
S. Iran-Pakistan (inc. mts of Baluchistan, Oman, Zagros and Taurus Mts)	3	0	4
N. Iran-Turkey (inc. Elburz and Caucasus Mts)	3-4	1	4
W. Burma-NE India	3	0	5
Indonesian Island Arc (inc. Sumatra, Java, East Indies)	3-4	2	3
Borneo	1-2	0	3

Source: Fookes *et al.* (1985)

Table 1.7. *Some characteristics of the circum-Pacific mountain chain.*

Mountain system	Seismicity	Vulcanicity	Erosion
North America			
Aleutians, Alaska Peninsula	5	2	3
Alaska-California (inc. Rockies, Coast Ranges, Cascade and Alaska Mts)	2-4	1	2-3
West Indies Island Arc	3-4	1-2	3
Mexico-Panama	3-5	2	3
South America			
Venezuelan Cordillera	2-3	1	3-4
Colombia-Chile (High Andes)	4-5	2	4
West Pacific island arcs			
NW Pacific (inc. Japan, Ryuku Archipelago, Sakhalin, Taiwan, Kuril Islands)	5	2	3-4
W. Pacific (inc. New Guinea, Philippines, Solomon Islands, New Hebrides)	2-5	2	3
SW Pacific (inc. New Zealand, Fiji)	2-3	2	3
Eastern Siberia			
Verkhoyansk Mts and Kamchatka Peninsula	1-5	1-2	2

Source: Fookes *et al.* (1985)

reaches (Mizuyama, 1984). According to Thompson (1964), the most obvious difference between mountains and other landscapes is the fact that mountains have significantly different climates at different levels and climate-related aspects of their landscapes are important. To some extent this is a function of relative relief. The intensity of contrasts which occur on 4000 m of relative relief is likely to be more impressive than those over 1000 m.

A summary of some of the essential characteristics of the world's Alpine fold mountain systems has been provided by Fookes *et al.* (1985) (Tables 1.6 and 1.7). The seismicity index, although subjective, is based on frequency and magnitude data provided by Lomnitz (1974) and Anon (1978). The low-risk areas (class 1) experience infrequent seismic events below magnitude of 5.9. Class 2 areas possess moderate seismic risk (maximum magnitude 6–6.9), class 3 are high risk areas (maximum magnitude 7–7.9), class 4 areas are severely at risk with occasional activity greater than magnitude 8 and class 5 areas experience frequent activity greater than magnitude 8. The vulcanicity classification is based on number of historically active volcanoes using data from Gutenberg and Richter (1954) and Anon (1978). Areas of class 0 possess no historically active volcanoes, class 1 areas possess a low density and class 2 areas possess a high density of volcanoes. The classification of erosion in tonnes per square kilometre per year (class 1, 0–10; class 2, 10–50; class 3, 50–100; class 4, 100–250, and class 5, over 250) is based on Strakhov (1967).

The data show that the mountains of the circum-Pacific belt are the most seismically active and contain the most volcanoes, which is clearly related to the tectonic framework discussed earlier. However, erosion rates are not necessarily higher. Indeed, the highest rates occur in the Himalayan range. This is because erosion rates are determined by other factors such as climate, vegetation cover, drainage networks and the availability of material to be transported. Rainfall amounts in mountainous areas are extremely variable, depending on relief and local topography. But some of the highest amounts are received in the Himalayas, especially in the eastern portions, dictated by monsoonal conditions. Also, the Himalayas are drained by large rivers, such as the Indus, Ganges, Tista and Brahmaputra, with well-integrated drainage nets. High erosion rates are also experienced in geologically older mountain systems (Table 1.8).

Thus, it appears that there are many distinctive elements in the geomorphology of mountains. This distinctiveness lies not in unique processes but in combinations of processes and landforms and the rate and intensity at which the processes operate. A composite landform association model for mountainous areas has been produced by Fookes *et al.* (1985) (Figure 1.5). It is based on terrain characteristics in east Nepal but, as a model, is applicable to most extra-tropical high mountains that have suffered extensive glaciation. The model also illustrates an

Table 1.8. *Some characteristics of old mountain systems.*

Mountain system	Seismicity	Vulcanicity	Erosion
Norway	1	2	2
Appalachians	1–2	0	2–4
Indochina and SW China	1–2	0	5
Central Asia (inc. Tien Shan, Altai, Pamir, Kunlun Mts)	2–4	0	?
Urals	?	0	1–2

Source: Fookes *et al.* (1985)

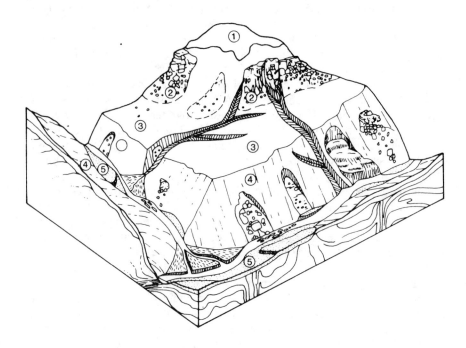

① High altitude glacial and periglacial

② Free rock face and associated debris slopes

③ Degraded middle slopes and ancient valley floors

④ Active lower slopes

⑤ Valley floors

Figure 1.5. Composite diagram of a mountain landscape.

Source: Fookes *et al.* (1985)

early state in the development of the landscape with active incision by the major rivers, the formation of river terraces and alluvial fans and the rapid extension of gully and mass movement systems.

Zone 1 is the zone of high-altitude glacial and periglacial activity. It may also be equated with the zone above the regional timberline. The slopes are mantled with highly variable glacial deposits or coarse weathering products. Denudation processes include glacial erosion, mechanical weathering, especially freeze-thaw activity, solifluction and instability in rock and snow masses. Landforms are those characteristic of glacial erosion and deposition such as cirques, angular ridges, peaks, U-shaped valleys, morainic ridges and till sheets. Ice sheets and ice fields may be present, as well as scree slopes and solifluction landforms. The extent of this zone will depend on altitude and latitude as well as erosional and climatic history. Landforms associated with zone 1 may be encountered at lower levels because of activity during the various colder phases of the Pleistocene. Glacial moraines and fluvio-glacial deposits are often abundant in the valley-floors of the larger rivers.

Zone 2 is a landscape composed of high-angle rock and coarse debris-covered slopes. It is essentially a landscape of weathering and mass movement. The most frequent types of mass movement are small rock-falls, wedge and toppling failures. Less frequent but involving larger volumes of material are rock avalanches and rock slides. Such movements may exceed by considerable amounts the average denudation rates. Landforms are either rock walls and cliffs or debris slopes of a variety of types. Debris slope types are common; coarse block fields or boulder screes below rock faces with relatively large joint spacing, scree or talus existing at angles in the range 33–38° and taluvium, created by the weathering of scree. Taluvium is basically a coarse mixture of grain sizes dominated by the gravel fraction. Debris slopes are usually being eroded by debris slides and avalanches.

Zone 3 slopes are gentler and possess a thicker soil cover. The topography represents ancient valley floors and degraded slopes, thus river terrace sediments and fan deposits are common. A wide range of slope materials is usually present. Slopes are generally stable but old landslides may be reactivated. Soil creep and surface wash may be extensive.

Many mountain landscapes are dominated by active valley slopes of zone 4. High rates of erosion occur on these slopes which are mantled with transported taluvium and colluvium overlying *in-situ* weathered rock. The thickness of the weathered material will vary locally with aspect and position and regionally with climate. Some slopes in tropical mountains will possess deep weathering profiles, whereas the depth on slopes in arid mountains will be considerably less. Chemical and mechanical weathering are important but the dominant geomorphological processes are mass movement and gullying. Mass movement processes include debris slides, rockslides and mudslides — activity being determined by rainfall intensity. Debris slides are commonest, with mudslides being confined mainly to fine-grained rocks and zones of deep weathering. Slope deposits that encourage the development of perched water tables may also be susceptible to mudslides. Rotational slides are less frequent. However, several different types of slide are usually encountered on any one slope, producing complex interactions. Landslides may provide the destruction of soil and vegetation cover to allow rilling and gullying to develop. Actively eroding gullies possess steep, irregular longitudinal profiles and generally V-shaped cross-sections with steep side-slopes. The density of the drainage network may be several times higher in zone 4 than in zone 3 (Fookes *et al.*, 1985). Rilling will occur on bare, unvegetated slopes. Gorges and high-angle rock slopes will also occur and the landforms and processes will be more similar to those in zone 2.

The valley floors of zone 5 consist of coarse alluvial materials deposited by either the main river or its many tributaries. Large debris loads and relatively steep valley floors lead to braiding, rapidly shifting channels and small terraces. The depositional processes of the main river are frequently interrupted by alluvial and debris fans emanating from side valleys and debris chutes. There can be considerable variation in fan shape and size related to catchment size and rock type (Sinha, 1975).

Studies in high mountains have often emphasised the role of short-lived but high-intensity events. In fact many workers have suggested that such high-magnitude/low frequency events are the formative events of medium- and long-term landscape evolution. The geomorphological importance of a given event is governed by the magnitude of the force or energy involved, the frequency with which it recurs, the processes during intervening intervals and the work performed during these intervals. 'During these events, the amount of work accomplished, and the way in which slope and channel units are intimately linked in a sediment transfer system, suggests that gross valley and slope forms are best considered as high magnitude response forms' (Brunsden and Jones, 1984, p. 383). There is no doubt that increasing relief enhances the potential for mass movements; therefore they play a relatively greater role in mountains.

Climate is also significant in this respect. There are more extreme heat and moisture conditions in mountains with both increased local intensities and greater spatial and temporal variations. There are also strong seasonal and diurnal fluctuations. In terms of surface runoff, steep channels tend to compress short-term hydrographs, and compression of discharge curves also seems to be a major factor in the higher sediment yields from glacierised mountains. The researcher in mountains can expect to obtain numerous observations of mass movement, changes of stream flow and glacier fluctuations. But, as Hewitt (1972) has stressed, research needs to be related to the larger questions of the mountain landscape. Brunsden and Jones (1984) have listed three areas where more information is required. These are: an assessment of the frequency and mean arrival times of events of different magnitude; an analysis of the formative events which actually do most work in initiating characteristic or persistent landforms; and a determination of the length of time required for a return to the previous or a new system state. These are some of the problems that are examined in the rest of this book.

Chapter 2

MOUNTAIN GEOECOLOGY

INTRODUCTION

The essential distinctiveness of mountain environments has been established in the previous chapter. Much of this distinctiveness is the result of the specific interaction between topography, climate, soils and vegetation; what can loosely be called the *geoecology* of mountains. This chapter seeks to establish the basic geoecological aspects of mountains as a foundation for the more detailed examination of the specific elements of mountains that forms the main part of the book. Any account of the geoecology of mountains must rely heavily on the remarkably extensive studies of Carl Troll and his associates. These studies have provided a firm basis for more specific studies of landscapes and processes. This chapter seeks, very deliberately, to establish the interactions between the various components in mountain environments — thus systematic treatment of climate, soil and vegetation is avoided. Good syntheses of these topics are already available in works such as those by Barry (1981), Price (1981) and Ives and Barry (1974).

A number of key issues consistently emerge in any geoecological analysis of mountains. These are generally related to the vertical zonation of environmental systems and the variation of these vertical zones with latitude. These interrelationships between geomorphological systems, vegetation, soils and climate are of paramount importance. Within these general themes, specific topics such as the significance of and controls on the upper timber line, the position of the present and Pleistocene snow lines and microclimatic influences have generated intensive study.

A GEOECOLOGICAL APPROACH

Geoecology has been defined as the science of the full and complex inter-relations between the organisms, or *biocoenosis*, and their environmental factors (Troll, 1972a). Troll (1970) has argued that there are two main aspects of geoecology. The first is the differentiation of the regionalisation of ecosystems and the spatial arrangement (landscape pattern) of ecotopes in a geographical region. This leads from global climatic, vegetation, soil and landscape belts to smaller and smaller landscape units within the general landscape hierarchy. The second is the analysis of single ecosystems functionally and quantitatively with regard to the full interrelations between biotic and non-biotic elements such as microclimate, rock material, soil type, water amount and movement, landforms and biocoenosis with plants, animals, microorganisms, and so on. The first approach can be categorised as the geographical approach and the second as the biological viewpoint. In essence, though, an understanding of specific landscape components can only be achieved by using both approaches and the way they interact.

O'Connor (1984), in the introduction to his paper on the stability and instability of ecological systems in New Zealand mountains, has summarised the situation very succinctly. Although mountains can be defined with real boundaries in space and time, it is more difficult to define ecosystems. Thus

it is very difficult to have a simple conceptual mountain community and its habitat defined so that it can correspond statistically with a genuinely homeostatic natural unit in the real world. It is likewise difficult to have such a conceptual ecosystem correspond with an actual land management unit, one that can be prudently managed without regard to its neighbouring units (O'Connor, 1984, p. 16).

Mountains are characterised by steep ecological gradients with intense mobility of water, material, energy and animals. Everything appears to be connected to everything else (Figures 2.1 and 2.2). Altitude and aspect control the zonation of mountain environments and ecosystem types function as coupled subsystems. This coupling is effected by continual or episodic transfers of matter or energy. O'Connor (1984) stresses that in mountain environments, it is important work with concepts such as landscape systems and to define the ecosystem as a system composed of physical-chemical-biologic processes within a space-time unit of any magnitude.

One of the main characteristics of mountain ecological systems is their fluctuating stability. Such systems are dynamic rather than static or stable. This dynamism was clearly seen in Chapter 1 and results from combinations of external influences such as tectonic activity, rejuvenation and climate fluctuations and internal changes such as weathering, mass movement, slope modification, fire, land use changes, and so on. In such

Figure 2.1. Contrast between braided channels on valley infill, forest-covered lower slopes and bare rocky upper slopes typical of alpine landscape. However, such a landscape is integrated by the interaction of geomorphological processes. (Photo: T.R. Slater.)

Figure 2.2. Typical geoecological mosaic present in high mountains. (Photo: T.R. Slater.)

situations concepts of ecological stability, as proposed by Gigon (1983) are extremely useful. Where disturbance factors external to the system are absent, little or no oscillation of ecological parameters indicates *constancy*. Large oscillations will produce *cyclicity*. Where disturbance factors are present but little or no oscillation occurs this would imply system *resistance* while large oscillations would indicate *resilience* or *elasticity*. The dynamic nature of mountains encourages recurrent stability over the long term in relatively immature systems which would otherwise become mature or even post-mature. Also, 'whether the type of stability embodied in such recurrent rejuvenation in mountain landscapes is termed *cyclicity* or *resilience* depends on whether or not the *disturbing factors* are judged to belong to the "normal household" of the ecological system' (O'Connor, 1984, p. 17).

The mountains of New Zealand provide an excellent 'natural laboratory' in which to examine many of these concepts. Instability has continually affected landforms, soils and vegetation (O'Connor, 1983). The mountains are young but essentially composed of old reworked sedimentary and metamorphic rocks. The main tectonic activity occurred from the Tertiary into the Pleistocene Periods and is still continuing. Thus a high-relief youthful landscape, coupled with high precipitation amounts, produces high erosion rates and extremely dynamic landscapes. Analysis of slopes and other geomorphic surfaces (see, for example, Molloy, 1964; Mosley, 1978; Chinn, 1981; Grant, 1981; Tonkin *et al.*, 1981) and river behaviour (O'Loughlin, 1969; Griffiths, 1981) indicates the episodic nature of landform evolution and sediment transport. This episodic characteristic has been analysed in terms of Gigon's stability types (Table 2.1). The landscape systems of the two high-rainfall regions (Westland and Northwest Canterbury), with periodic rock avalanches, debris avalanches and debris slides, are interpreted as showing cyclic stability or cyclicity. The drier mountain system of Central Otago is interpreted as possessing general constancy although some cyclicity of periglacial activity caused by climatic oscillation may be superimposed on the general constancy. The Central Canterbury fold mountains may be classed as possessing either cyclic stability or natural endogenous instability. Endogenous instability may have been present even with apparently mature soils and biotic successions and before the advent of man. Some elements of the landscape show an inclination to change irreversibly. The general conclusion is that regional differences in ecological stability of the landscape reflect differences in both the frequency and pace of soil-biotic succession (O'Connor, 1984). These, in turn, are influenced by geomorphological and geological processes.

Geoecological principles, outlined above, are evident in attempts to define high-mountain landscapes. Three criteria have been used to determine the lower limit of such landscapes; the upper timber line; the high-mountain landform complex; and the lower limit of cryonival denudation.

Table 2.1. *Nature of instability in New Zealand mountains.*

Region	Westland	Northwest Canterbury	Central Canterbury	Central Otago
Geology and geomorphology	Mica-schist fold mountains with glaciated valleys and fluvially dissected tributaries	Greywacke fold mountains with glaciated valleys	Greywacke fold mountains with post-glacial deposits	Mica-schist block mountains with tectonic basins infilled with post-glacial deposits
Climate	Superhumid	Humid	Sub-humid	Semi-arid
Altitudinal zonation of vegetation				
Alpine	Herb-field	Fell-field and herb-field	Fell-field and herb-field	Cushion-field and herb-field
Subapline	Scrub and grassland	Tall tussock grassland and scrub	Tall tussock grassland and scrub	Tall tussock grassland and scrub
Montane	Evergreen Rata-Kamahi rain forest	Evergreen Podocarp and Silver Beech forest	Evergreen Podocarp and Mountain Beech forest	Evergreen Podocarp forest and grassland
Lowland	Podocarp rain forest	Podocarp and Beech forest	Podocarp and Mountain Beech forest	Low tussock grassland, scrub and woodland
Principal erosional processes	Ice and snow avalanches Glaciation Fluvial dissection Debris slides Debris avalanches (generally frequent and shallow)	Ice and snow avalanches Rockfalls Rock avalanches Debris flows (generally less frequent and shallow)	Rock avalanches Debris flows Screes (infrequently active)	Periglacial solifluction Local landslides (very infrequent, deep)
Inferred stability type	Cyclicity	Cyclicity	Cyclicity or natural endogenous instability	Constancy with local cyclicity

Source: O'Connor (1980)

Table 2.2. *Relationships between the altitudes of the timber line and Pleistocene snow line.*

	Actual timberline (m)	Pleistocene snow line (m)
Northern Urals	750	750
Central Urals	1000	1100
Southern Urals	1300	1250
Alps, northern border	1500–1600	1200–1300
Central Alps	2250	2000
Pyrenees (west, northern border)	1600	1500
Pyrenees (east, maximum)	2300	2300
Great Atlas (central part)	3250	3000–3200
Rila Mts (Bulgaria)	2000–2100	2200
Pirin Mts (Bulgaria)	2100–2200	2300
Mount Olympus	2200	2300
Altai Mountains, northwest	1500	1500
Altai Mountains, southeast	2500	2500
Western Anatolia	2100–2200	2400–2500
Armenia	2700	3000–3200
Elburz (Iran)	2700	3500

Source: Troll (1973b)

The *upper timber line* is a suitable criterion for extra-tropical mountain systems where seasonal change of temperature and duration of snow cover are effective. The definition of the upper timber line in tropical mountains is more difficult largely because seasonal changes of temperature are smaller and less significant ecologically than diurnal variations. The upper timber line is such an important boundary, both ecologically and geomorphologically, that it is examined in greater detail later.

The *high-mountain landform complex* encompasses a variety of landforms that exist in all high mountains between the Equator and the poles and which are regarded as being the result of glacial erosion during the cold periods of the Pleistocene Age. The lower limit is customarily taken to be equivalent to the Pleistocene snow line which itself is defined as the lowest level at which cirques occur. Interestingly, the snow line and the upper timber line ascend and descend together, and within temperate zones they almost coincide (Table 2.2). In more arid mountains distances between timber lines and Pleistocene snow lines become greater (400–800 m) because tree growth is not possible at high altitudes due to dryness and frigidity. In extreme oceanic climates of the temperate belt the Pleistocene snow line may have been depressed below the current timber line.

The *cryonival belt* can be regarded as the belt between the timber line and the snow line. Frost activity is strong and patterned ground — for example, polygons, stripes, garlands, structure soils — is common. The lower limit of solifluction in temperate zones corresponds approximately to the timber line. In Scandinavia and the Urals this limit is approximately 1250 m and in the European Alps it is in the range 1500–2200 m. In continental systems it is higher, being 4000 m in Iran and 5000 m in Tibet. In the tropical Puna of northern Bolivia it is 4700 m and on equatorial Mount Kenya it is 3950 m.

These characteristics enable the high-mountain environment to be divided into three zones:

(i) The upper zone or nival belt above the limit of perpetual snow.
(ii) The intermediate zone between the snow line and the limit where a more or less dense plant cover with a soil layer begins. It is a zone of frost debris, patterned soils and open pioneer vegetation.
(iii) The lowest zone where plant cover struggles with frost action in the soil. It reaches down to the timberline with alpine meadows, dwarf shrubs, tussocks etc. forming a nearly closed carpet.

A somewhat similar approach has been adopted by Stablein (1984) in a geoecological synthesis of altitudinal zonation in the mountains of Greenland. In the mountains of mid-latitudes the forest belt up to the treeline is mainly influenced by fluvial processes, whereas above the treeline lies the alpine tundra with signs of recent cryonival activity. Higher still is the rocky frost-debris zone and the nival belt where active glacial processes dominate. But conditions in polar regions are different because alpine relief rises from sea level and glaciers extend to sea level in fjords. Apart from the ice sheets and glaciers, only the tundra belt and frost debris belt can be differentiated. Altitudinal boundaries change from region to region according to relief conditions so that 'no altitude can be fixed for a general climatic upper boundary of the tundra region' (Stablein, 1984, p. 319).

Investigations in Greenland (see, for example, Poser, 1936; Paterson, 1951; Malaurie, 1960; Washburn, 1965; 1967; 1969; Nichols, 1969; Stablein, 1977a; 1982) have raised the fundamental question whether there are any geomorphological belts at all in artic-alpine mountains. The general answer to this appears to be negative, although relative zones can be recognised dominated by meso-scale relief influences. Near glaciers, zones can be differentiated on the basis of energy transfer, mass budgets and mechanisms of movement. From high to low elevations the following zones may be distinguishable (Stablein, 1984):

(i) the dry snow zone where no melting occurs even in summer with mean annual air temperature less than $-25°C$;
(ii) the percolation zone, where meltwater percolates and refreezes in the snow pack;
(iii) the wet snow zone, where all the snow of one year melts and percolates during the summer;
(iv) the superimposed ice zone, where refrozen meltwater is exposed at the surface;

(v) the ice ablation zone, where firn and glacier ice melts below the equilibrium line at heights from 1000 m to more than 1500 m.

On individual slopes and in simple valley systems, zonation of processes and landforms can be discerned but based more on local climatic parameters, conditioned by geoecological, edaphic and orographic factors, rather than on altitudinal limits. This can be seen in the catena profile of arctic-alpine geomorphodynamics (Figure 2.3). Four main valley-side slope zones can be discerned (Stablein, 1984):

(i) the belt of peaks, plateaux and saddle sites on which frost weathering, solution weathering, slope wash, cryoturbation and wind deflation zones are common;
(ii) the upper slope belt where frost weathering, rockfall and debris creep occur;
(iii) the mid-slope belt on which slope wash, solifluction, nivation, and slope dissection are active;
(iv) the lower belt, often with terraces, where cryoturbation, frost heaving and wind action occur.

The dynamics of the system can be explained by variations in soil properties (Stablein, 1977b; 1979). In the peak belt, bedrock surfaces are pitted by weathering pans, solution hollows and tafoni. In more sheltered areas, frost weathered debris accumulates and there is sufficient material and moisture for freeze-thaw cycles to produce patterned ground. Needle ice and turf exfoliation may also occur. Slope wash is active on the upper slopes with little vegetation, whereas where the slopes are steeper than 25° intense rill and gully formation occurs.

An interesting sub-division of frost-slopes has been suggested by Wirthmann (1976). A *cryogenic rock fall slope* with a steep wall and debris slope indicates that denudation does not extend above the upper break of slope, whereas a *compensating frost-slope* shows evidence of denudation by surface runoff and solifluction (See Figure 2.3). Snow drifts and nivation processes are important and solifluction is active below snow drifts due to percolating snow melt. On lower slopes, generally covered with dense tundra vegetation, solifluction is periodic or episodic.

SYNTHETIC ALPINE SLOPE MODEL

The synthetic alpine slope model (Burns and Tonkin, 1982) illustrates similar concepts to those just described but uses soil-geomorphic models to interpret the spatial distribution and relative development of soils in alpine geomorphic provinces. Many soil-geomorphic models have been proposed (Gerrard, 1981), some specifically for mountain regions (see, for example, Richmond, 1962; Birkeland, 1967; Parsons, 1978; Tonkin

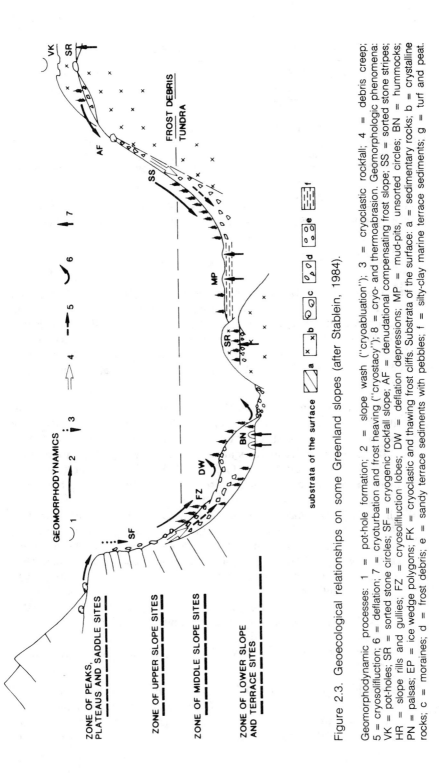

Figure 2.3. Geoecological relationships on some Greenland slopes (after Stablein, 1984).

Geomorphodynamic processes: 1 = pot-hole formation; 2 = slope wash ("cryoabluation"); 3 = cryoclastic rockfall; 4 = debris creep; 5 = cryosolifluction; 6 = deflation; 7 = cryoturbation and frost heaving ("cryostacy"); 8 = cryo- and thermoabrasion. Geomorphologic phenomena: VK = pot-holes; SR = sorted stone circles; SF = cryogenic rockfall slope; AF = denudational compensating frost slope; SS = sorted stone stripes; HR = slope rills and gullies; FZ = crysolifluction lobes; DW = deflation depressions; MP = mud-pits, unsorted circles; BN = hummocks; PN = palsas; EP = ice wedge polygons; FK = cryoclastic and thawing frost cliffs. Substrata of the surface: a = sedimentary rocks; b = crystalline rocks; c = moraines; d = frost debris; e = sandy terrace sediments with pebbles; f = silty-clay marine terrace sediments; g = turf and peat.

Table 2.3. *Soil-geomorphic relationships in the Alpine tundra zone.*

Alpine province	Dominant geomorphic processes	Soil variability control by state factors	K-cycle stable phase, period of duration*
Ridge-top	periglacial	spatial > temporal	long/medium
Valley-side	gravity some glacial	temporal > spatial	short
Valley-bottom	glacial fluvial gravity	temporal > spatial	short/medium

*'Long': greater than 15 000 years; 'medium': 5000 to 15 000 years; 'short': less than 5000 years.
Source: Burns and Tonkin (1982)

et al., 1981). The synthetic alpine slope model is based on the K-cycle of Butler (1959) where a K-cycle involves an unstable period of erosion and deposition followed by a period of stability when soils form. Burns and Tonkin (1982) divide the alpine tundra zone into three geomorphic provinces (Table 2.3). The classification was developed for the southern Rocky Mountains but it should be applicable to other alpine areas. The provinces are:

(i) The *ridge-top tundra province* which consists of broad interfluves that have not been glaciated in late Pleistocene times and can be interpreted as a persistent zone which has, in recent times, been characterised by relative stability. Conceptually this analysis is similar to that proposed by Gigon (1983) and the synthesis of New Zealand mountains presented earlier. The stability of the landscape is reflected in mature soils with deep profiles.

(ii) The *valley-side tundra province* is located on the steep slopes of the valley-side walls. This province is the direct analogy of the erosion, alternating and deposition zones of K-cycles as it has been undergoing dynamic change throughout the last 10 000 years. Soils reflect this instability, being predominantly young, thin soils intermixed with rock outcrops.

(iii) The *valley-bottom tundra province* is primarily present in cirque floors but can be present further downvalley. This province is equivalent to the major deposition zones of K-cycles and the soils have developed on a variety of glacial, fluvial and slope derived materials.

The length of the various stable phases is reflected in the 'maturity' of the soils: short phases (less than 5000 years) are estimated from glacial soils that have not reached steady state characteristics; medium-length phases 5000–15 000 years) are characterised by soils that have minimum

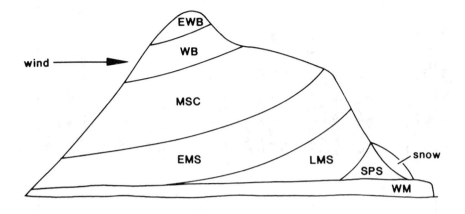

Figure 2.4. Schematic representation of the synthetic alpine model (after Burns and Tonkin, 1982). **Note:** abbreviations are explained in Table 2.4.

steady-state requirements; and long periods (greater than 15 000 years) are based on the ages of well-developed glacial soils.

The operation of the model can be assessed by examining the ridge-top tundra province in greater detail (Figure 2.4 and Table 2.4). Seven microenvironmental sites have been defined based on edaphic–topographic–snow cover relationships. The soil terminology follows that of Soil Survey Staff (1975). Soils differ mostly in A horizon characteristics, thickness of aeolian deposits, moisture regimes and mean annual temperature.

(i) The *extremely windblown* (EWB) sites are located on the crests of the drainage divides. Vegetation is sparse and is primarily cushion plants. EWB soils (90 per cent Dystic Cryochrept, 10 per cent Typic Cryumbrept) are poorly developed and well-drained with thin sandy A horizons over thin cambic B horizons.

(ii) The *windblown* (WB) sites are found from the tops to 30 per cent down the slopes. They are dry and have mixed vegetation of cushion plants, sedges and herbs. WB soils (80 per cent Dystic Cryochrept, 20 per cent Typic Cryumbrept) are similar to EWB soils except that the cambic B horizons are thicker and patches of loess occur. Some loess is mixed in the A horizons which are thicker than EWB soils.

(iii) *Minimal soil cover* (MSC) sites occur in cols and on large plateaux. They have more snow cover than WB sites and possess thick turf vegetation. MSC soils (80 per cent Pergelic Cryumbrept, 20 per cent Dystic Pergelic Cryochrept) have thick, fine-textured A horizons of aeolian origin overlying cambic B horizons. They are the best-developed soils with, except for WM soils, A horizons containing the highest organic matter content.

Table 2.4. Relationships of the ridge-top tundra province of the synthetic alpine slope model (after Burns and Tonkin, 1982).

Microenvironmental sites	Approximate snow-free days per year*	Slope angle (degrees)	Mean annual soil temperature (°C at 50 cm)	Associated indicator plant species	Braun-Blanquet groupings of vegetation (Komarkova and Webber (1978))
Extremely windblown (EWB)	> 300	0–8	+ 0.5	Paronychia pulvinata Silene acaulis	Association Sileno-Paronchietum
Windblown (WB)	225–300	0–10	+ 1.4	Carex rupestris Dryas Octopetala Trifolium dasyphyllum	Association Potentillo-Caricetum; Ass. Trifilietum dasyphyllum; Ass. Eritricho-Aretiodis
Minimal snow cover (MSC)	150–200	0–10	− 0.1	Kobresia myosuriodes Carex elynoides Kobresia myosuriodes with Acomastylus rossii	Alliance Caricion foeneo-elynoidis; Ass. Selaginello densae-Kobresietum myosuroidis
Early-melting snowbank (EMS)	100–150	5–15	+ 0.5	Acomastylus rossii Trifolium parvii Deschampsia caespitosa Vaccinium spp.; Salix spp.	Order Trifolio-Deschampsietalia
Late-melting snowbank (LMS)	50–100	10–40†	+ 3.4	Sibbaldia procumbens Carex pyrenaica Juncus drummondii	Order Sibbaldio-Caricetalia pyrenaicae
Semipermanent snowbank (SPS)	0	5–40	0	no vegetation	no vegetation
Wet meadow (WM)	about 100	0–5	− 0.1	Carex scopulorum Pedicularis groenlandica Salix spp. (wet)	Class Scheuchzeno-Caricetea fuscae; Class Betulo-adenstyletea; Class Montio-Cardaminetea.

*Burns (1980) and May (1973).
†Backslopes.

(iv) *Early melting snowbank* (EMS) sites are found 30–90 per cent of the way down slopes, usually on gentle slopes. Such sites possess a great variety of vegetation. EMS soils (60 per cent Typic Cryumbrept, 30 per cent Pachic Cryumbrept, 10 per cent Dystic Cryochrept) are moderately well drained with the thickest A horizons of any soils in the sequence. Loess occurs in the A horizons but is not as thick as in the soils of MSC sites.

(v) *Late-melting snowbank* (LMS) sites are present 50–90 per cent of the way down slopes in leeward nivation hollows. The snowbanks normally melt out in early July or August. Vegetation is sparse and the sites are generally rocky. LMS soils (100 per cent Dystic Cryochrept) are poorly developed, moderately well-drained soils overlying weakly developed cambic B horizons.

(vi) *Semi-permanent snowbank* (SPS) sites are found 60–90 per cent of the way down slopes in nivation hollows with little vegetation cover because snowbanks rarely melt completely. SPS soils (Lithic Cryothent, headwall; Pergelic Cryoboralf or Pergelic Cryochrept, nivation hollow) are poorly developed. The headwall soil is the least-developed soil and is simply a thin sandy A horizon over a C horizon.

(vii) *Wet meadow* (WM) sites are characterised by bog vegetation and occur below snowbank sites in depressions and on turf-banked terraces and lobes at the base of the slopes. WM soils exhibit a variety of characteristics but are poorly drained with either A and/or O horizons of variable thickness overlying gleyed and mottled B and C horizons. They have low mean annual temperatures with possible permafrost and contain large amounts of silt and clay which have been moved downslope by water action.

The three examples, examined in some detail here, have shown consistent relationships between geomorphological processes, soils, vegetation, climate and landforms and provide strong justification for a geoecological approach to mountain environments. One of the most significant geoecological boundaries is the position of the upper timber line. This line marks the boundary between high alpine landscapes dominated by nival, cryonival and mass movement processes and the lower forested zones where fluvial processes, soil creep and throughflow achieve greater importance. The factors which influence the position and variability of the timber line are also the factors which influence the workings of the geomorphological systems. These factors are now examined.

THE UPPER TIMBER LINE

The nature and position of the upper timber line in different climatic zones has been reviewed by Troll (1973a). The timber line reaches its maximum elevation on the high plateaux in the arid zones of the marginal tropics. Thus in eastern Tibet the forests of western China meet the highland steppes at about 4700 m and in northern Chile and western Bolivia, on the dry Puna belt of the Cordillera de Los Andes, stunted trees occur in open stands among the tussock grasslands at nearly 5000 m. In latitudes with seasonal climates, the thermic summer conditions determine the limitations of tree growth. It is difficult to be precise about the climatic conditions that are directly relevant but tree growth appears to be controlled by the temperature of the summer season or perhaps of the warmest month, the duration of the frost-free period in the growing season and the accumulated degree days in summer.

The specific nature of the timber line will be determined by local topoclimatic differences such as the accumulation of winter snow in hollows by wind drifting or deflation and consequent exposure on ridge tops. Such conditions produce wind-deformed trees, krummholz, puckerbush and snow glades. Patterns such as ribbon forest and snow glade have been described by Billings (1969) in the Medicine Bow Mountains, Wyoming, and Glacier National Park, Montana. Mountain tops and sharp ridges, even if below the tree line, are often treeless. Because of these different controls it is necessary to examine different parts of the world separately.

Humid tropical mountains

The main differences in the upper timber line in the Andes and the mountains of East Africa, Indonesia and New Zealand are created by the diverse seasonal and diurnal variations of temperature. Tropical mountains lack the contrast of a cold winter and a warm summer. In the tierra fría of the equatorial Andes at 3000–4000 m, mean January and July temperatures exhibit a variation of less than 2°C. Daily temperature variations easily surpass such annual variations. A further distinction is that such mountains only experience a thin and short-lived seasonal snow cover during the rainiest periods and then only high above the timber line. Thus the decisive factor in determining the upper limit of tree growth in tropical mountains must be lack of heat.

The timber line is normally abrupt and is usually composed of dense evergreen forest with dozens of broad-leaved trees. Single species may form a narrow belt of fringing woodland such as *Polylepis* in eastern Peru and Bolivia, *Hagenia abyssinica* and *Hypericum leucoptychodes* in East Africa, *Philippia longiflora* on Mount Ruwenzori, *Erica arborea* in Ethiopia and *Leptospermum javanicum* in Indonesia. The upper timber line in humid tropical mountains usually climbs higher in the valleys than on exposed ridges, in contrast to the Alps, Rocky Mountains and northwest Himalayas and other boreal mountain systems. This difference is largely accounted for by depth and nature of snow drift and temperature inversions. Thus, in New Zealand the floors of valleys at high altitudes usually support grasslands rather than forest because of temperature inversions (Wardle, 1973).

Arid mountains

High mountains in arid zones are generally islands of humidity in a sea of dryness. Thus there are two timber lines, a lower as well as an upper. At a certain height the upper limit of aridity is reached and forests commence. Forest then occurs until the upper timber line is reached, which results in a girdle of forests around the mountain systems. Arid mountains often occur in the transition zone between humid areas and an arid core. Thus lower forest limits descend to the humid zones on one side of the mountains and rise on the other side in the direction of the arid core. The lower limit of forests often rises more steeply than the upper limit and the girdle forest belt will gradually lessen in width and eventually disappear. Troll (1973a) has illustrated this phenomenon with a transect from the Punjab across the north-west Himalayas to the Karakorams. The lower and upper forest limits on the Pir Punjal range between the Punjab and Kashmir basins on the Nanga Parbat and Rakaposhi massifs are 2700 m and 3800 m, respectively. In the main range of the Hunza Karakoram the altitudes are 3400 m and 3900 m, whereas in the following Luphar Group the forest belts peter out at about 4000 m. The geobotanical character and ecological types of upper timber lines in arid belts differ according to climatic zonation. Girdle forests often coincide with very distinct cloud belts.

Messerli (1973) has examined the factors governing the upper tree line in the Ahaggar and Tibesti Mountains of the Sahara. The upper tree line appears to be either thermally or hygrally conditioned. An upper thermal line is indicated in a transect from the highest northern to highest southern part (Mouskorbe 3376 m to Emi Koussi 3415 m) by an increase of the highest level for acacia from north to south of about 300 m, and precipitation at about 2000 m is five to ten times higher than in the valleys. A hygral level is indicated by the following:

(i) The acacia is a representative of a frost-sensitive tropical savanna vegetation.
(ii) The difference between the east and west flanks of the mountain range on Emi Koussi shows a clear hygral dependency. In open spaces, often below 2000 m, acacias are often stunted. The highest examples are found in gorges and the top level appears to be dependent more on relief than altitude.
(iii) Precipitation does not increase linearly with height. A precipitation maximum occurs 300–500 m below the summit with precipitation decreasing rapidly above this height.

(iv) The high radiation intensity and continual strong winds in this fully arid mountain range cause an increased evaporation, in spite of relatively low temperature, which leads to an extremely low soil-moisture content.

(v) Species comparable to those of the Mediterranean–North Africa areas appear only rarely.

Messerli (1973) concludes that a comparative tree line has ceased to exist for hygral reasons and that there is no theoretical tree line. The tree line cuts out above the fully arid belt in the same way as the snow line and perhaps the solifluction line (Figure 2.5). These results have important implications for geomorphological systems in such arid mountains and there are grounds for suggesting that such mountains are fundamentally different to those in humid areas.

Mountains of the Mediterranean and Middle East

Mountains in the Mediterranean and the Middle East also possess complicated three-dimensional arrangements of vegetation. Three climatic subzones can be recognised:

 (i) Northern sub-Mediterranean subzones with mountains experiencing winter and summer rains, and the forests are of the boreal type with conifers and deciduous trees. The timber line is often composed of beech trees such as *Fagus silvatica* or *orientalis* in northern Spain and the Italian Apennines, sometimes with white fir, *Abies alba.*

 (ii) The fully Mediterranean subzone, with specific Mediterranean trees such as *Quercus tozza* (pyrenaica) in the Spanish Nevada, *Pinus leucodermis* on Mount Olympus (2200–2500 m) and evergreen holly oak (*Quercus ilex*) in the westernmost parts of the Great Atlas of Morocco (2900 m). Similar relationships, but different species, occur over 7000 km east in Afghanistan.

(iii) The southern subbelt, the Mediterranean–Iranian mountain zone, where summer dryness extends higher and species of *Juniperus* are important.

The southern hemisphere

The situation in the southern hemisphere is somewhat simpler as there is a much closer vegetation affinity between diverse mountain ranges. In New Zealand, two evergreen species of *Northofagus* form level, abrupt timber lines, in contrast to irregular timber lines where mixed forest gives way via subalpine scrub to alpine grassland. The altitude of the timber line is related to summer mean temperatures. For *Northofagus* to survive there must be enough time between germination and the end of the first

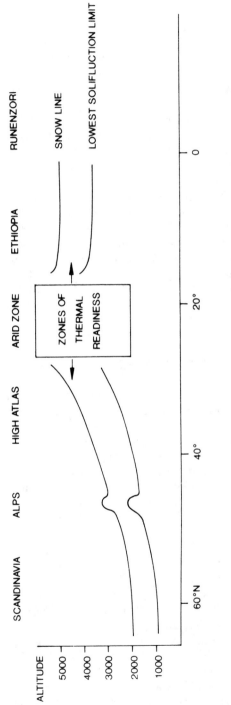

Figure 2.5. Relationships between snow line and lowest solifluction limit through the arid mountains of North Africa (after Messerli, 1973).

growing season for the cycle of shoot growth, bud formation and hardening to be completed. Above the timber line, seedlings do not have sufficient time to harden and they dry out when cold weather begins. Also the inability of beech seedlings to tolerate strong light above the timber line explains the abrupt forest limit.

On coastal mountains in New Zealand at latitude 41°S the tree limit is 1200 m, rising to 1500 m inland. Forest belts tend to be absent from cirques and from areas where there is little or no soil over ultrabasic or hard quartzose rocks. Avalanche and wind-throw also influence the pattern of the timber line and on the drier mountains there has been extensive deforestation by fire since the advent of man about 1000 years ago. There is also evidence that the absence of *Northofagus* is due to its failure to recover from ground lost during the last glaciation. This raises the important issue, stressed by Ives (1978), that current timber lines may not reflect current climates. Some timber lines may be higher than present-day climate would warrant because they are related to previously more optimum climatic conditions. In other instances, the timber line will have been depressed by anthropogenic influences. Thus comparisons of timber lines throughout the world must be viewed carefully, but these problems do not alter the fact that the timber line is an important geomorphological boundary.

ZONATION OF MOUNTAIN SYSTEMS

The strong relationship between geomorphological, pedological and ecological systems within mountains has already been established. This interdependence allows some of these factors to be used to distinguish spatial zonations of interacting systems. Such systems vary in three directions; a change with altitude; a change from north to south; and a change from east to west. Such variation is seen best in the zonation of vegetation patterns. A number of specific examples will demonstrate this.

Tropical Mountains

The vertical zonation of vegetation in the mountains of the humid tropics has been well established (Figure 2.4). The typical sequence is from lowland tropical forest, through submontane, montane, subalpine and eventually to the alpine zone of grasses and shrubs. The constantly warm-temperate frost-free zone, called the tierra templada, corresponds with the lower montane zone. Above this zone is the tierra fría, with cloud forests caused by humidity gathered in clouds formed by upslope winds. The cloud forest is rich in epiphytes and is sometimes called moss forest. The tierra helada zone, beyond the tree line, is the zone of páramos, with tussock-like bunchgrass, woolly rosetted herbs, hard

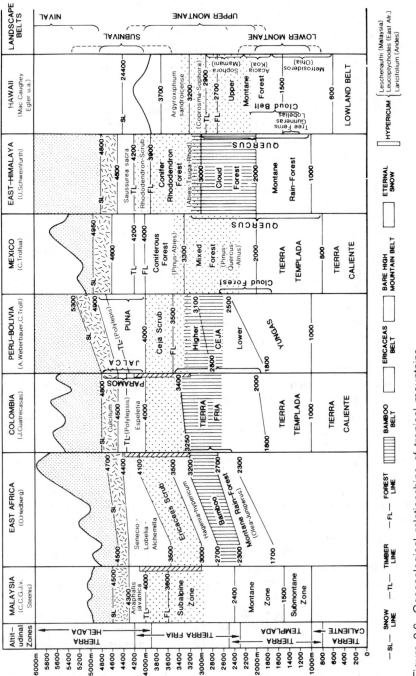

Figure 2.6. Geoecological relationships of tropical mountains.

cushion plants and shrubs with dense foliage of rolled leaves. In the upper part of the tierra helada the grassland gradually disappears into the frost-desert. As Figure 2.6 shows, there are remarkable consistencies and similarities between mountains across the humid tropical zone. This enables very general statements to be made concerning the operation of the geomorphological systems. However, local variations, superimposed on this general pattern, govern the detailed operation of such systems within any one mountain pass or on any one mountain.

Almost every type of tropical mountain environment can be found in Venezuela. Monasterio and Sarmiento (1984) have shown how the elements of vegetation zonation and the precise altitudinal limits between them vary because of rainfall amounts and its annual distribution (Figure 2.7). On the driest slopes facing the Chama valley, the vegetation sequence commences with a cactus shrub and ends with a dry type of páramo that appears at 2500 m. In the wetter areas of the same inner part of the valley the majority of the slopes are covered by cloud forests that rise to over 3300 m. The cloud forest is followed by a wet type of páramo, replaced at greater heights by the Desert Páramo, the periglacial desert and above 4700 m by the nival zone.

The bedrock of the slopes is often overlooked in the type of geo-ecological zonations that have been examined in this chapter. But the role of geology in determining the overall landscape form was stressed in the previous chapter. In the Venezuelan Andes, geology determines topography and soils. Metamorphic rocks, mostly gneisses and schists, produce massive relief; hard sedimentary rocks, such as conglomerates and sandstones, produce a contrasted relief of crests and monoclines; whereas softer sedimentary rocks, such as shales and limestones, produce a gentler topography where deeper soils develop.

In Bolivia, it is possible to assign plant communities to a three-dimensional temperature and humidity model based on the vegetation zonation. This zonation will be modified by local topoclimatic conditions (Figure 2.8). In the Charazani Valley, the northern and southern slopes differ greatly in their thermal budget and moisture conditions (Lauer 1984a). The drier and warmer side of the valley between 3000 m and 3900 m is characterised by an association of shrubs, with the main species being *Satureja boliviana*. The opposite, cooler and more humid slopes at the same altitude possess *Baccharis pentlandii* associations. On the drier side of the valley at lower elevations *Mutisia* shrubs are dominant with cacti and dry grass of *Vulpia* sp. in the driest areas. These slopes are exposed to an upvalley wind during the day. On the strongly sunlit slopes at medium elevation *Calceolaria parviflora* occurs. It is only at higher elevations where the upslope winds condense to form fog banks, usually above 3600 m, that Satureja dominates the vegetation associations. It is usually in association with *Chuquiraga*, characteristic of wetter areas.

On the opposite side of the valley, which is in shadow during most of

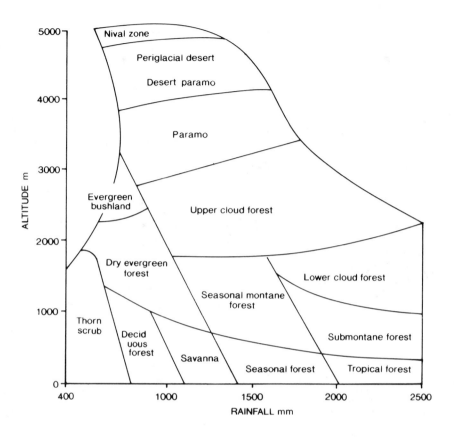

Figure 2.7. Vegetational zonation in the Chama valley, Venezuela.
Source: Monasterio and Sarmiento (1984)

the winter, in the *Baccharis pentlandii* zone, the steep slopes are grass-covered. *Eryngium* societies associated with *Puya* are found scattered in this zone, with *Puya* especially on rocky sites. In the humid upper valley portions, remnants of *Polylepsis* probably represent a mesophytic bush-type forest line at an elevation between 3400 m and 3600 m.

The entire bush belt, on both valley sides, demonstrates a dependence on decreasing temperature with increasing altitude as well as a dependence on moisture. Above 3900 m the bush belts of both slopes merge into the grass belt with *Aciachne pulvinata* dominant.

Similar relationships exist in the mountains of Central America and Mexico (Hastenrath, 1966; Lauer, 1973). The vegetation of the tierra caliente offers major contrasts between the wetter areas, such as the north coast of Honduras or north-western Nicaragua, and the more arid parts of south-eastern Guatemala, north-western El Salvador, the interior

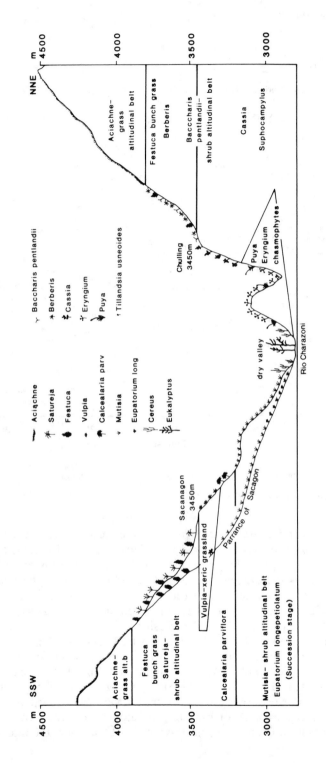

Figure 2.8. Vegetation zonation in the Charazani Valley of Bolivia.

Source: Lauer, 1984a.

basins of Honduras and the zone from eastern El Salvador through southern Honduras into northwestern Nicaragua. Within the tierra caliente zone there is a lack of vertical variation of vegetation which might be traced over large areas.

At higher elevations the vegetation of the tierra caliente passes into the oak-pine forest of the tierra templada. Oaks tend to occur at lower elevations and there is a gradual transition at higher levels to almost pure pine forest. This character is found from Nicaragua through El Salvador and Honduras to Guatemala and also in Mexico. In many places the oak belt is affected by human activity. The upper tree line in the Cordillera de Talamanca of Costa Rica occurs at about 3100 m, formed of *Quercus costaricensis* but the tree line is not reached in Nicaragua, Honduras or El Salvador. Tree limits on recent volcanoes have edaphic origins. Volcanoes near Guatemala City possess pine forests up to their peaks, except for recently active volcanoes such as Mounts Fuego and Acatenango. The upper tree line on Mount Tajumulco (4210 m) in western Guatemala appears to be 3800–3900 m and on the Mexican Meseta it is about 4000 m.

Western Cordillerra of North America

It is difficult to establish vegetation patterns for a large mountain chain such as the Western Cordillerra which extends from arid zones in the extreme south and west to northern coastal mountains clothed in dense forests. Two areas that have been examined in some detail are the White Mountains and Basin and Range areas of California and the Front Range of the Rocky Mountains in Colorado. Basin and range topography consists of subparallel mountain ranges, with piedmont plains and debris-filled basins. The southern and southeastern sections are arid all year round. In the alpine environment of the highest mountains, aridity decreases to three months a year with nival conditions lasting for up to eight months (Hollerman, 1973b). The geoecological pattern in Death Valley has been established by Hunt (1966). In general four zones of vegetation can be recognised. A creosote bush zone up to about 1200 m is followed by shadscale semidesert and desert up to 2000 m, then pinyon woodland followed by subalpine forest generally above 3000 m.

A similar sequence occurs in the White Mountains of California. The lower limit of the pinyon is the so-called 'dry timber line'. The local height of the lower timber line is controlled by factors such as cold air settling and bedrock. In the western Panamint Range woodland is common on dolomite bedrock but is sparse or absent on shale or quartzite. The subalpine belt extends from 2900–3050 m to 3400–3500 m. The typical subalpine community is an open forest or woodland of bristlecone pine (*Pinus aristata*) and limber pine (*Pinus flexilis*). Ground cover of trees rarely exceeds 16 per cent. The pattern of bristlecone forest and treeless shrub appears to be controlled by slope exposure and bedrock. The bristlecone forest prefers the north- and west-facing slopes, while sagebrush cover is maximum on south- and east-facing slopes.

At an altitude of 3090 m on White Mountain, below freezing average temperatures exist for six months with 210 frost-change days and 75 ice days. A high proportion of the ground within the subalpine belt is covered by angular rock fragments. Chemical alteration is minor and the clay content of soils is low (LaMarche, 1968). A considerable proportion of the sand and silt fraction is aeolian or volcanic ash (Marchland, 1970). Only the summit uplands of the White Mountains exceed the upper timber line. The average air temperature is below freezing for at least eight months, nivation hollows and stone pavements occur and some snowdrifts may persist through the summer.

A considerable amount of work has been conducted in a number of ares of the central and northern parts of the Western Cordillerra. One of these areas is the Indian Peaks region, Colorado, where a series of studies has been undertaken by the Institute of Arctic and Alpine Research, Boulder, Colorado. The results of some of these studies are examined in Chapter 9. Vertical zonation of vegetation has been well established and also shown to influence geomorphological systems. Orders of magnitude differences in surficial process rates were found when the data were stratified by plant community (Frank and Thorn 1985). The alpine vegetation has been mapped (Komarkova and Webber, 1978) and the geoecology admirably summarised by Ives (1980).

The Himalayas

The zonation of ecosystems in the Himalayas follows the principles established in previous examples but on a grander scale. The Himalayan mountain system extends for about 3000 km from north-west to south-east and its climatic and vegetation belts can be explained by transitions from south to north, from east to west and from lowland to mountain top. This reinforces the concept of 'central–peripheral', 'planetarium' and 'hypsometric' change. The southern slopes and ranges are exposed to the rain-bearing winds and are densely forested. Forests change on the Tibetan side of the mountains to cold high-mountain steppe lands. In the same direction the alpine tree line rises from 3400–3800 m in the outer southern ranges to 4400–4600 m in Tibet (Troll 1972b). The zone of perpetual snow begins at 4500–4600 m in the south and at a maximum of 6400 m in Tibet. It is only just over thirty years since the first reasonably detailed vegetation map of the Himalaya was published (Schweinfurth, 1957; 1981). Before the publication of this map, the 'Himalaya were a little known maze of ranges, gorges, peaks and valleys with a barely discernible topographic order' (Schweinfurth, 1984, p. 339).

Schweinfurth (1984) has made the distinction, in a north–south transition, between Outer, Inner and Tibetan Himalaya. The Inner Himalaya character is imposed by local conditions of topography and climate. Thus Bhutan with its deep valleys is Inner Himalayan in character. In the north-west, the moist coniferous forest is well developed around the valley of Kashmir and is also Inner Himalayan in character. Sikkim displays this differentiation well. Lower Sikkim represents the wet part that is Outer Himalayan; Upper Sikkim, with the high valleys of Lachen and Lachung, is Inner Himalayan; and the high valley of Lhonak is Tibetan Himalayan.

In a south-east to north-west direction, in the foothills zone, there is a transition from near-tropical rainforest in Assam to semi-desert conditions in parts of the north-west. Thus, in Assam and Bengal as far as Bhutan, the lowlands are covered with tropical evergreen rainforests. This gradually changes westwards in Sikkim and Nepal to tropical humid deciduous forests (wet Sal forests of *Shorea robusta*). From west Nepal to the Sutlej a drier, semi-humid type of deciduous Sal forest occurs. The Punjab is much drier and its natural vegetation is a subtropical semi-arid thorn steppe and in the valleys of the extreme north-west towards the Hindu Kush and the Karakoram semi-desert conditions give way above 2000 m to central Asiatic *Artemesia* steppe.

Troll (1972b) has proposed a subdivision of the Himalayas into seven natural sections:

(1) The Tsangpo Himalayas, the mountain ranges and valleys of southeastern Tibet.
(2) The Assam Himalayas.
(3) The Sikkim Himalayas, from Bhutan through Sikkim to eastern and central Nepal.
(4) The Garhwal Himalaya, from western Nepal through Kumuon and Garhwal to the Sutlej.
(5) The Punjab Himalayas, from the Sutlej to the Indus.
(6) The Indus Himalaya.
(7) The Tibetan Himalaya.

In all these regions there is a distinct vertical gradation of climate and vegetation (Figure 2.9). In the east, tropical rainforest is followed by evergreen lower montane forest, evergreen upper montane forest, subalpine *Rhododendron* forest and at higher levels by wet alpine scrub and meadows. In the western parts of the central Himalayas, the tropical deciduous forest gives way to *Pinus roxburghii* forest, followed by mixed oak and coniferous forest, subalpine forest of *Betula utilis* and moist alpine scrub and meadows. In the drier north-west the sequence ranges from thorn steppe along the foothills, into evergreen sclerophyllous forest, *Pinus roxburghii* forest, mixed oak and coniferous forest, subalpine forest of *Betula utilis* and moist alpine scrub and meadows. As conditions become even drier the sequence ranges from the semi-desert of the valleys, through *Artemisia* steppe and steppe forest of *Pinus*

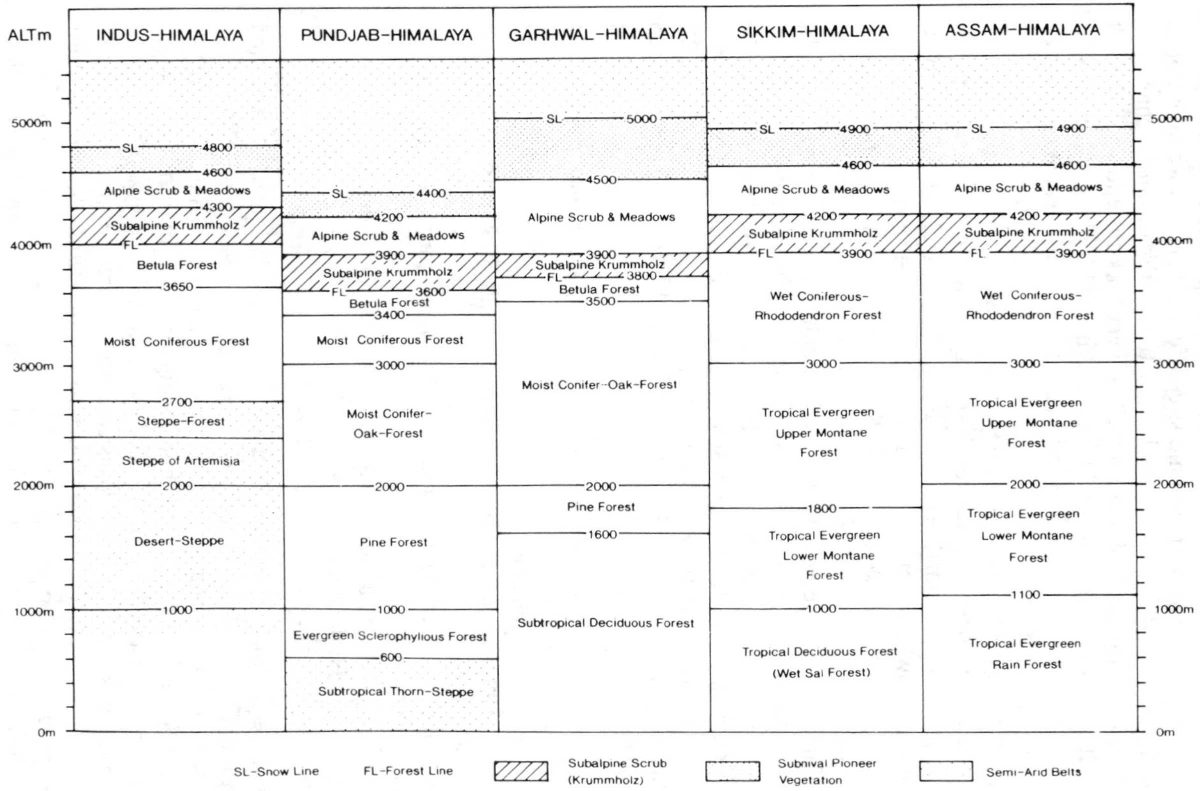

Figure 2.9. Vegetation zonation in different parts of the Himalayas.
Source: Troll, 1972b.

Table 2.5. *Radiation balance in different vegetation zones of the Caucasus Mountains.*

High-mountain belt	Radiation balance (kcal cm^{-2} yr^{-1})	Evaporation (mm)	Radiation index of dryness
Steppe	55–50	300	2
Broad-leaved forests	50–45	400–700	0.9–1.2
Dark coniferous forests	45–40	400–700	0.9–1.2
Subalpine meadows	40–30	400–250	0.8–0.6
Alpine meadows	30–20	250–150	0.5

Source: Zimina *et al.* (1973)

Table 2.6. *Biomass relationships in the Caucasus Mountains.*

Characteristic of the phytomass	Horn beam-oak forests	Beech forests	Dark coniferous forests	Subalpine meadows	Alpine meadows
Total biomass (c ha^{-1})	6,400	7,900	6,510	820	630
Structure (%)					
Production of green mass	9.7	0.8	0.6	19.5	9.3
Perennial mass above the ground	82.0	85.0	94.1	0.1	0.5
Underground mass	5.7	13.0	4.3	78.1	88.1
Perennial litter and detritus	2.6	1.2	1.0	2.4	2.1

Source: Zimina *et al.* (1973)

gerandiana and *Quercus baloot* into the western Himalayan forest, subalpine forest of *Betula utilis* and moist alpine scrub and meadows. In the far north-west there is no moist vegetation type and dry vegetation types of the semi-desert valley floors lead into *Artemisia* steppe and into the alpine steppe of the Tibetan plateau.

A number of interesting variations to this general pattern occur, created by local topography and geomorphological processes. Local wind systems develop in the larger and deeper valleys, resulting in a dry climate and dry vegetation on the valley floor and lower slopes, in contrast to wet vegetation at higher elevations in the belt of cloud forest. There is also a contrast between north- and south-facing slopes. Slopes prone to frequent snow avalanches have their vegetation patterns altered by avalanche tracks. Conifers are replaced by alders, birches and willows.

Caucasus Mountains

The mountains of the Caucasus range exhibit similar characteristics (Isakov *et al.*, 1972; Zimina *et al.*, 1973). Detailed studies have established relationships between radiation balance, evaporation, phytomass and zoomass (Tables 2.5 and 2.6). Steppe vegetation is associated with the plains and piedmont areas of the northern and eastern Caucasus with less than 400 mm of precipitation. The next zones include a variety of broadleaved trees with the lowest mountain belt dominated by hornbeam-oak forests. The warm season lasts for seven or eight months and there is a combination of radiation heat and atmospheric moisture optimum for development of vegetation. Total phytomass of up to 6400 c ha^{-1} yields a high annual production of green mass. The middle forest belt is dominated by beech forests, total phytomass is large but green mass productivity is low. The top forest belt is occupied by coniferous trees. The warm season lasts only about four months but there is abundant rain. Phytomass is large but is broken down very slowly. The warm season lasts only three or three-and-a-half months in the belt of subalpine meadows but temperatures are quite high. In the alpine meadow zone, the warm season is short (between six and eight weeks), the summer temperature of air and soil is low, and the radiation balance is much lower than in any of the other mountain zones.

Japan and Taiwan

The vertical vegetation zones on Japanese mountains are usually called the piedmont or hilly zone, the montane zone, subalpine and alpine zones (Table 2.7). Zones correspond in general terms with the warm temperate evergreen broad-leaved forest, the cool temperate deciduous broad-leaved forest, the subarctic evergreen needle-leaved forest, and the arctic evergreen needle-leaved scrub zones (Numata, 1972). As in other mountain systems, there are local variations superimposed on these patterns. This is especially true of the volcanic mountains. The mountain flora of the high volcanoes is generally immature and there is an incomplete vegetation zonation. Volcanic processes, such as solfatara activity, also have a great effect on vegetation type.

Vertical vegetation zonation in Taiwan is partly similar to that of Japan and partly to that of the central and eastern Himalayas (Table 2.8). The piedmont zone is the laurel-leaved forest zone consisting of the lower Shiion zone and the upper Cyclobalanopsion zone. This is similar to the Japanese hilly zone. The montane zone, rather than being deciduous broad-leaved forests as in Japan, is an evergreen needle-leaved zone dominated by *Chamaecyparion taiwanensis*. The subalpine zone, up to the forest limit, is covered by evergreen needle-leaved forests of *Abies*

Table 2.7. *Vegetation zonation in Japanese mountains.*

Zone		Physiognomy	Association or alliance
Alpine	upper	meadow, desert	Arcterion-Loiseleurietum Dicentro-Violetum crassae
	lower	evergreen needle-leaved scrub	Vaccinio-Pinetum pumilae
Subalpine	upper	deciduous broad-leaved scrub	Alno-Betuletum ermanii Nanoquercetum
	lower	evergreen needle-leaved forest	Abietum mariesii Tsugetum diversifoliae
Montane	upper	deciduous broad-leaved forest	Fagion crenatae
	lower	dec. br.-leaved or evergreen nd.-leaved forest	Quercetum crispulae Tsugion sieboldii
Hilly	upper	dec. br.-leaved, evergr. nd.-leaved or evergr. br.-leaved forest	Illicio-Abietum firmae Cyclobalanospion Carpinetum japonicae
	lower	evergreen broad-leaved forest	Shiion sieboldii Machilion thunbergii

Source: Numata (1972)

Table 2.8. *Zonation of vegetation in the mountains of Taiwan.*

Zone		Physiognomy	Association or alliance
Alpine	upper	alpine desert	*Epilobium nankotassanensis-Picris ahwiana* alliance, *Festuca orina* alliance, *Brachypodium kawakamii* alliance
	lower	evergreen scrub	Rhododendreto-Juniperetum squamatae
Subalpine		evergreen needle-leaved forest	Abietum kawakamii
Montane		evergreen needle-leaved forest	Chamaecyparion taiwanensis
Hilly	upper	evergreen broad-leaved forest	Cyclobanopsion paucidentatae
	lower	evergreen broad-leaved forest	Shiion stipitatae

Source: Numata (1972)

kawakamii. The lower alpine zone is occupied by the shrubs of *Juniperus squamata* and *Rhododendron pseudochrysanthum*. This continues to the upper alpine desert zone. Thus the piedmont zone is similar to that of Japan whereas the higher zones are similar to those in the humid Himalayas.

CONCLUSIONS

This chapter has tried to demonstrate the interdependence of the processes and landforms that are characteristic of mountain environments. It has also shown that these relationships create recognisable zones or spatial systems that can be recognised on mountains world-wide. It is now necessary to examine some of these processes and landforms in greater detail. In doing so it is necessary to treat many of these processes and landforms separately but as far as possible the linkages between the various system elements are stressed. If these linkages are not stressed explicitly, it is nevertheless assumed implicitly that they occur.

Chapter 3

WEATHERING AND MASS MOVEMENT

Mountains are extreme, high-energy geomorphological systems. Physical weathering is intense but chemical weathering is far from inactive. Snow patch weathering and erosion achieve their greatest potency in high mountains. Major changes of climate create the possibility that weathering residues of former climates still exist on many of the mountain slopes. Steep slopes and high relative relief also ensure that mass movement has the potential for tremendous landscape modification as well as being a major hazard. But for movement to occur the rock must be weakened by weathering.

WEATHERING

It is customary to divide weathering into physical weathering (disintegration), the mechanical breakdown of rock without any contributory chemical alteration; chemical weathering (decomposition) which often involves irreversible chemical change; and biological weathering, which is effected by the growth or movement of plants and animals. Although the various processes can be differentiated in this way, the various weathering processes often act together. Biological weathering, in particular, rarely acts alone and can be subsumed under the other types. In high mountains most attention has been focused on physical weathering, especially frost and ice action, but the effects of chemical weathering must not be underestimated.

Physical weathering

Physical weathering is concerned with processes that lead to the brittle fracturing of the rock. Failure occurs when stresses, between and within grains of the rock, created by the weathering agencies, exceed the strength of the rock. The main activating agencies are unloading, thermal processes (thermoclasty), and the group of processes that involve the growth and expansion of material in the cracks and pores of the rock, namely ice (gelifraction) and various salts (haloclasty).

Unloading is an extremely important process in shaping the relief of high mountains, especially those mountains which have been heavily glaciated and are also currently tectonically active. Rock masses that have been deeply buried or been subjected to tectonic stresses acquire a considerable amount of 'locked-in' strain energy. Some of this strain will be released when confining pressures are reduced and will result in the production of joints or cracks in the rock. If the release of pressure is fast enough catastrophic failure of large rock masses can result.

The possibility of unloading being initiated by erosion has been suggested by many workers. The retreat of glaciers from deeply incised valleys has permitted the fracturing and uparching of rock on valley sides and floors (Lewis, 1954; Harland, 1957; Gage, 1966). This phenomena is treated in greater detail in Chapter 6. Most unloading phenomena have been reported in igneous rocks, especially granite, but unloading has also been reported on other rock types.

There is little doubt that considerable amounts of strain energy are locked up in rocks as a result of tectonic forces. There is also the possibility that pressure created by the movement of lithospheric plates will be transmitted to and locked in the rocks. However, the mechanisms of simple unloading have been questioned (Twidale, 1972; 1973; 1982). Twidale argues that expansive stress during unloading will be taken up by existing joints or by grain-boundary sliding. However, if some of the rock is in a state of compression then there will be few open joints to take up the release of pressure. Horizontal stresses in near-surface rocks are generally greater than vertical stresses (Hast, 1967; Ranalli, 1975) and rock at the base of valley slopes may be under the greatest compression (Sturgl and Scheidegger, 1967). The hypothesis that offers the best explanation is that involving lateral compression, induced by horizontal stresses and the creation of stress patterns near the surface when vertical loading is decreased by erosion. Whatever the mechanism, seismotectonic and unloading phenomena are responsible for mechanically destabilising rock masses and producing material which is unsound, jointed and microfissured (Bonnefond, 1977; Bousquet and Pechoux, 1977; 1980; Coltorti *et al.*, 1983).

Frost weathering is an extremely important process in mountains and is largely responsible for the production of coarse angular material which often accumulates in vast scree and fan deposits. Freeze-thaw activity in soils and sediments is also responsible for patterned ground, solifluction and other mass movement processes. But there is still considerable uncertainty concerning its action. This uncertainty is caused by confusion over its relationship with climatic factors and by the paucity of detailed data on the climatic regimes of high mountains. Some information is available from freeze-thaw weathering studies in Spitsbergen (Rapp, 1960b), Norway (Hall, 1980; McGreevy and Whalley, 1982), the Alps (Francou, 1982), central Andes (Francou, 1984), the Rocky Mountains (Thorn, 1980), and the Karakorams (Whalley *et al.*, 1984). Although attention is rightly focused on alpine mountains, frost weathering should not be ignored as a process on tropical mountains. The higher East African mountains experience a number of frost cycles each year and frost is not uncommon on the high mountains of Papua New Guinea (Brown and Powell, 1974; Loffler, 1977).

Several different kinds of frost weathering can be distinguished. Frost scaling is due to the formation of thin ice plates parallel to the rock surface. Frost splitting of previously non-cracked rock is a rock-bursting process, as distinct from frost wedging, which exploits pre-existing cracks and joints. Little is known about frost bursting but it may relate to the expansive action of rigid absorbed non-freezable water (McGreevy, 1981; Fahey, 1983; Fahey and Dagesse, 1984; Mugridge and Harvey, 1983).

Water freezing at $0°C$ in a closed system increases in volume by about 9 per cent and a maximum pressure of 2115 kgf cm^{-2} is possible at $-22°C$. Such pressure is more than capable of fracturing even the strongest rocks but will only occur in a completely closed system. An additional force will also be present, that related to the growth of ice crystals. When ice crystals form in large pores, water is withdrawn from smaller pores leading to a build-up of pressure as the crystal grows. Discussion of maximum likely pressures may be irrelevant because there is a considerable body of information to show that it is not necessarily the intensity of cold that is important but the number of times that the temperatures crosses the $0°C$ threshold (McGreevy and Whalley, 1982), although there is some confusion over the importance of short-term cycle frequency (Dahl, 1966a; 1966b; Ives, 1966). Fahey (1973) has charted 238 diurnal freeze-thaw cycles over a twenty-two-month period at 2600 m in the Colorado Front Range. The number of cycles declined to 89 at 3750 m. Frequent temperature shifts across freezing point may also be especially significant in tropical and subtropical high mountains (Hewitt, 1968).

Hewitt (1968) has attempted to differentiate the important phases of the freeze-thaw cycle. He makes the initial distinction between frost shifts, which merely represent crossings of the $0°C$ boundary, and a freeze-thaw cycle which involves three crossings of the boundary, at the beginning, middle and end of a cycle. He argues that analysis of freeze-thaw cycles requires consideration of:

(a) absolute or mean number of frost shifts in given periods;
(b) frequency of cycles, comprising
    (i) wavelength (e.g. daily, annual), and

(ii) recurrence intervals or density of cycles;
(c) intensity of freeze-thaw, depending on
    (i) amplitude of temperature change, and
    (ii) slope of temperature curve;
(d) scale relations of freezing and thawing phases; and
(e) the problem of what constitutes the 'effective' temperature shift for
    a given process and the best measure of it.

Terminology for cycle length usually follows that used by Rapp (1960b). Thus, there are short freeze-thaw cycles with four or more shifts in twenty-four hours; diurnal freeze-thaw cycles approximately twenty-four hours long; several-day freeze-thaw cycles with one melt and one freeze-thaw cycle in more than thirty-six hours; and monthly and annual freeze-thaw cycles.

These factors are important because they will govern the nature and intensity of the resulting processes. Thus the number of frost shifts is a measure of variability of the environment and the energy exchange involved in the freezing and melting of the water. The length of the cycle allows comparison between environments and with other phenomena that operate cyclically, such as precipitation. The rate of temperature change affects the intensity of the process. In processes such as frost bursting, the larger the temperature change the more intense the process. The efficiency of other processes such as frost heaving in soil may depend on a slower rate of temperature change to allow soil water to migrate to the freezing front.

Hewitt (1968) has used this terminology to examine the freeze-thaw environment of the Karakoram Mountains. At a height of 10200 ft in the Braldu valley frequent and quite large diurnal frost cycles dominate the records. Prolonged periods of 'ice days' are also likely to occur. Valley station records reveal the incidence of frost cycles up to 11 000 ft. At 4000–5000 ft (Gilgit), the freeze-thaw zone is just beginning with daily and several day cycles in mid-winter. About 7000–8000 ft (Skardu) seems to be the limit below which ice days are uncommon and frequent daily frost shifts occur from November to February inclusive. The zone at 10000–11000 ft is one with between one and four months of ice days. At high altitudes (above 17 000 ft) thaw days are rare. This demonstrates the strong seasonality in temperature environment at all altitudes.

Hewitt's work demonstrates the need to use realistic temperature data in experimental studies of frost shattering. This can be illustrated by considering some of the work which has been conducted on frost action at or below rock surfaces around snow patches. Data have been provided by Thorn (1979a; 1980) and Thorn and Hall (1980) which seems to indicate that temperatures of $-5°C$ are necessary for freezing to occur. Similar temperatures have been used by Fakuda (1971; 1972), and Lautridou (1971) employed a temperature of $-5°C$ over a ten-hour period for his experiments. Dunn and Hudec (1966) noted that in rocks in which water actually froze, freezing occurred suddenly between $-2°C$ and $-7°C$. But, as discussed earlier, the effectiveness of frost action will also depend on the rate and duration of freezing and the frequency of frost cycles. Battle (1960) proposed that the freezing rate must equal or exceed $0.1°C$ min^{-1}, a rate which McGreevy (1981) has suggested is unlikely to occur. The freezing process will also be affected by the rock type. Tourenq (1970) found that in rocks possessing a large percentage of pores with radii greater than 0.03 μm, freezing was initiated at temperatures between $-2°C$ and $-4°C$ and that freezing occurred between $-10°C$ and $-20°C$ in rocks in which all the pores had radii less than 0.007 μm.

Although there can be no doubt that frost weathering is an important process in mountains this brief review has demonstrated a number of uncertainties concerning the factors governing its action. Thus, one must concur with Lautridou (1988) that the current understanding of frost cracking is limited and requires the support of both physical studies and theoretical modelling.

Insolation weathering (thermoclasty) must also be considered a likely process in high mountains with clear skies and large diurnal radiation fluxes. In the Hunza Valley of the Karakorams rock temperatures can exceed air temperatures by as much as $24°C$ with actual rock temperatures around $40°C$ even at 4250 m (Whalley *et al.*, 1984). The process is less significant in tropical mountains where vegetation cover prevents excessive temperature changes.

Salt weathering (haloclasty) is also likely to be an important process in mountains, and has been noted in most high mountains such as the Karakorams (Hewitt, 1968; Goudie, 1984). Weathering by salt crystals involves the thermal expansion of salt crystals, the hydration of salts and the growth of salt crystals. Hydration forces created by anhydrous salts may approach those of frost action. There is also the possibility that interaction between salt and frost weathering may be important but the results of research are contradictory. McGreevy (1982) and Litvan (1980) consider that frost is more efficient with low concentrations of salts in water, whereas Williams and Robinson (1981) believe in greater efficiency with more concentrated solutions. Clearly more work is required using standardised methods before this problem can be resolved.

Chemical weathering

Chemical weathering is extremely important in tropical mountains but must also not be thought insignificant in alpine mountains. Rapp (1960a) has demonstrated that solution was one of the most important processes of removal in the Karkevagge Mountains and solute studies of glacial meltwater streams have tended to substantiate this (see Chapter 6). Between $2°C$ and $15°C$ chemical weathering processes show little variation in intensity and this, coupled with the fact that snow and its meltwaters are often quite acidic (Williams, 1949; Clément and Vaudour, 1968) is sufficient to demonstrate that chemical weathering is an important process in alpine mountains. Whalley *et al.* (1984) have shown that surface temperatures at 4000 m in the Karakorams may easily reach 30°C under sunny conditions and 20°C under cloud. Such temperatures, combined with moisture, are very favourable to chemical weathering. Snow patches are often sites of increased chemical activity. Hall (1975) has demonstrated the importance of chemical weathering at a nivation site in Norway and Thorn (1975) noted a two- to four-fold increase in weathering rind thickness at snow patch sites in Colorado. Significant increases in the solute load of meltwaters emanating from snow packs have also been noted (Thorn, 1976). In temperate alpine environments chemical weathering processes may be intense (Reynolds, 1971) and the weathering processes are similar to those in other temperate environments (Dixon *et al.*, 1984).

Detailed studies in the Marmot Creek Experimental Watershed (2113 m) on the east slopes of the Rocky Mountains in Alberta have shown yields of geochemical components in the range 12.4–409.2 t $km^2 yr^{-1}$ (Singh and Kalra, 1984). The main constituents were CA^{++}, Mg^{++}, HCO_3^{-1} with appreciable amounts of NA^+, SO_4^{--} and SiO_2. These results were different from those obtained by Lewis and Grant (1979; 1980) from Como Creek Watershed near the Continental Divide in Colorado, largely because of differences in soil and parent material. This stresses the danger of attempting to generalise on the basis of very little information.

Although chemical weathering is important in alpine mountains, the production of a chemically weathered profile is rare. This is in contrast to many tropical mountains, where variable thicknesses of weathered material occur. The greatest thicknesses occur on broad crests and relict surfaces. Active mass movement on steep slopes keeps the weathered mantle thin. Haantjens and Bleeker (1970) have observed an altitudinal zonation of weathering on the mountains of Papua New Guinea with the proportion of mature weathering decreasing with increasing altitude. Mature weathering was rare above 1800 m and virtually absent above 3000 m. Where inconsistencies occur they are usually related to relict surfaces. The survival of deep weathering profiles at high altitudes is favoured in Papua New Guinea by an absence of slope wash.

Weathering profiles on extra-tropical mountains, such as the 5 metres of weathered dolerite described by Caine (1983) on Mount Barrow and Ben Lomond, Tasmania, can usually be explained by climatic change. The weathered horizons analysed by Caine (1983) were dominated by kaolinite and have been ascribed to the Tertiary Period, when temperatures up to 12°C higher than today were experienced (Kvasov and Verbitsky, 1981).

WEATHERING-RELATED LANDFORMS AND PROCESSES

Mountain permafrost

Permafrost, of one form or another, is a conspicuous feature of high mountains in middle and low latitudes. This is generally known as 'alpine permafrost' and is somewhat different from permafrost in mountain areas in high latitudes, which is regarded as the same as polar permafrost. The main areas of alpine permafrost are in the European Alps, Scandinavia, Iceland, the North American Cordillerra and south Alaska, the Caucasus, Sikhola Alin Range, the Tibetan Plateau, Hindu Kush, Pamirs and Tien Shan. Permafrost also occurs at 4140 m on Mauna Kea, Hawaii (Woodcock, 1974), on the summits of Mount Taisetsu and Mount Fuji, Japan (Fujii and Higuchi, 1972) and at high elevations in the Andes (Catalano, 1972; Corte, 1978). It has also been noted near the summit of Citlatepetl in Mexico (Lorenzo, 1969) and it presumably occurs on the highest peaks in New Zealand. The permafrost on Mauna Kea, Hawaii, is interesting because the mean air temperature appears to be well above freezing. However the permafrost is at least 10 m thick and its survival is associated with local trapping of radiationally cooled air, a low angle of incidence of sunlight and a very dry atmosphere.

A detailed examination of its distribution in western North America demonstrates the ways its lower limit is related to latitude (see Table 3.1 and Figure 3.1). There have also been some reports from the eastern United States. In New Hampshire it has been noted on Mount Washington, which has a mean annual air temperature of $-2.8°C$ (Antevs, 1932; Schafer and Hartshorn, 1965; Howe, 1971; Ives, 1974). It may also occur on Mount Katahdin, Maine, and on other New England summits.

The relatively simple pattern noted above obscures a great amount of local variation in response to local climatic influences. Also, Harris and Brown (1982) have argued that the lower limit of permafrost does not steadily decrease with latitude but undulates (Figure 3.2). There is a major contrast between the two sides of mountain ranges which run at right angles to the prevailing movement of air masses. This is most strikingly seen in western North America, where temperature gradients and high snowfall amounts on the west of the divide prevent permafrost occurrence in south-west British Columbia, except perhaps on the very highest mountains (Mathews, 1955), whereas it is continuous at lower levels on the Rocky Mountains east of the continental divide (Harris and Brown, 1978; 1982; Harris, 1986).

Different classifications of mountain permafrost have been proposed. Many workers differentiate between continuous, discontinuous and sporadic permafrost. However, Gorbunov (1978) has used the terms 'perennial', 'seasonal' and 'short-term permafrost'. Cheng (Cheng and Wang, 1982) has differentiated unstable from stable permafrost zones in China, based on the relationship between mean annual ground temperature

Table 3.1. Information on permafrost in North American mountains.

Map location	Locality	Altitude (m)	N Latitude	Source	Comments
1	Cassier, British Columbia	1370	59°17'	Brown (1969: 28)	Mining excavations
2	Jasper National Park, Alberta	2150	52°50'	Luckman and Crockett (1978: 545)	Ice-cemented rock glaciers
3	Banff, Alberta	2347	51°50'	Oglivie and Baptie (1967: 744)	Frozen peat reported as permafrost
4	Banff, Alberta	2655	51°50'	Scotter (1975: 93)	Ice in the ground, reported as permafrost
5	80 km SW of Calgary	2224	51°00'	Harris and Brown (1978: 387)	Ground temperature measurements
6	Beartooth Mts, Wyoming	3230	45°00'	Johnson and Billings 1962: 121)	Bogs with ice, reported as permafrost
7	Beartooth Mts, Wyoming	2950	44°53'	Pierce (1961: 155)	Ice in peat, reported as permafrost
8	Yellowstone National Park, Wyoming	2600	44°50'	Pierce (1979: 7)	Ice in peat, reported as permafrost
9	Yellowstone National Park, Wyoming	2400	44°50'	Pierce (1979: 7)	Ice in peat, reported as permafrost
10	Absaroka Mountains, Wyoming	2700	44°38'	Potter (1972)	Ice-cemented rock glacier
11	Niwot Ridge, Colorado	3500	40°04'	Ives and Fahey (1971)	Ground temperature measurements
		3500	40°04'	Ives (1974: 187)	Ground temperature measurements
12	Colorado Front Range	3500	40°05'	White (1976: 83)	Ice-cemented rock glaciers
13	McClellan Mountains, Colorado Front Range	4000	39°50'	Weiser (1875: 77)	Permanent ice in mine in bedrock
14	Sangre de Cristo Range, Colorado	2800	37°40'	Johnson (1967: 218)	Ice-cemented rock stream
15	San Juan Mountains, Colorado	3850	37°30'	Hyers (1980: 102)	Temperature extrapolation. Mean annual air temperature −2.2°C at wind-swept crest

Table 3.1. contd.

Map location	Locality	Altitude (m)	N Latitude	Source	Comments
16	San Juan Mountains, Colorado	3500	37°30'	Howe (1909: 33)	Ice-cemented rock glaciers
	San Juan Mountains, Colorado	3500	37°30'	Spencer (1900)	Ice-cemented rock glaciers
	San Juan Mountains, Colorado	3500	37°30'	White (oral comm., 1981)	Ice-cemented rock glaciers
	San Juan Mountains, Colorado	3500	37°30'	Brown (1925: 465)	Ice-cemented rock glaciers and permanent ice in mine in bedrock
17	Tesuque Peak, New Mexico	3720	35°47'	Retzer (1965: 38, 39)	Alpine turf, frozen; reported as permafrost
18	White Mountains, Arizona	3475	33°50'	Merrit and Pewe (1977: 41)	Temperature extrapolation. Mean annual air temperature −1.1°C at wind-swept crest
19	Istaccihuatl, Mexico	4700	19°30'	Lorenzo (1969: 168)	Active block fields
20	Volcanic Mountains, central Mexico	4800	19°05'	Heine (1977: 170)	Near modern alpine snowline
21	Volcanic Mountains, central Mexico	4500	19°05'	Gorbunov (1978: 284)	Frozen ground reported as permafrost
22	North Cascade Range, Washington	2200	48°13'	Libby (1968: 318)	Ice-cemented rock glacier
23	Head of Yosemite, Sierra Nevada, California	3300	38°00'	Kesseli (1941: 205)	Ice-cemented rock glacier
24	Mount Whitney, California	4300	36°35'	Retzer (1965: 38)	Mean annual air temperature extrapolation at wind-swept crest

Source: Pewe (1983)

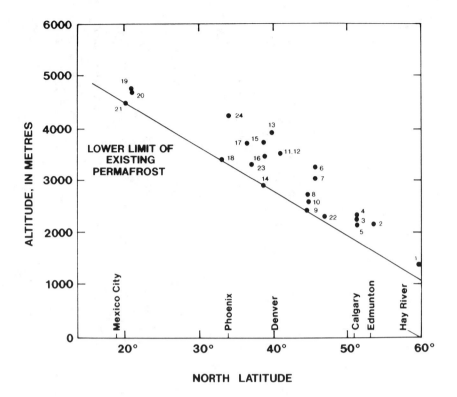

Figure 3.1. Altitudinal distribution of permafrost in North America.
Source: Pewe (1983). **Note:** Numbers refer to Table 3.1.

and mean annual air temperature.

The extent of the various zones depends on local relationships between climate and topography. In the Canadian Rocky Mountains the lower limit of continuous permafrost lies in the range 2180–2575 m, with the zone of discontinuous permafrost 100 m lower. The pattern is largely related to the depth of winter snowfall as thick snow insulates the ground. In the Front Range of the Colorado Rocky Mountains the lower limit of continuous permafrost at 40°N is 3750 m (Ives, 1974; Ives and Fahey, 1971). The lower limit of discontinuous permafrost is 3500 m. Based on freezing and thawing indices, continuous permafrost might extend down to 3600 m on south-facing slopes and down to 3550 m on north-facing slopes. Discontinuous permafrost might extend down to 3300 m or 3200 m on similar slopes.

Discontinuous permafrost up to 50 m thick occurs in association with the Gruben rock glacier at 2800 m in the Swiss Alps close to the Grubengletscher. But it is likely that this was associated with Little Ice

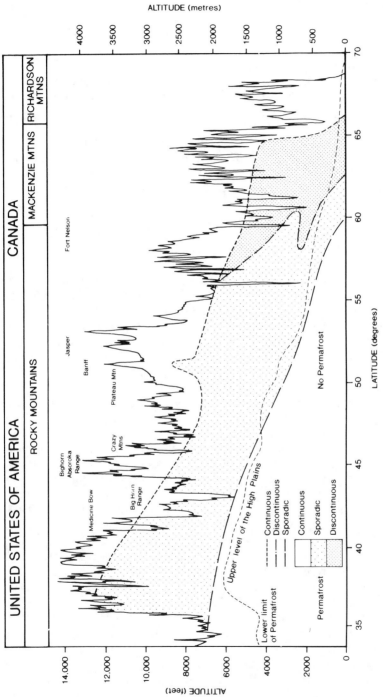

Figure 3.2. Distribution of permafrost zones along the eastern ranges of the Rocky Mountains between latitudes 70°N and 35°N. *Source*: Harris (1985).

Age temperatures, which were 0.5–1.0°C colder. In the areas of the Alps with high precipitation amounts (greater than 2000 mm), the glacier equilibrium line descends below 2600 m which is close to the lower limit of discontinuous permafrost. Therefore permafrost is restricted to rock outcrops, summits, crests and rock walls. Continuous permafrost is found only above 3500 m. In the drier interior with precipitation 1000 m or less the equilibrium line is 3000 m or above, and the lower limit of discontinuous permafrost is at 2400 m. Therefore there is a larger potential zone for permafrost.

There is a general feeling that permafrost has been considerably underestimated in treatments of the physical geography of mountains. This may also imply that features and processes associated with permafrost, such as rock glaciers, ice wedges and active cryoplanation terraces, have also been underestimated.

Nivation

The term 'nivation' was introduced by Matthes (1900) in his study of the glaciation of the Bighorn Mountains, Wyoming, and refers to the processes associated with snow patches. Since then it has been used frequently and somewhat indiscriminately, as the extensive review by Thorn (1988) has revealed. It is also a term that is very difficult to define and is 'a useful word that explains nothing' (Tricart, 1970, p. 100). Thorn (1988) has shown how it is used in two different senses. It is used as a collective noun to identify an assemblage of weathering and transport processes that are intensified by a late-lying snow patch. The processes are widespread in periglacial areas and the usefulness of the term is based on supposition that the action of the processes is intensified. Nivation is also used to describe landforms that have been substantially modified by the processes defined in the first usage although it is not believed that the landforms were actually initiated by snow patches.

It was thought by some early workers (for example, Ekblaw, 1918; Lewis, 1936; 1939; McCabe, 1939) that snow patches were static, but gradually other workers (for example, Chamberlin and Chamberlin, 1911; Dyson, 1937; 1938) were producing evidence of snow creep and abrasion. More recent studies have documented rates of snow creep and have shown that very high pressures can be created by such creep (Costin *et al.*, 1964; 1973; Mathews and Mackay, 1963; 1975). Solifluction and overland flow are also important processes associated with snow patches and must be considered part of nivation. Thorn (1976; 1979b) has shown that transport rates by overland flow are between twenty and thirty times greater within a colluvium-mantled snow patch site than on a nearby snow-free surface. A few studies have attempted to provide comparative data on the relative importance of solifluction and overland flow. In the

Shirouma-dake region of the Japanese Alps, Iwata (1980) found that frost creep and gelifluction comprised more than 60 per cent of total mass transfer on rubble slopes. Also the absolute total of material moved is about 1.5 times greater on slopes affected by nivation than on those not experiencing nivation.

This very brief review of nivation has shown that understanding of the processes involved is rudimentary and an unambiguous definition of nivation impossible. This reinforces Thorn's (1988) plea that nivation, as a term, should be abandoned and that research should be focused on the nature of the relevant processes, with weathering and transport being separated into two process groupings.

Cryoplanation

Cryoplanation is the name given to the production of an erosional surface by processes involving freeze-thaw activity and periglacial transport processes. Such cryoplanation or altiplanation terraces and cryopediments have been described from most of the alpine mountain regions of the world. They have also been described in areas not currently experiencing periglacial conditions (Gerrard, 1989b). But cryoplanation, as a term, is faced with problems similar to those of nivation. Indeed nivation has often been invoked as the mechanism by which the rock scarp retreats to produce the terrace or pediment (Reger and Pewe, 1976).

The inclination of cryoplanation terraces varies between 2° and 10° with the riser varying in steepness from 40° to 90°. A variety of processes, solifluction, sheetwash and piping, transport the products of weathering across the developing terrace. Cryopediments develop at the foot of valley sides and generally occur only at one level, whereas cryoplanation terraces often occur in stepped sequences. Little is known about the precise climatic requirements for the development of cryoplanation terraces but optimum formation seems to occur from 300 m to 500 m below the snow line to a little above it (Richter *et al.*, 1963). Cryoplanation terraces are rare on mountain sides steeper than 20° and in relatively soft rocks without initial irregularities such as in the Richardson Mountains and the Brooks range of North America (Priesnitz, 1988).

Stratified slope deposits

Stratified slope deposits are common in the periglacial zones of mountains and are generally of two types: grèzes litées and éboulis ordonnés (stratified screes). It is important to differentiate between the types because the terms have sometimes been used indiscriminately. Grèzes litées are low-angled deposits occurring at the base of slopes, whereas a stratified scree is a scree with alternations of coarse and fine layers, inclined parallel to the ground surface and having little or no sorting from top to bottom. The stratification in grèzes litées may not be parallel to the slope surface. Conditions necessary for the formation of stratified slope deposits appear to be frequent temperature oscillations around 0°C, the existence of sufficient water to sort the material and absence or extreme scarcity of vegetation. These conditions generally span the zone between the lower limit of perennial snow and the timber line. Thus stratified screes are conspicuous on long east-facing slopes of the central Peruvian Andes between 4700 m and 5200 m (Dollfus, 1964). In some mountains, such as the eastern Pyrenees, two series of stratified slope deposits can be found, one of Pleistocene age and one of more recent age.

DeWolf (1988) has identified three zones in which stratified slope deposits are likely to occur:

(i) Permafrost zones where north-facing slopes do not thaw and where only south- or south-west-facing slopes can produce the gelifracts and water necessary for mobilisation.

(ii) Zones without permafrost but with seasonal frozen ground, where a wider range of slope aspects exists for the production of stratified deposits.

(iii) Zones with no frozen ground where only snow is available to provide the water necessary for the transport and bedding of the frost-shattered material.

Processes involved include slope wash, solifluction and limited amounts of subsurface flow and the resulting stratification is influenced by the grain size of the material, the shape of the gelifracts and the water-retention capacity and internal cohesion coefficient of the sediment. The development of bedding results from a repetition of discontinuous phenomena which are highly variable in space and time (DeWolf, 1988).

Rock glaciers

Rock glaciers are features of the periglacial belt in all the high mountain systems. They have been identified in north America (Kesseli, 1941; Johnson, 1967; Libby, 1968; White, 1976; Luckman and Crockett, 1978; Johnson, 1975; 1978), the European Alps (Vietoris, 1972; Barsch, 1969), Scandinavia (Barsch and Treter, 1976), the Himalayas (Mayewski *et al.*, 1981), the Andes (Corte, 1976), New Zealand (Jeanneret, 1975), Antarctica (Mayewski and Hassinger, 1980), Svalbard (Swett *et al.*, 1980), Tasmania (Caine, 1983) and Iceland. Although they have been identified as landslides (Patton, 1910) and fossil glaciers (Brown, 1925) they are essentially lobate- or tongue-shaped bodies of frozen debris with interstitial ice and ice lenses which move downslope or downvalley by deformation of the ice contained within them (Barsch, 1988).

Rock glaciers display a steep front (35–45°), steep side slopes 10 m or more high and a surface relief of arcuate ridges and furrows generally aligned perpendicular to the direction of flow. The detailed morphology of rock glaciers has been described by a number of workers (see, for example, Wahrhaftig and Cox, 1959; White, 1976). Rock glaciers were identified by Spencer (1900) but knowledge of the mechanics of internal flow and the temporal/spatial patterns of surface movements is still incomplete. However, it is clear that their movement is largely governed by their ice content and the slope of the land. They can be extremely large, up to $10^6 m^3$, with a clastic debris content of 40–50 per cent and ice content of 50–60 per cent.

The available evidence suggests that the ice is disseminated throughout the main mass of the rock glacier, and some workers (for example, Potter, 1972; White, 1976) have suggested that there is no central ice core. It is also clear that some debris-covered glaciers, which are common in semi-arid regions with high radiation amounts, such as the Andes and Himalayas, have been misinterpreted as rock glaciers. But debris-covered glaciers are quite distinct from rock glaciers.

A distinction has sometimes been made between tongue-shaped rock glaciers and lobate rock glaciers. Harris (1979, 1981a, 1981b) has argued that tongue-shaped rock glaciers occur below ice glaciers and are characteristic of the discontinuous alpine permafrost zone in maritime climates, whereas lobate rock glaciers generally occur below talus slopes and are indicators for cold regions with continuous permafrost. This interesting idea needs further testing but there is little doubt that active rock glaciers need a full periglacial climate, which raises the interesting question of their age. Some workers believe they are very young. Thus Boesch (1961) has suggested that they have formed in the Swiss Alps since AD 1500 and Osborn (1975) has argued for a maximum age of 375 years for those in the Banff National Park, Canada. In contrast, Haeberli (1979) believes that the Gruben rock glacier in Switzerland is more than 3500 years old, and extremely large active rock glaciers must be essentially Holocene features with ages between 5000 and 10 000 years.

On the basis of an extensive review of the literature, Barsch (1988) has produced a number of generalisations:

(i) Active rock glaciers are best developed in continental and semi-arid climates.

(ii) In very humid environments, locations suitable for rock glaciers are normally glacierised and the periglacial belt is very narrow.

(iii) Within a mountain area, rockglaciers are formed where permafrost is possible and clastic debris is concentrated.

(iv) Debris concentration is due to glacial action, weathering and rockfall. Therefore, rock glaciers occur below or at the side of substantial glacial moraines or below talus slopes.

With reference to the last point, Giardino *et al.* (1984) have shown how snow avalanches chutes and talus cones have provided material for the twenty-three rock glaciers identified in Mount Mestas, Colorado.

Notwithstanding the validity of these generalisations, Barsch (1988) has stressed that more information is needed to resolve problems such as those concerned with summer and winter movement of active rock glaciers and retardation and acceleration in response to fluctuations in climate. In addition, little is known about the ice–debris mix in different environments, the thickness and volume of active rock glaciers and the nature of the interface between the rock glacier and the ground surface. Rock glaciers may appear to be insignificant features in the broader mountain landscape but they can account for 15–20 per cent of periglacial mass transport.

MASS MOVEMENT

Rockfalls

Rockfalls are extremely important processes on mountain slopes, as demonstrated in the extensive research by Luckman (1976) Gray (1972; 1973) and Gardner (1969a; 1969b; 1970a; 1977) in the Canadian Rocky Mountains. Rockfalls occur when an individual block or series of blocks becomes detached from a cliff face. If release is from joint surfaces, some sliding may occur before the free fall phase. Rockfalls contribute considerable volumes of material to lower slopes in mountains and the rate and intensity of such events governs the long-term evolution of the cliff face. This topic is considered in greater detail in Chapter 5. All that is attempted here is to establish some of the general principles.

A distinction has been made between primary falls, which are those just released from the rock face; and secondary falls, which result from the transport of previously released material which has rested on ledges. In cliff faces dominated by interbedded sedimentaries, differential rock strengths often lead to the formation of numerous ledges and secondary rock falls will be the major means of moving material from the top to bottom of the slope. The commonest causes of small rockfalls are high rainfall, freeze-thaw and dessication weathering. Several workers have demonstrated the enlargement of joints by the freezing of water and frost bursting (Bjerrum and Jorstad, 1968; Gardner, 1970a) and the role of cleft water pressure has been emphasised by Bjerrum and Jorstad (1963a; 1963b; 1968). Larger falls may be triggered by earthquakes, unloading or erosional undercutting. Primary rock falls are more likely to occur in gullies (Whalley, 1974) and are possibly related to joints created by the release of stresses which, where they intersect rock walls, produce V- or X-shaped forms (Scheidegger, 1963; 1970; Gerber and Scheidegger, 1969; 1973; 1975).

Many studies have noted that the highest frequency of small rockfalls coincides with the increase in freeze-thaw cycles in spring and autumn, with diurnal variations corresponding to midday temperature maxima. Church *et al.* (1979) have described how rockfall activity commenced at 0600 hours, peaked around the solar noon and then declined. A similar pattern was noted by Gardner (1967; 1970a). Francou (1982) has compared the frequency of freeze-thaw cycles and rockfall cycles in Combe de Laurichard. Freeze-thaw cycles were most frequent in spring, from the end of February to the end of May on south-facing slopes and from the end of March to the middle of June on north-facing slopes. The maximum period of rockfall activity was not temporally related. Rockfalls tended to occur during periods of effective transport by snow avalanching and melting of snow. On the south-facing slopes most activity was from late April to the middle of June, and on north-facing slopes activity was retarded by a month and continued into late September. This stresses that it is important to make the distinction between the release of rock and its transport downslope.

Failure occurs in a number of ways. *Plane failure* occurs along joints or bedding planes inclined toward the cliff face. *Wedge failure* occurs where rock fails along two intersecting discontinuities. *Slab failures* are common on steep faces and occur where there is a conspicuous development of vertical discontinuities parallel to the rock face. Unloading is often the cause of such discontinuities. Slab failure may be a special case of *toppling failure*. Toppling failures are common where the rock is thin relative to the slope height. Four types of toppling failure can be recognised. *Flexural toppling* occurs when near vertical discontinuities exist and high columns bend forward and fail. *Block toppling* requires the sliding of equal-shaped blocks lower down the slope, causing a lack of support and the toppling of other blocks higher up the slope. *Block flexure* is a relatively continuous movement of long columns involving small accumulated displacements on numerous cross joints. *Secondary toppling* is where failure is initiated by an undercutting agent.

Rockslides

Rockslides involve the downward and usually rapid movement of detached portions of bedrock sliding on some form of discontinuity. Most rockslides break up very rapidly into separate masses of rock and move as a rock avalanche. Thus, this type of mass movement is now more often called a 'flowslide' (Rouse, 1984). Rockslides often involve vast quantities of material and are very significant agents of landscape change. This can be seen in Cruden's (1976) analysis of major rockslides

Table 3.2. *Characteristics of major rockslides in the Rocky Mountains of Canada.*

Slide	Material	Surface of rupture			
		Width (km)	Length (km)	Angle (degrees)	Coefficient of friction
Beaver flats N	Limestone	0–40	0.10–0.79	26	0.32
Beaver flats S	Limestone	0.48	0.43	40	0.25
Jonas Creek N	Quartzite	0.31–0.40	0.81	30	0.27
Jonas Creek S	Quartzite	0.50	0.61	28	0.37
Maligne Lake	Siltstone Limestone	0.98	1.56–1.67	25–40	0.16
Medicine Lake	Limestone Dolomite	1.64	0.43	35–48	0.26
Mt Kitchener	Limestone Dolomite	1.93	0.16–0.40	30	0.21

Source: Cruden (1976).

in the Rocky Mountains of Canada (Table 3.2). The distinctive feature of large rockslides is the high efficiency of transport. This can be assessed by calculating the ratio of the maximum height dropped to the maximum distance travelled. This is also known as the equivalent coefficient of friction (see Table 3.2). Small rockslides tend to have equivalent coefficients of friction between 0.5 and 0.6 whereas large slides have values between 0.1 and 0.3. Comparisons can be made with the low values characteristic of extremely mobile volcanic avalanches (see Chapter 7). Such low values have led to a great deal of speculation concerning the mechanisms involved. Mechanisms proposed include fluidization (Kent, 1966), air layer lubrication (Shreve, 1966; 1968) and cohesionless grain flow (Heim, 1932; Hsu, 1975).

Irrespective of the manner of the movement, large rockslides can be truly catastrophic. The slide which buried the village of Yungay in the Peruvian Andes in 1970 reached a speed of 480 km hr^{-1}. The Said-marreh slide in the Zagros Mountains of Iran was probably the world's largest slide (Harrison and Falcon, 1937). It involved an area 15 km long, 5 km wide and about 300 m thick. The total volume involved was about 20 km^3 and covered an area of 166 km^2 to an average depth of about 130 m with a maximum depth of 300 m. Many other examples, such as Turtle Mountain, Frank, Alberta, and Heim and Goldau in Switzerland, could be cited as examples of the scale and power of large mountain rockslides. Evidence from the Tyrolean Alps and central Nepal indicates that, under exceptional circumstances, rock can be fused by friction-generated heat along the sliding plane of large rockslides (Heuberger *et al.*, 1984). The landslides concerned are the Koefels, in the

Oetz Valley, Tyrol, and Langtang, Nepal, with displaced volumes probably of 2–3 km³ and over 10 km³ respectively.

Debris flows, mudflows and debris slides

There are a large number of mass movement processes that can be subsumed under the general heading of debris flows, mudflows and debris slides. The terminology of this section is left deliberately vague because of the uncritical use of the terms 'debris flows', 'mudflows' and 'debris slides' in the literature. In many cases this is because it is not easy to be specific about the mode of movement. In addition, one movement mode often changes into another. Movements occur across talus, alluvial fans and soil-covered slopes and range from movements largely involving snow, through slush or west snow avalanches, to movements involving water and debris. Landforms and deposits produced by snow avalanches have been well documented (Rapp, 1959; Davis, 1962; Peev, 1966; Gardner, 1970b; Luckman, 1971; 1972; 1977; Corner, 1980) but direct observations of avalanches and their immediate erosional effects are rare. Snow avalanches may abrade bedrock surfaces (Matthes, 1938), may form impact depressions at the slope foot (Corner, 1980), and entrain, transport and deposit material (Caine, 1969; Luckman, 1978a). But most literature on snow avalanches implies that avalanches have little direct contact with the ground surface. Conditions for avalanche erosion depend on the type of avalanche, the presence or absence of basal ice layers, the free water content of the snow and the presence of snow-free and thawed debris surfaces (Gardner, 1983a). There is little doubt that wet snow or slush avalanches, sometimes called 'slushflows' (Washburn and Goldthwait, 1958) are more potent geomorphologically.

Debris flows not involving snow are usually the result of either snow melt or torrential rainfall (Rapp, 1985). Rapp (1960a) has documented numerous debris flows at Karkevagge following 107 mm of rain in twenty-four hours. Two smaller storms in July 1972 (20 mm in thirty minutes followed three hours later by 15 mm in thirty minutes) produced similar effects on 15–20° slopes in the Tarfula area of Lappland (Rapp, 1975). Approximately 50 000 m³ of debris was involved. Rainstorms in July 1972 (31 mm in ten hours) in the Longyear Valley, Spitzbergen, triggered eighty debris flows (Larsson, 1982; Rapp, 1975). More failures occurred in August 1984 after 55 mm of rainfall in twenty-four hours (Rapp and Nyberg, 1981). Most failures occurred where lateral concentration of drainage occurred on the slope or at the break of slope between steep sections and lower gradients. The failure surface was either the frost table or a bedrock surface. Observations suggest that the movements started as debris slides, then stopped temporarily in the narrow portions of the slide scar and then continued as a debris flow.

Alpine mudflows have been well documented and have been described in North America (Sharp, 1942; Fryxell and Horberg, 1943; Winder, 1965; Curry, 1966; Broscoe and Thompson, 1969), the Andes (Sutton, 1933; Dollfus, 1960), the European Alps (Stiny, 1910; Tricart, 1957), the Himalayas (Hewitt, 1968; Brunsden *et al.*, 1981), Central Asia (Vinogradov, 1969), New Zealand (Brundall, 1966) and Scandinavia (Rapp, 1960a). Many mudflows are small but collectively can be extremely important transporting agents. Rapp (1960a) has estimated that ten mudflows in Karkevagge, Scandinavia, moved about 1300 m³ of material. Dollfus (1960) has recorded volumes up to 100 000 m³ and Stiny (1910) has estimated maximum volumes of 300 000 m³ for mudflows in the European Alps. Owens (1972) has suggested that alpine mudflows may be differentiated on the basis of size, one group consisting of small features which modify local slope form above the timber line and another comprised of large features which are significant in constructing large fans upon which most of them occur. Their dynamics and magnitude-frequency characteristics may also be different.

Debris flows are not restricted to alpine environments but occur in all mountains on steep slopes when the water build-up on the slopes overcomes the resistance of the soil and vegetation cover. Thus Temple and Rapp (1972) have described numerous debris flows that occurred in the Uluguru Mountains of Tanzania following torrential rain. In this instance the resistance of the landscape had been reduced by land-use changes. Similar flows repeatedly occur on the lower and middle slopes of the Himalayas as a result of deforestation and monsoonal rainfall (Starkel, 1972). Debris flows and slides are so conspicuous on steep mountain slopes that it is likely that they are the main mechanism for long-term slope evolution.

Some mass movement activity seems to be related to annual climatic cycles. Rapp's (1960a) study of the incidence of earthslides and rockfalls along railway lines showed that they were most frequent during months when thawing of ice in the ground was at a maximum. Thus, rockfalls on the Voss–Granvin line were concentrated in April, as were those on the Voss–Bergen line, but with a secondary peak in October. However, movements on the Riskgransen–Narvik line showed peaks in June and September. A double peak in spring and autumn was also found by Bjerrum and Jorstad (1968), who argued that frost action was at a maximum in these seasons because of continual change of air temperatures about freezing point.

A summary of landslides, rockfalls and mudflows in Iceland for the period 1958–70 inclusive has been provided by Jonsson (1974). Although movements occurred in all months, there are peaks of activity in February and September–October, which suggests that a relationship might exist with annual climatic patterns (Figure 3.3). Analysis of summary climatic statistics does not immediately suggest reasons for these peaks, probably because of the inappropriateness of standard climatic parameters. Frost activity is strongest in the period from

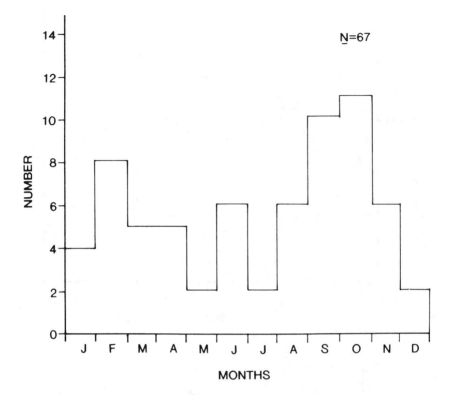

Figure 3.3. Frequency of landslides throughout the year in Iceland for the period 1958-1970 inclusive using data from Jonsson (1974).

January to March, while September and October are the wettest months and have the highest precipitation intensities (Figure 3.4). These observations suggest that heavy rainfall, perhaps also associated with the sudden thawing of snow, is the factor which triggers many mass movements. Many slides occurred in the north-west peninsula on 20 November 1958, following 53–56 mm of rain, with warm westerly winds raising the temperature to 10°C, thawing much of the lying snow. In the same area, extensive slipping occurred on 13–14 November 1961 following a ten-day period in which 102 mm of rain fell and also on 21 October 1962 following 96 mm of rain in less than twenty-four hours. Similar relationships can be established in other parts of Iceland; they reinforce an eyewitness account of the effects of heavy rain in the spring of 1922 in Austerdaler, north Iceland, when forty screes were active at the same time. Typical channels created by these movements are shown in Figures 3.5 and 3.6.

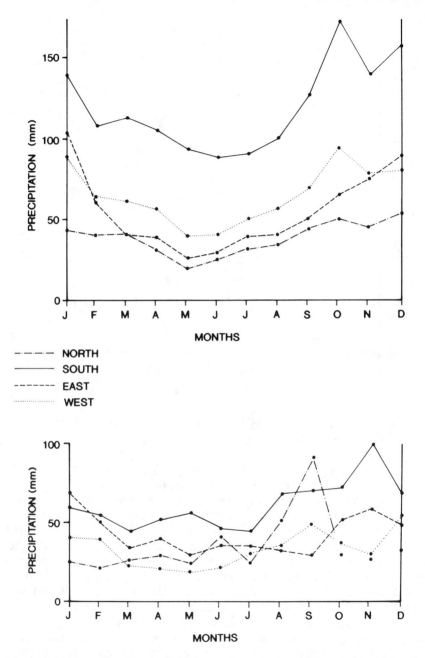

Figure 3.4. Mean monthly precipitation and maximum precipitation in 24 hours for stations representative of different parts of Iceland for the period 1931–66 inclusive.

Source: published climatological data.

Figure 3.5. Debris flow tracks on a scree cone in north Iceland. (Photo: author.)

Figure 3.6. Erosion of coarse scree by intermittent debris flow activity in north Iceland. (Photo: author.)

Skinflow

There are a number of mass movement types that are specifically related to the characteristics of seasonally thawed ground. Skinflows, sometimes called 'active layer slope failures' (French, 1976) and 'active layer detachments' (Strangl *et al.*, 1982), are shallow in comparison to their length and are merely the detachment of a thin veneer of vegetation and mineral soil over a planar surface. Skinflows involve behaviour like a viscous fluid and generally develop on steep slopes. They can be explained using standard geotechnical slope stability analyses without invoking excess porewater pressures (C. Harris, 1981).

Active layer skin slide is a term used by Mackay and Mathews (1973) to describe the downslope gliding of the active layer as a discrete unit with a minimum of internal disturbance. There is often confusion with the skinflow movements just described since terms such as 'block glide' (Brown *et al.*, 1981), 'detachment failures' (Hughes, 1972) and 'active layer detachment slide' have also been used. Movement generally occurs along the permafrost table but the underlying cause appears to be an excess of pore-water pressures at the base of the active layer. Many of the movements appear to be triggered by rain.

Slow mass movement processes

The relatively fast-moving and often spectacular mass movements discussed above are conspicuous and it is easy to assume that they are the major landforming and transporting agencies in mountain landscapes. However, slow-moving processes are ubiquitous and may be moving larger masses of material, albeit at a much slower rate. Such movements involve the slow flowing and creep of a variety of types of material. Solifluction or gelifluction is the most noticeable. Gelifluction is solifluction associated with frozen ground. Suitable conditions for gelifluction occur where the downward percolation of water through the soil is limited and where the melting of segregated ice lenses provides excess water which reduces internal friction and cohesion in the soil. Gelifluction is influenced by soil moisture, slope gradient and soil grain-size distribution (Washburn, 1979) and vegetation cover. Gelifluction can produce a number of small-scale slope features such as lobes, terraces and ploughing blocks.

Rates of gelifluction occur in the range 1–12 cm yr^{-1} in Spitzbergen (Jahn, 1960), 0.9–4 cm yr^{-1} in Sweden (Rapp, 1960a, Rudberg, 1962; 1964), 1.0 cm yr^{-1} in the French Alps (Pissart, 1964), 0.4–4.3 cm yr^{-1} in the Colorado Rockies (Benedict, 1970) 1–6 cm yr^{-1} in Norway (Harris, 1972) and 0.6–3.5 cm yr^{-1} in Canada (Price, 1973). Differences in rates were largely the result of varying slope angles.

Creep and shift of materials also occur on talus. Talus shift involves movement in sudden jerks and is extremely variable from one part of the talus to another. Talus shift rates vary considerably — for example, in the range 6–111 cm yr^{-1} in the Canadian Rocky Mountains (Gardner, 1983b), and up to 22 cm yr^{-1} in northern Scandinavia (Rapp, 1960a). The spasmodic nature of movement is shown by observations in the Colorado Front Range, where there was no shift at all for eight years and then sudden movement of 8–10 cm yr^{-1} for the next three years (White, 1981). White (1981) has also measured movements of 1–2 cm yr^{-1} of blocks in block slopes and block streams.

This chapter has demonstrated that intense activity is characteristic of most mountain slopes. This activity often interacts with fluvial processes, either directly or indirectly, and is largely responsible for supplying most mountain streams with abundant load. Fluvial processes are examined in the next chapter.

Chapter 4

MOUNTAIN HYDROLOGY AND RIVER PROCESSES

INTRODUCTION

The hydrology of mountain areas, although governed by the same set of factors, is markedly different from that of lowland areas. It is the combination of these factors that produces distinctive associations. This distinctive hydrology is reflected in specific channel and valley characteristics. This is especially the case where the mountains are greatly influenced by snowmelt and glacier melt processes. In other mountains ephemeral streamflow possesses distinct characteristics (Wertz, 1966). Snowmelt runoff has unique characteristics which place it midway between glacial and temperate hydrological regimes. The risk to peak flow and subsequent recession is more dramatic than in glacier-fed catchments. High altitude may elongate both hydrograph limbs because seasonal warming spreads sequentially up the mountain side.

Snowpacks are also an important temporary store for meltwater (Fitzgibbon and Dunne, 1981). The effect of mountain climates is to increase the range of moisture availability through time, producing a concentrated annual hydrograph. Steep channels also compress short-term hydrographs. Thus 'the alpine and high-mountain areas of the world play an extremely important and distinctive role in the hydrological processes of the planet and in the regional hydrology of all continents' (Roots and Glen, 1982, p. v). But it is also true that high-mountain areas present difficult problems for hydrological observation and measurement. This is one of the reasons why hydrological maps of all the high-mountain areas of the world are being compiled as part of the *World Atlas of Snow and Ice Resources* at the Institute of Geography of the Academy of Sciences of the USSR.

It is in alpine regions where meteorological, glaciological, periglacial and hydrological phenomena have the most intimate and complex interaction and variability on short spatial and temporal scales. Barsch and Caine (1984) have made the distinction between 'Alp type relief' dominated by glacial erosion and 'Rocky Mountain type relief' with a lesser impact of glacial erosion and retaining significant summit and interfluve areas of low relief.

The presence, absence or percentage of river basins occupied by glaciers is related to the mass balance of areas of snow and ice, which depends on the size, slope and orientation of the basin and interactions with insolation and precipitation patterns. In maritime mountainous areas in higher latitudes, glaciers frequently extend to sea level. In the arid continental interiors of the tropics permanent snow and ice is found only at very high altitudes. There is also a general decrease in height of the snow line and permafrost limits on north–south mountain ranges (Harris, 1985; Harris and Brown, 1982) and east–west environmental gradients across mountain ranges such as the Andes and Rockies. Contrasts, such as these, manifest themselves in streamflow characteristics. In the Tien Shan and Pamirs most streamflow (1200–1600 mm) is formed on the windward slopes of the Pskemskiy, Chatkalsky, Gissarskiy and other ranges of the western exposures most accessible to moist air masses (Dreyer *et al.*, 1982). On slopes with a northern exposure, streamflow decreases to 800-1000 mm and to 50–200 mm on east-facing slopes. The north-eastern slopes of the Karakoram are in a rain shadow area and possess low runoff amounts. In a similar way, the north-western and northern slopes of the Hindu Kush possess lower streamflow than the southern and south-western slopes. Thus great regional contrasts can be expected in the hydrology of mountain regions.

The distinctiveness of mountain hydrology is best developed at certain scales. Hydrologic processes at the micro scale are affected by local terrain factors such as vegetation, bedrock geology and soil type. Variations in slope angle and aspect act through the radiation balance to affect precipitation storage and slope drainage. At the macro scale, mountain river basins are dominated by the effects of past glacial, periglacial and fluvio-glacial activity. It is 'the discrimination of the mesoscale spatial entities, such as the ubac and adret, with their characteristic energy, mass and moisture balances, and the understanding of ways in which each of these environmental systems receives, transforms and releases energy, moisture and mass which are the two most difficult challenges in alpine hydrology' (Slaymaker, 1974, p. 133).

The essential components of mountain hydrology are the high radiant energy fluxes, high moisture fluxes in the form of rain and snow, discharge hydrographs influenced by snowmelt and glacier melt processes, high available relief, low-percentage ground cover by trees, and other vegetation and loose, poorly developed soils and regolith. The spatial variability of direct incoming radiation is particularly complex in alpine environments (see, for example, Lee, 1964; Garnier and Ohmura, 1968; Williams *et al.*, 1972). Latitude, time and topography combine to make calculations of incoming radiant fluxes extremely complicated (c.f. Hay, 1971). The complexity of mountain river basin hydrology can be simplified by identifying broad systems within which there is some uniformity of response to the primary inputs. Slaymaker (1974) has noted ten such systems, listed in Table 4.1, together with the main processes operating in them over four time-scales. This synthesis provides a useful framework for much of the subsequent discussion.

THE HEAT BALANCE

The driving force of mountain hydrology is the heat balance. This not only determines the operation of many geomorphological and pedological processes but also determines the nature of the vegetation cover (Isard, 1983; Williams *et al.*, 1972). The heat balance can be expressed in a number of ways but the basic heat balance equation is of the form

$$R_N = S + H + LE$$

Where R_N is the net radiation heat transfer, S is the soil heat transfer, H is the sensible heat transfer, and LE is the gain or loss of heat by evaporation or condensation. Slope angle and aspect are major factors in determining the spatial variability of the energy flux. Frank and Lee (1966) have shown that radiation intensity over a year at 50°N differs by a factor of four between north-facing and south-facing 45° slopes. Net radiation can be estimated from the following equation:

$$R_N = G(1 - \alpha) + \epsilon_a \sigma T_a^4 - \epsilon_e \sigma T_e^4$$

where G is global radiation, α is albedo, ϵ_a is emissivity of the atmosphere, ϵ_e emissivity of the surface, α is Stefan's constant (5.67 × 10^{-8} W m^{-2} K^{-4}), T_a is the radiation temperature of the lower atmosphere, and T_e that of the surface. The albedo and transmissivity of the various surfaces found in mountains vary greatly and have a major influence on the energy balance. Thus the energy balance of most mountain areas is extremely variable unless the surface is completely covered in fresh snow. The albedo of a glacier varies from 20 per cent for dirty glacier ice to 46 per cent for clean ice. However, the albedo for fresh snow lying on a glacier can be as high as 95 per cent. This means that in the autumn period when a glacier may possess fresh snow at its upper end but dirty, debris-covered ice near its snout, a highly variable energy reflecting surface exists. Albedo of snow can be estimated from empirical relationships that have been established between it and the age of the snow cover (Male and Granger, 1979; US Army Corps of Engineers, 1956).

Transmissivity also varies. It has been shown that 10 per cent of incident radiation not reflected will reach 40 cm in glacier ice but will only reach 12 cm in dry snow and 4 cm in wet snow. Alpine lakes, influenced

Table 4.1. Summary of major hydrologic processes operating in mesoscale alpine systems.

	> 10³ years	10–10³ years	10⁻¹–10 years	< 10⁻¹ years (c. 1 month)
1) Glaciers	Water storage term in hydrological cycle	Glacier retreat and advance	Ablation, accumulation	Physics of glacier motion, surging
2) Snowpacks	–	–	Snow metamorphism firn ice	Melt, storage, accumulation
3) Alpine lakes	Glacial scour and lake formation, sedimentation and meadow formation (see 6)	Sedimentation	Temperature stratification, draining events, ice formation and break-up, sedimentation regime	Seiches, daily level changes, sediment movement into and through lake, evaporation
4) Mountain streams	Glacial erosion	Downcutting, supply creep	Discharge regime, sediment supply (talus)	Runoff concentration, 'tumbling' flow, rapid response to precipitation and evaporation
5) Morainic mounds	Glacial deposition	Degradation of morainic slopes	Chemical weathering, frost action, vegetational change	Infiltration, interflow
6) Alpine and subalpine meadows; valley bottoms	Sedimentation in glacial lakes (see 3)	Dissection by streams, flood plain development, soil formation	Periodic inundation jökulhlaups (see 1), vegetational change	Precipitation, infiltration, interflow, baseflow
7) Alpine and subalpine meadows; adret slope	Postglacial soil development	Slope degradation, channel dissection	Vegetational change, mass wasting	Precipitation, infiltration, interflow
8) Alpine and subalpine treed slopes, adret	Postglacial soil development, soil creep, gullying	Slope degradation, channel dissection	Frost action, mass wasting	Interception, precipitation, infiltration, interflow

Table 4.1. *contd.*

	> 10^3 years	$10-10^3$ years	$10^{-1}-10$ years	< 10^{-1} years (c. 1 month)
9) Alpine and subalpine treed slopes, ubac and ridge top	Postglacial soil development, slides and flows of earth and mud	Slope degradation	Frost action, mass wasting	Overland flow, precipitation, interception, infiltration, interflow
10) Alpine barren	Glacial erosion postglacial rockfall and talus, slope formation	Slope degradation	Frost action, mass wasting	Overland flow, precipitation

Source: Slaymaker (1974)

by glaciers, have an albedo in the range of 15–20 per cent, which is higher than for lakes in unglacierised basins (5–10 per cent) because of the generally high sediment content of the lakes. Bare rock surfaces have albedos in the range 5–15 per cent, and high-mountain grassland 12–25 per cent.

SNOW ACCUMULATION AND SNOWPACK CHARACTERISTICS

The distribution of snow accumulation in mountain regions is one of the most important controls on mountain river hydrology. Several works (such as Caine, 1976; Thorn, 1978) have stressed the importance of the range and magnitude of snow inputs to mountain systems. Thorn (1978) has argued that significant geomorphological process thresholds are related to snowfall, snowpack surface, depth, mobility and duration. Accurate information of snow accumulation rates and the spatial variability of snowpack depth is often the weakest link in the hydrology of alpine regions (Figure 4.1). Snow accumulation is governed by relief and altitude, variations in freezing and melting levels, turbulence and especially interception by vegetation (Clark *et al.*, 1985). However, marked fluctuations in accumulation rates in the St Elias Mountains, Yukon, were not altogether explained by elevation, topography, continentality or exposure (Marcus and Ragle, 1970). Accumulation rates were stable in the 1950s, increased in the early 1960s and were at a minimum in the period 1964–6. It is also necessary to estimate orographic increase in precipitation and the change in form of precipitation with altitude (Bagchi, 1982). Ming-ko Woo (1972) has noted how many river basins in British Columbia experience simultaneous snow accumulation at high altitudes and rain-on-snow melt at lower altitudes. Because of the changing position of the freezing level with different storms it is necessary to budget the snow packs by altitudinal zones in order to predict basin streamflow and snow storage. Some indication of this variability in Finland has been provided by Kuusisto (1984). Drifting of snow after it has settled is also a major complicating factor (Tabler, 1975; Daly, 1984).

Landsat imagery provides a reasonably practical method of determining snow-line altitude (Meier and Evans, 1975) and snow cover (Tarar, 1982) which can then be used in a snowmelt-runoff model (Rango and Martinec, 1979). The areal extent of the snow cover can be detected and used as a measure of snow storage although the water equivalent of the snow cannot be estimated.

SNOW MELT

The variability of snow accumulations makes accurate information on

Figure 4.1. Late-lying snow patches provide concentrations of water flow throughout the summer on mountain slopes. (Photo: author.)

snowmelt processes difficult to obtain (Moore and Owens, 1984; Obled and Harder, 1979). In general, three main approaches have been adopted to estimate snowmelt; the energy budget approach, the multiple regression approach and the temperature index approach.

The sources of heat involved in the melting of snow are net radiation, sensible heat transfer from air to snow, latent heat of vaporisation by condensation from the air, conductive heat transfer, and the heat content of rainwater. The energy available for melting on a horizontal snow or ice surface of unit area over a unit of time is a combination of the following components (Rothlisberger and Lang, 1987):

$$Q_M = Q_{NR} + Q_S + Q_L + Q_P + Q_G$$

where Q_{NR} is net radiation, Q_S is sensible heat, Q_L is latent heat of condensation or evaporation, Q_P is heat provided from precipitation, Q_G is heat from heat conduction in the snowpack, and Q_M is heat used for melting or gained from refreezing of meltwater. The melt rate M is then

$$M = Q_M/S$$

where S is the latent heat of melting. A similar relationship has been produced by Rantz (1964).

The multiple regression approach involves correlating snowmelt with available meteorological data and has been used by Garstka *et al.* (1959). The third approach is to use a temperature value such as mean or maximum temperature or even degree days as predictors of snowmelt. A more geomorphological approach, involving factors of basin topography and vegetation cover, has been suggested by Hendrick *et al.* (1971).

INFILTRATION, THROUGHFLOW AND OVERLAND FLOW

The relationship between soil moisture, infiltration and overland flow in mountains can be extremely complex. In general the combination of steep slopes, thin soils with low infiltration capacities, low tree cover and rapid snowmelt leads to high rates of overland flow. An interesting study was conducted by Haupt (1967) at about 2000 m near Reno, Nevada. He demonstrated that the nature of soil ice greatly influenced overland flow rates. Porous, concrete frost reduced infiltration rates and induced overland flow whereas needle ice (stalactite soil frost) increased infiltration. Rapidly melting snow over soil containing dense frost accelerated runoff. Overland flow can thus be generated extremely quickly. Also the relative importance of overland flow and throughflow can vary considerably, both temporally and spatially, as the nature of soil ice and freeze-thaw cycles fluctuate rapidly. This is in addition to the inherent spatial variability imposed by soil and regolith characteristics. This variability will have a major influence on mountain stream hydrographs.

Slaymaker (1974) has demonstrated the speed with which snowmelt can occur. In a period of eleven days an average of 5 cm per day of snow was removed from the alpine meadows of Miller Creek, British Columbia, and the snow line rose by 550 m on the adret slopes and by 350 m on the ubac slopes. The significance of this rapid change is that overland flow appears to be concentrated in a 30–40 m band below the snow line, a zone which retreats upslope following the retreat of the snow line. The concept of contributing basin areas is especially important in alpine basins. During spring and summer, when the snow melts, the contributing area increases accordingly, but will reverse quite dramatically if weather conditions change.

DISCHARGE REGIMES

Streamflow regimes in mountain areas are often highly irregular but usually exhibit a number of distinctive patterns depending on the relative importance and seasonality of factors such as precipitation, snowmelt and glacier melt. Thus, in a transect across the Western Cordillera of Canada, the alpine coastal areas of British Columbia will experience all three influences on discharge hydrographs, but in the more continental parts of the Rocky Mountains only snowmelt and glacier melt peaks are important.

A typical seasonal pattern can be examined using stream flow regimes for the Upper Indus basin the Karakoram Mountains (Ferguson, 1984). The Karakorams contribute about 25 per cent of the total flow of the Indus from about 15 per cent of its catchment area. The total annual runoff and its monthly distribution are shown in Figure 4.2. Discharge starts to increase in March at the lower stations as snowmelt commences, but not until May and June in the Upper Hunza and Upper Indus. Between 40 per cent and 70 per cent of the total runoff is in July and August, when discharges are between fifteen and forty times those in March. A detailed discharge regime is shown in Figure 4.3 which, as well as demonstrating the general pattern, also exhibits considerable variability. In some cases secondary peaks in the autumn are the result of heavy rainstorms. The timing of the peak discharges indicates an icemelt origin and all the major tributaries of the Hunza River are fed by glaciers. Monsoon rainfall is rather insignificant in this region. Suspended sediment concentration is also shown to establish the link between streamflow patterns and those of sediment removal.

Genetic components of streamflow for the Hindu Kush, Karakoram, Pamir and Tien Shan mountains have been established by Dreyer *et al.* (1982). The method is based on the regular changes in streamflow structure during the year. Rivers are fed by groundwater from November to February. Snowmelt begins in the lower parts of the basins in March, and groundwater discharge increases to a maximum in July–August.

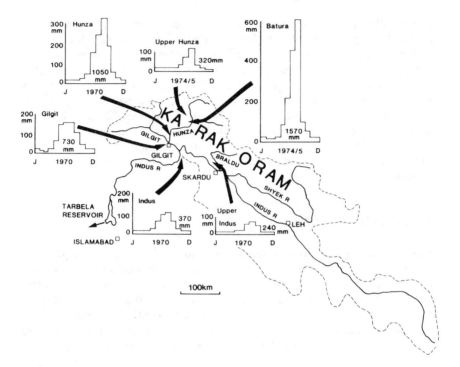

Figure 4.2. River flow characteristics in the Karakoram.

Source: Ferguson (1984).

Until September rivers are fed by groundwater, icemelt, snowmelt and precipitation. Snowmelt has finished by October in the high mountains, and rivers are fed solely by groundwater. In low-mountain areas the hydrograph is occasionally complicated by one- or two-day rain floods.

The share of rain in feeding the rivers in the Hindu Kush and Karakorams is less than 5 per cent and less than 10 per cent in the Tien Shan and Pamirs. An increase in the rainwater component occurs on the boundary with the Himalayas, where monsoon rains are important. Rivers in the lower parts of the southern slope of the Himalayas obtain up to 60–80 per cent of streamflow from monsoon rains. Most of the rivers have a groundflow component of 30–40 per cent, with 50–70 per cent snow and ice component. This snow and ice component increases with altitude especially in zones of local and continuous permafrost. The ice component for the rivers of the Pamirs and Tien Shan decreases downstream from 60–70 per cent near the glacier tongues to 20–25 per cent where the rivers leave the mountains.

Major rainfall-induced floods in mountains often occur in the second half of the ablation period when winter snow cover and its retention capacity are at a minimum and when internal drainage systems of glaciers are well developed (Rothlisberger and Lang, 1987). Extreme

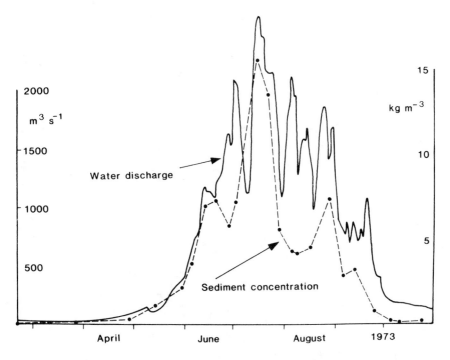

Figure 4.3. Discharge characteristics of the Upper Hunza River.
Source: Ferguson (1984).

floods can occur when a period of maximum meltwater from the lower part of a glacier combines with an intense rainstorm in late afternoon or early evening, such as occurred on 17 September 1960 in the Oetzal Alps. A similar extreme flood event has been described for the Hintereisferner region of the European Alps by Rudolph (1962). Spatial differences in average runoff are related to many of the factors discussed earlier, especially differences in the percentage of each catchment covered by permanent snow and ice. Thus, the discharge of the Upper Hunza River draining 5000 km^2 on relatively low mountains north of the main Karakoram chain and without major glaciers is doubled by the confluence of the short river draining the 300 km^2 Batura glacier and its surrounding peaks. The discharge of the Hunza River is then increased by a number of major glacier systems to produce a mean runoff for the whole 13 000 km^2 catchment of about 900 mm yr^{-1}. This can be contrasted with the similar-sized but less glacierised basin of the Gilgit River to the west, which produces 20 per cent less streamflow. The Upper Indus to the east drains four times the combined area of the Gilgit and Hunza rivers but has only about 30 per cent more discharge because the high runoff from the large glaciers of the central and eastern Karakoram is counterbalanced by low runoff in Ladakh and on the Tibetan Plateau.

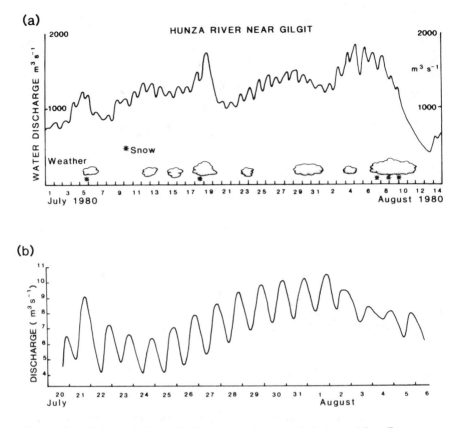

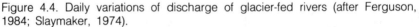

Figure 4.4. Daily variations of discharge of glacier-fed rivers (after Ferguson, 1984; Slaymaker, 1974).

The Karakoram rivers, and alpine rivers in general, exhibit marked diurnal fluctuations (Figure 4.4). In Miller Creek, British Columbia (Figure 4.4b) minimum discharge is recorded around 0900–1000 hours and the maximum around 1900–2000 hours. The River Gilgit also exhibits a lagged diurnal cycle interrupted by sharp recessions when snowfall or cloud cover halts glacier ablation (Figure 4.4a). These characteristics can be explained by examining the nature and timing of glacier runoff.

GLACIER RUNOFF

Glacier discharge is dominated by meltwater runoff. Precipitation usually has a negative influence on glacier runoff because incoming solar radiation is reduced, and if precipitation is in the form of snow a higher albedo is created. Glacier runoff variations generally exhibit a reverse pattern to a rain-dominated runoff regime. Thus, in large mountain basins, only partially glacierised, the upper parts will experience a meltwater runoff regime, whereas in the lower parts of the basin runoff will be dominated by rainfall. This counterbalancing effect of precipitation and melt processes in glacier basins reduces the variability of annual flow. Kasser (1959) has noted that a minimum variation in annual runoff in the European Alps occurs in river basins with 30–40 per cent glacier cover, while during the main ablation period in August the minimum variation is associated with a glacier cover of 30–60 per cent (Rothlisberger and Lang, 1987).

In the area occupied by Glacier AX010 in Shorong Himal, Nepal, it has been estimated that about 55 per cent of total precipitation during June to September is in solid form (Higuchi *et al.* 1982). At the snowline on the maritime glaciers in the periphery of the Tibetan Plateau in China, 70 per cent of annual precipitation is solid (Li, 1980). Rain on glaciers is discharged almost immediately just after the rainfall.

The hydrological relationships of glacierised basins can only be understood following detailed observations over a number of years. The benefits to be gained from such study are exemplified in work carried out on the Peyto Glacier basin on the eastern slope of the Rocky Mountains as part of the International Hydrological Decade (Young and Stanley, 1977).

Glacier outflow hydrographs are composed of a peaked diurnal quickflow component superimposed on a flatter delayed-flow component. The volume of delayed flow often exceeds the volume of the quickflow but the absolute and relative volumes of the two components will vary over the ablation season according to changes in input conditions and routing conditions as affected by the expansion and contraction of the capacity of drainage channels. Thus outflow hydrographs vary in size and slope from day to day over the ablation season (Elliston, 1973). The base flow consists of groundwater runoff, runoff from storage zones within the ice, runoff from the firn water aquifier and regular drainage from lakes. The quickflow component consists of meltwater from the lowest parts of the glacier basin, draining supra- and subglacially; and meltwater from the snow-free part of the glacier, which drains along short connections to the main subglacial drainage.

The various drainage paths in a glacier are shown in Figure 4.5. If the glacier surface is covered in snow and firn, water will gradually percolate through this layer until it reaches the glacier ice. Several studies have shown that appreciable storage of water occurs in this snow and firn layer (see, for example, Oerter and Moser, 1982; Schommer, 1976; 1978; Lang *et al.*, 1977; 1979) and that changes in this storage may partly account for long-term runoff variations. Meltwater on the ice surface usually disappears into the glacier via crevasses and moulins and it is unusual for large surface streams to reach the glacier snout. Individual passageways within the glacier join together to form a partially integrated

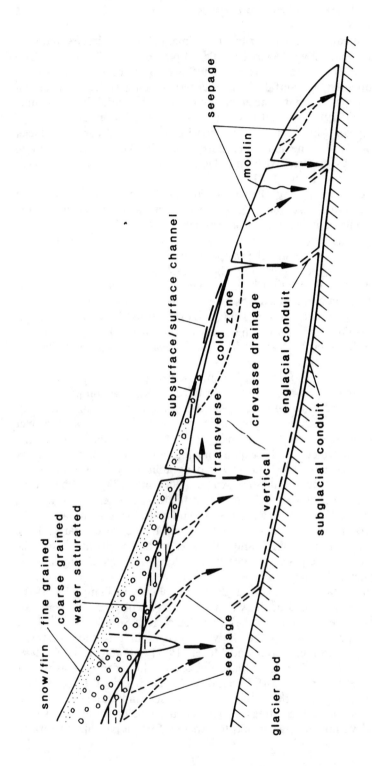

Figure 4.5. Drainage paths in a typical glacier.

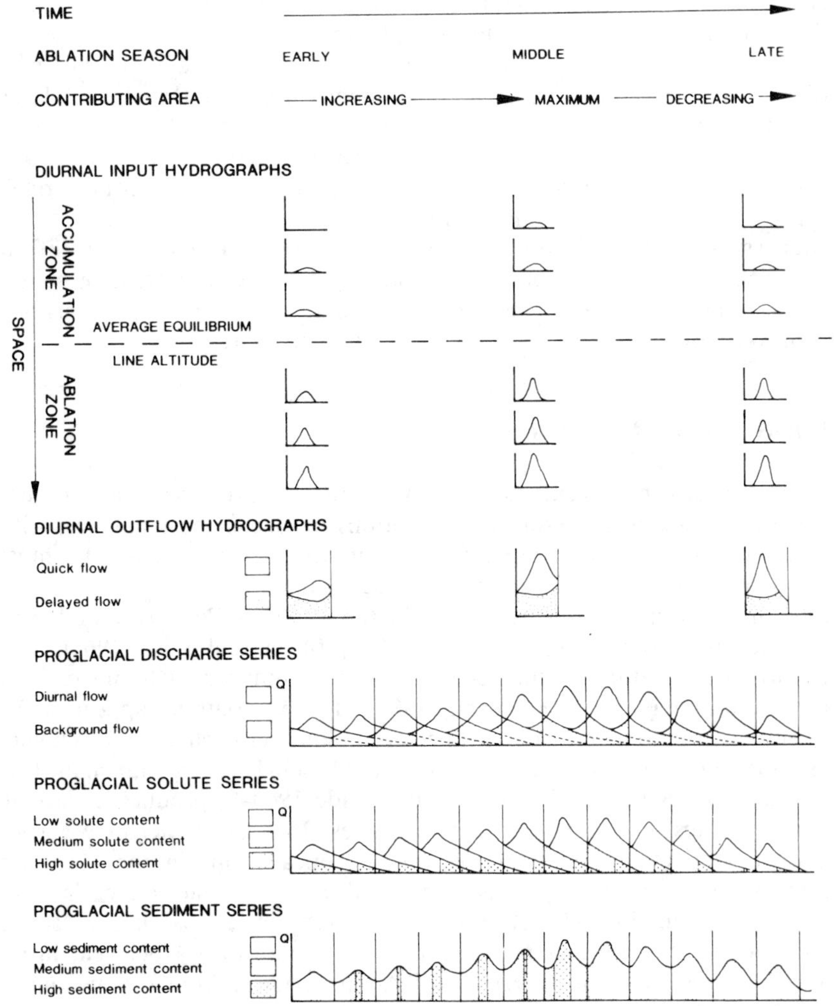

Figure 4.6. Hydrograph models for glacier-fed streams.
Source: Fenn (1987).

system which eventually exits in a small number of subglacial tunnels.

Daily hydrographs vary systematically throughout the ablation season as the glacier drainage system develops. During the winter most of the water passages in glaciers decrease in size, and the drainage system is incapable of transmitting the water supplied by fresh melting in the spring. This results in high pressures which may affect the movement of the glacier, enabling new passages to develop and the enlargement of the glacier drainage system (Haefeli, 1970; Muller and Iken, 1973; Iken *et al.*, 1983, Hooke *et al.*, 1985). There often occurs what has been called the 'spring event', which includes high water levels close to the glacier surface in moulins, the sudden drainage of surface water at the same time as an ice velocity peak, a simultaneous uplift of the glacier, large sediment loads with a high percentage of fines, and sudden changes in the sub- and proglacial drainage pattern (Rothlisberger and Lang, 1987). The various hydrograph characteristics have been summarised in 'model' form by Fenn (1987) (Figure 4.6). The models attempt to combine changes in the diurnal input and outflow hydrographs and the quickflow and delayed flow components of the proglacial discharge time series. Models have also been developed for solute and sediment transport. Transport of solutes and sediments is discussed later.

LONG-TERM VARIATIONS

There can also be considerable variations in discharge from year to year which are usually the result of fluctuations in glacier mass balance, with weather conditions and ablation rates in the summer being the most significant. Variations in mass net budgets and their effects on runoff rates have been extensively studied in the Western Cordillera of North America (see, for example, Collier, 1957; Meier and Post, 1962). The runoff in the Batura catchment in the Karakorams was 840 mm in 1974 and 1230 mm in 1975 (Ferguson, 1984). These variations appear to be essentially random and are distinct from long-term changes in climate and the mass balance of glaciers. In the Aletschgletscher, maximum loss of glacier mass occurred in the warm decade 1940–9, producing extreme ablation rates and maximum runoff values. In general, negative glacier mass balances have characterised the European Alps in the twentieth century. In the Swiss Alps there were 1817.6 km² of glacier area in 1876 and 1342.2 km² in 1973 (Kasser, 1981). This glacier loss has caused a reduction of the annual specific runoff of 35 mm or 3.4 per cent in an average year. The general retreat of Alpine glaciers from the maximum of the Little Ice Age was halted in 1965 but increasing mean air temperature resulting from the 'greenhouse' effect is likely to trigger another phase of glacier reduction.

OUTBURST FLOODS

Several studies have demonstrated that major discharge events occasionally occur unrelated to the factors discussed above. In glacierised basins glacier outburst floods, débâcles or jökulhlaups, can produce a dramatic increase in discharge. One of the commonest ways in which such floods occur is if a main valley is occupied by a glacier while the tributaries are ice-free. Water is then trapped in the tributaries and is often released rapidly through or under the ice dam. In the Karakorams, rapid glacial advances have frequently trapped river flow, creating lakes which have been suddenly released with catastrophic effects. There have been at least seven such incidents in the Shimstal Valley since 1884. In that year an ice dam burst leading to a 3 m rise in the river level. Similar events were experienced in 1893 and 1905, while in 1906 the river rose by over 15 m above its normal summer flood level in a similar situation.

Hewitt (1982) has noted that thirty-five destructive outburst floods have been recorded in the past 200 years in the Karakorams, thirty glaciers are known to have advanced across major headwater streams of the Indus and Yarkand rivers, and there is evidence of large lakes ponded by eighteen of these glaciers. In addition, a further thirty-seven glaciers interfere with the flow of trunk streams in a potentially dangerous way.

Sometimes, ice-dammed lakes form due to glacier sliding, involving the rapid forward movement of ice and the blocking of rivers. Movements of such kind have been observed in the Andes but not the European Alps. Lakes may also be impounded in the angle between two confluent glaciers. A retreating glacier can create situations in which catastrophic floods can occur. As the ice melts, small lakes may develop on the ice tongue which may coalesce to form supra- or proglacial lakes. Eventually frontal lakes will form in depressions in the outwash area. About 230 such lakes are found in the Cordillerra Blanca of Peru, some of which have emptied violently with great damage and loss of life. This series of outbursts began only after a widespread glacier recession set in the 1920s (Lliboutry *et al.*, 1977). Outbursts can also originate in water pockets formed within or beneath a glacier. Such a catastrophic outburst occurred in 1892 from the Tête Rousse glacier in the French Alps. It appears that pockets of meltwater became trapped within the ice at a point where the glacier was passing over a rock sill. The pressure of water which gradually accumulated (about 200 000 m^3) was such that the ice barrier gave way and a violent débâcle occurred (Vivian, 1974; 1979).

Climatic and glacier fluctuations play a major role in the formation of such floods. It has been observed that glacier débâcles tend to occur in late summer and autumn. This appears especially true of Washington State, USA (Richardson, 1968). Liestol (1955) has argued that this timing is because of the way in which the physical processes operate. The sudden release of ponded water caused by landslides can have a similar effect. In the 1850s the Phungurh landslide near Sarat created a floodwave on the Indus which raised the water level by about 9 m in less than ten hours and the discharge may well have exceeded 10^6 m^3 s^{-1}. The blocking of rivers by mass movement can be very short-lived — for example, the mudflow which in 1974 blocked the Hunza River in the vicinity of the Batura glacier for only one hour. Nevertheless, the floodwave that must have been created would have been noticeable on the hydrograph and must have resulted in considerably increased sediment yield.

FLUVIAL CHARACTERISTICS

River channels in mountain regions differ in many important respects from lowland channels. As Bathurst (1987a) has stressed, historically most research into sediment transfer processes in rivers has concentrated on sand- and silt-based rivers of lowland areas and it is only recently, partly in response to development pressures, that a similar level of interest has been shown in the gravel- and boulder-bed rivers of mountain regions. Channels of mountain rivers are usually steeper with a greater bed roughness and either have resistant non-alluvial boundaries or very narrow and discontinuous floodplains. These characteristics have been shown by many workers (for example, Kellerhals, 1972; Russell, 1972; Caine and Mool, 1981) to have a major influence on hydraulic performance. Direct input of sediment from hillslopes is thus facilitated and sediment transport by rivers is often related to the episodic nature of slope input. Width of channels incised into glacial or periglacial debris is closely related to upstream discharge regime. However, in basins modified by glacial action, channel slopes may be considered to be independent of both discharge and drainage area. As an example, the downstream hydraulic geometry of the Green and Birkenhead Basins, British Columbia, was found not to conform to the usual relationships because of the effect of recent glaciation (Ponton, 1972). Equilibrium had not been established in the channel systems and the rapid decrease in channel slope implied that grade had not been reached.

Two major factors have to be considered when examining the nature of mountain rivers; the wide range of sediment sizes and the spatial and temporal variability of sediment sources. The bed material of mountain streams commonly varies from sand to gravel, cobbles and boulders often greater than a metre in diameter. This means that sediment transport is affected by bed armouring and packing effects. Also, in the highly seasonal flow regimes of many mountain areas, there are short periods of very intense bedload transport separated by long periods of very little transport. Miller (1958) has observed that a large fraction of bed material is immobile even at bankfull stages.

As mentioned earlier, magnitude–frequency relationships may be different for mountain streams. In some areas bankfull discharge appears to occur less frequently than in lowland regions, and high-magnitude infrequent floods appear to be especially significant in shaping the channels. However, it is difficult to generalise as differences occur from mountain region to mountain region. Also, the majority of the detailed hydrological studies have been conducted in alpine regions which either possess appreciable snow and ice cover or have experienced glacial conditions in the past. Relationships in tropical mountains have been studied less intensively but it is clear that steepland channels, wherever they occur, behave differently from lowland channels. In order to assess the ways in which mountain streams differ from lowland rivers, some basic characteristics of river flow need to be examined.

CHARACTERISTICS OF WATER FLOW

Flow in open channels can be described as steady or unsteady, uniform or non-uniform (varied). Steady flow occurs when the velocity does not change in magnitude or direction. Flow is uniform when velocity is constant with distance along the channel. These definitions also imply that with steady flow there is constant stream depth over time, whereas with unsteady flow there is changing depth. For uniform flow, depth must be constant over distance. Flow can be classified in terms of viscous forces defined by the Reynolds number Re:

$$Re = \frac{\varrho V^2/L}{\mu V/L^2} = \frac{\varrho VL}{\mu} = \frac{VR}{\eta}$$

where ϱ is density of water, V is mean velocity of flow, L is a length dimension which in this instance is the hydraulic radius R, μ is dynamic viscosity and η is kinematic viscosity where $\eta = \mu/\varrho$. Laminar flow occurs when Reynolds number is low (less than 500) and viscous forces dominate. When the Reynolds number is high (greater than 2000) flow becomes turbulent.

Flow can also be classified by using the Froude number Fr:

$$Fr = \frac{V}{gD}$$

where V is mean velocity, g is the gravitational constant, and D is water depth. If $Fr < 1$ velocity is low and flow is subcritical, tranquil or in the lower flow regime. If $Fr = 1$, flow is critical or transitional. When $Fr > 1$ velocity is high and flow is supercritical. The Reynolds and Froude numbers can be combined to describe flow characteristics (Table 4.2).

Sudden changes in depth can lead to a change in flow regime. One such change is known as a hydraulic drop where there is a change from a gentle to a steeper slope (Figure 4.7a). Flow changes from subcritical to supercritical as depth suddenly decreases. A hydraulic jump occurs when there is a change from a steep slope to a gentler one with the sudden change in depth and velocity causing turbulence (Figure 4.7b). A hydraulic jump can also be caused by an obstruction in the channel (Figure 4.7c).

This synthesis can be used to examine the nature of mountain streams and to assess how they differ from lowland streams. Water flow in mountain streams is often characterised by frequent hydraulic jumps,

Table 4.2 *Differentiation of flow regimes.*

Regime	Froude number	Reynolds number
Subcritical-laminar	< 1	< 500
Supercritical-laminar	> 1	< 500
Subcritical-transitional	< 1	500–2000
Supercritical-transitional	> 1	500–2000
Critical	1	any value
Subcritical-turbulent	< 1	> 2000
Supercritical-turbulent	> 1	> 2000

Source: Morisawa (1985).

large velocity fluctuations, irregular and ill-defined flow areas and entrainment of air into the flowing water (Kellerhals, 1972). This type of water movement has been called 'tumbling flow' (Peterson and Mohanty, 1960). Energy dissipation is less dependent on boundary friction and channel slope has much less influence on hydrological relationships than in lowland rivers. The effect of channel slope decreases with increasing channel roughness as proportionally more of the flow occurs in pools and jumps, Also the pool sequence seems to exist during very high discharges in the steeper channels (Day, 1972). Most tumbling flow channels are self-formed through degradation into debris on valley floors. The steeper reaches are rougher than flatter ones and larger channels are rougher than smaller ones (Kellerhals, 1970).

SEDIMENT TRANSPORT

The load carried by natural streams can be separated into three components:

(i) the bed load, usually the larger particles which are moved by sliding, rolling or bounding (saltation) across the bed;
(ii) the wash load, comprising finer particles (less than 0.64 mm in diameter) which are moved readily in suspension;
(iii) the dissolved load, consisting of material transported in solution.

The relative proportion of bedload to suspended load appears to vary with the coarseness of the bed material. In sand bed and gravel bed rivers the proportion tends to vary between 5 per cent and 20 per cent, whereas this value may rise to 50 per cent for boulder-bed steams. Some studies (Milhous and Klingeman, 1973; Hayward, 1979) have reported that bedload transport is sometimes in excess of 75 per cent of total sediment discharge. In most alpine glacierised catchments bedload is greater than 30 per cent of total sediment load.

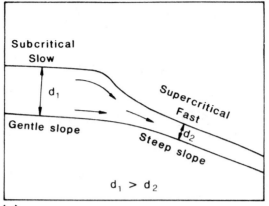

(a)

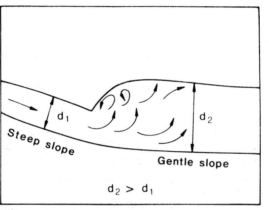

(b)

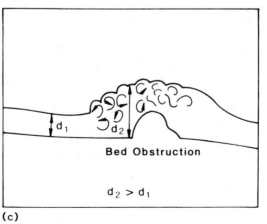

(c)

Figure 4.7. Various flow characteristics of mountain streams.
Source: Morisawa (1985).

Table 4.3. *Important variables in determining bed-material transport.*

Flow properties	Fluid properties	Sediment properties	Other properties
Discharge	Kinematic viscosity	Density	Gravity
Velocity	Density	Size	Plan-form geometry
Flow depth	Temperature	Sorting	
Width	Wash load concentration	Fall velocity	
Slope			
Resistance			

Source: Knighton (1984).

Bed load

One of the most distinctive elements of mountain streams is the movement of bed load. The nature and rate of sediment transport depends on
the relationships between flow characteristics and channel sediments
(Table 4.3). Mountain rivers differ from lowland rivers in both sediment
and flow properties. It might be expected that there is a continuous spectrum of river types as the median grain size increases from the sand
range to the gravel and cobble range. But this is not so and there appears
to be a gap with very few rivers possessing median sizes in the very
coarse sand to gravel range (Parker and Peterson, 1980). There may also
be a gap between gravel rivers and rivers dominated by cobbles and
boulders. Although gravel-bed rivers occur in lowland areas, they seem
to be especially characteristic or steeplands and mountains.

The major difference between sand and gravel rivers is that sand beds
are active beds with flows that are low enough to render the bed
immobile corresponding to nearly dry channels. Dunes are common in
sand-bed rivers and are a major contribution to bedform resistance but
are rare in gravel-bed streams. The beds of gravel rivers appear to be
mobile only at the higher flows. Bar structures formed at high flows in
sand rivers can be reworked during low flows but bars in gravel rivers
are formed exclusively at flood stages and are static at all flows but the
highest. Streams with mostly gravel or coarser material generally possess
a coarse surface pavement not to be confused with armour, which is a
coarse layer that never moves (Parker *et al.*, 1982). Laboratory
experiments have demonstrated that such pavements are mobile features
and that they act to decrease the inherent difference in mobility between
large and small grains by overrepresenting the percentage of large grains
exposed to the flow. The genesis of such pavements appears comparatively simple (Parker and Klingeman, 1982). Each time a large grain is
dislodged it leaves a 'hole' of comparable size into which smaller grains

fall, gradually work their way below the pavement and reduce their probability of re-erosion. This is a vertical winnowing process not to be confused with the development of armour, which is due to horizontal winnowing. This vertical winnowing process operates to create equal mobility of particles and to enhance the amount of coarse grains available for transport just sufficient to counter their lower intrinsic mobility.

Relationships between sediment transport, discharge and channel pattern in steep gravel rivers are extremely complicated and difficult to understand because mountain streams are difficult to measure (Bathurst *et al.*, 1986). Sawada *et al.* (1983) have established four categories of sediment transport:

(a) sediment transport with neither destruction of armoured bed nor erosion of the channel side;
(b) sediment transport with destruction of the armoured bed;
(c) sediment transport which accompanies channel side erosion;
(d) sediment production and transport due to channel course variation.

Material on the bed cannot be mobilised before the pavement is mobilised and the annual duration of flows high enough to move the pavement is generally very small. Parker and Peterson (1980) have estimated that in Oak Creek, Oregon, with median grain size 60 mm, the bed is active for only eleven days per year. But during such periods nearly all the sizes are in motion (Hollingshead, 1971). But even at bankfull conditions the bed-shear stress exceeds the critical mobilising stress only slightly, and the low-shear stresses also prevent all but the finest portion of bed material from being carried in suspension.

The threshold shear stress for initial motion of gravel in the beds of steep mountain streams is more difficult to determine than that for sand-bed streams. The standard means of calculating the critical flow conditions for the initiation of bed material movement is the Shields (1936) equation:

$$\frac{\tau_c}{(\varrho_s - \varrho)gD} = \tau_{*c}$$

where τ_c is the critical shear stress, ϱ_s is bed-material density, ϱ is water density, g is acceleration due to gravity, D is bed-material particle diameter, and τ_{*c} is the Shields parameter. For gravels and coarser materials with uniform size distributions, the parameter is usually given a value in the range 0.04–0.06. However, for bed material with non-uniform size distributions, the value of the parameter varies as inter-action between the different particle sizes affects the critical conditions for each size fraction. Thus, alternative methods of calculating the critical conditions for particle movement in steep channels with coarse, non-uniform bed material have had to be derived (see, for example,

Carson and Griffiths, 1985; Bathurst, 1987b; Wiberg and Smith, 1987).

In non-uniform sediments, smaller particles are often protected by larger particles and require higher flows to initiate movement than would be necessary for uniform sediments of the same size. White and Day (1982) have found that the stability of a particle depends on the position of its size within the overall size distribution, relative to a critical diameter — which is usually that size for which 50 per cent of the particles are finer (D_{50}). Andrews (1983) has found that for the range

$$0.3 < D_i/D_{50} < 4.2$$

$$\tau_{*ci} = \frac{0.0834 \, D_i^{-0.872}}{D_{50}}$$

where τ_{*ci} is the average critical Shields parameter for particles of size D_i in the surface or armour layer, i is the size fraction, and D_{50} refers to the subsurface or parent material.

The comparatively shallow depths and high velocities characteristic of steep mountain streams complicate estimates of critical shear stress because the presence of clasts that are large with respect to the flow depth causes the velocity profile to deviate considerably from its typical logarithmic form (Wiberg and Smith, 1987). The drag force created by flow around large clasts causes the momentum of the flow to be reduced in the near-bed zone thus decreasing the mean flow velocity. Also, as mentioned earlier, methods based on shear stress are inappropriate for boulder-bed channels.

Bed roughness can be defined in a variety of ways. Bathurst (1978) has used the ratio k/d, where k is the height of the roughness and d is the depth of flow. Lowland, alluvial rivers with low roughness generally have values less than 0.3; there is then a transitional region with roughness values 0.3 to 1.0. Large-scale roughness implies values greater than 1.0 which is when boulders emerge above the water surface. It appears that the behaviour of mountain streams is governed by the spacing of boulders that jut through the flow. An alternative way of combining roughness and bed-load transport is to use the ratio of the critical depth of particle entrainment (d_c) and particle diameter (D). Entrainment thresholds, relations between friction and roughness and flow regime begin to change as d_c/D becomes less than 15 or channel slope exceeds 0.5–2 per cent. Three domains of relative submergence have been defined (Bathurst, 1978). Bed roughness is small-scale if d_c/D exceeds 15 and it is possible to utilise the usual relations between friction and roughness. At an intermediate scale, with values between 4 and 25, a substantial portion of the flow is below the tops of protruding particles, velocities are much lower and rising wakes of low-velocity water are created by protruding particles. Large-scale roughness elements ($d_c/D < 4$) individually affect flow and sediment transport in complex ways depending on their shape, spacing and location.

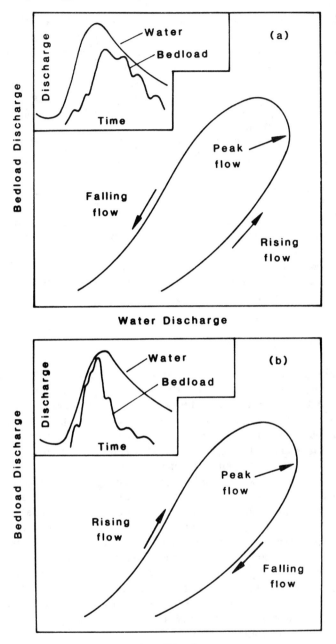

Figure 4.8. Relationships between bed-load transport and different stages of a flood hydrograph.

Source: Bathurst (1987a).

It was mentioned earlier that bed-load transport is often non-uniform and unsteady in mountain rivers. This has clearly been shown by Bathurst (1987a) and is related to the availability of sediment to transport at the different stages of a flood hydrograph (Figure 4.8). Transport may be inhibited, during the rising limb, by the presence of an armour layer. This layer may break up near the peak flow, releasing material for transport (Figure 4.8a). If the rising limb is able to entrain sediment which has accumulated in the channel or which has recently become available by a landslide, the opposite may occur (Figure 4.8b). Rapid removal of sediment might mean that transport rates on the falling limb are low.

Bathurst (1987a) has also noted seasonal variations dependent on the source of the high flows. Rivers dominated by snow-melt regimes often carry high bed loads in the early meltwater flows because of the availability of sediment produced during the winter. Reduction of supply in the summer means that similar flows later in the year may carry less bed load. Also a snow-melt flood is likely to carry less sediment than a rainfall flood of the same magnitude because the latter will be supplied by the erosion from overland flow.

Suspended sediment transport

The distinctiveness of suspended sediment transport in mountain rivers results from the interaction of the abundant availability and highly variable nature of erodible sediments and characteristic flow regimes. But, apart from a few intensively studied glacierised basins, there are few detailed measurements of suspended sediment load. Also, the information that does exist must often be treated carefully because of different measurement techniques and the problems of sampling.

Ferguson (1984) has provided some interesting information concerning suspended sediment transport on Karakoram rivers. The trans-Karakoram headwaters of the Indus in Xijang and Ladakh provide relatively little sediment compared with the main Karakoram range. These differences appear to be related to relative relief and percentage glacial cover. The mean yield per unit area of $4800 \text{ t km}^{-2} \text{ yr}^{-1}$ for the Hunza River is twenty-five times the world average. It is also high when data from other mountain regions are considered. Only two other mountain regions, the Nepal Himalayas and Southern Alps, New Zealand, plot as high as the western Karakorams. Both these regions are similar to the Karakorams in tectonic activity and extent of glacial cover but each receive much greater annual rainfall and snowfall.

Gurnell (1987) has analysed suspended sediment transport of forty-three drainage basins with glaciers, not all of which are in mountain regions. The data presented in Figure 4.9 are subdivided into glaciated mountain catchments, subarctic mountain catchments and arctic

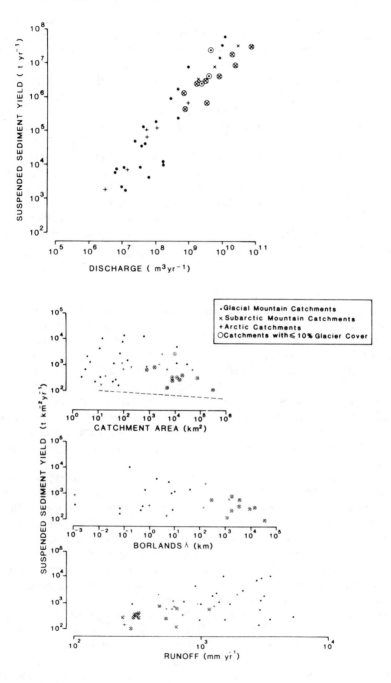

Figure 4.9. Suspended sediment relationships.
Source: Gurnell (1987).

catchments. Drainage basins with less than 10 per cent glacier cover are indicated. There is a strong relation between total suspended sediment yield and discharge volume, which is not surprising since the former is calculated from suspended sediment concentration and discharge. There is a great range in suspended sediment yield and discharge volume from less than 700 t yr^{-1} and 1×10^6 m^3 yr^{-1}, respectively (Hilda Glacier, Hammer and Smith, 1983) to over 63×10^6 t yr^{-1} (Hunza River, Ferguson, 1984) and over 82×10^9 m^3 yr^{-1} (Yukon River, Guymon, 1974).

Relationships between suspended sediment yield and some possible explanatory factors are also shown in Figure 4.9b. The relationship obtained by Walling and Kleo (1979) between suspended sediment yield and catchment area from their world suspended sediment data base is shown by a dotted line. All the glacier catchments plot above the line. Borland (1961) derived an index (λ) for estimating sediment yield from glacier catchments in Alaska:

$$\lambda = (A_T L_G)/A_G$$

where A_T is the total catchment area, A_G is the glacier cover of the catchment, and L_G is the length of river from the glacier snout to the monitoring site. But the data analysed by Gurnell (1987) indicates a poor fit for this relationship (Figure 4.9c). Figure 4.9d provides the relationship between suspended sediment yield and runoff. There are no real indications of differences in the relationships between catchments except that glacier mountain catchments possess higher runoff rates. Suspended sediment load and runoff are often linearly correlated. There may be a real causal link in this relationship since more runoff implies larger runoff source areas, higher river levels and the tapping of more extensive sediment stores.

Suspended sediment load in mountain rivers fluctuates over several time-scales. Most information concerns fluctuations over annual, seasonal and daily time-scales but it is possible to speculate on fluctuations over longer time spans. Over 10^2–10^3 years there can be marked variations, both high and low, related to abnormal years. Lakes may burst, landslides may affect rivers and there may be glacial advances and retreats. Some of these events create long-term trends superimposed on more random events. Over 10^4–10^5 years fluctuations will be associated with major climatic changes and glacial–deglacial cycles.

Ferguson (1984) has noted an almost twofold year-to-year range in runoff at six gauging sites on Karakoram rivers, and a three- to seven-fold range in annual suspended sediment yield, but variations in load at different stations are synchronous, suggesting a common cause. The connecting factors appears to be yearly variations in glacier mass balance.

The largest fluctuations of suspended sediment transport occur at the seasonal time-scale. This is well shown in Figure 4.3 for the Hunza River, referred to earlier. An approximately twentyfold increase in discharge from March to July is associated with a greater rise in suspended sediment concentration from 0.01–0.1 kg m^{-3} in winter to peaks of over 10 kg m^{-3} in summer. Between 97 and 99 per cent of annual load is carried between 1 June and 30 September. Winter movement of suspended sediment is limited by restricted availability of fine sediment to winter runoff and not by the transporting capacity of the rivers. The increase in sediment concentrations in the period from March to July reflects the tapping of fresh sediment sources as snow and ice melt occur at progressively higher elevations and as swollen rivers flood previously dry parts of the valley floors.

Sediment availability also accounts for a clockwise seasonal hysteresis loop in suspended sediment. A given discharge tends to be associated with a higher sediment concentration in spring or early summer than in later summer or autumn. This is a consistent feature of many Karakoram rivers and seems to be related to an initial depletion of the finer grain sizes from sediment source areas. Also, deposition of silt and sand occurs at channel margins during falling river stages to provide stores of material for movement the following year. This storage of sediment is examined in greater detail in the section on basin sediment budgets.

There are also fluctuations in suspended sediment movement over diurnal cycles, especially in glacierised basins. Discharge generally lags behind suspended sediment concentration and there is a similar exhaustion effect. Many workers have also noted sudden, apparently random changes of suspended sediment concentration, often without a major change in discharge. These anomalies may be generated by sudden increases in sediment inputs to the rivers, such as slope failures. Sudden drops in concentration are more difficult to explain but may be related to the sudden initiation or extension of sediment sinks.

Excessive sediment transport is associated with glacial meltwater outburst events. Because of the nature of such events there are very few suspended sediment measurements. Two such outbursts were noted from the Tsidjiore Nouve glacier during 1981 (Beecroft, 1983; Gurnell, 1987). The first outburst resulted in 19 per cent of the season's suspended sediment load being transported by less than 5 per cent of the discharge value and the second event transported 35 per cent of the load in 15 per cent of the discharge volume. Thus these two events accounted for over 50 per cent of the suspended sediment transport during the three-month monitoring period. The first outburst was more effective in transporting sediment because it was able to tap the sediment usually available at the start of the meltwater season.

The amount of suspended sediment transported by mountain rivers will depend on how the sediment is delivered to the rivers. In non-glacier basins sediment delivery is mainly through surface runoff, channel incision, bank erosion and mass movement activity. In glacierised basins

Table 4.4 *General trends in suspended sediment concentration in glacierised catchments.*

Location	Trend
Supraglacial, englacial, subglacial,	Suspended sediment concentration generally lower in supraglacial than in englacial, subglacial or ice-marginal streams.
ice-marginal streams	There is a possible trend towards increasing suspended sediment concentration towards the glacier snout, which has been demonstrated for supraglacial streams.
Tributary streams from the glacier snout	Diurnal hysteresis in the suspended sediment concentration-discharge relationship.
	Major and minor single and multiple tributary suspended sediment flushes: minor to major meltwater outbursts.
	Rationalization of glacial meltwater drainage system during ablation season leading to diversion of majority of flow and suspended sediment transport to one (or more) increasingly dominant stream(s).
Main proglacial river	Diurnal, sub-seasonal and seasonal hysteresis in the suspended sediment concentration — discharge relationship.
	Major differences in the level of suspended sediment transport between years (presumably as a result of differences in the sediment supply).
	Great impact of meltwater outbursts on suspended sediment transport.
	Flushes of sediment from glacial and proglacial sediment sources. Reduction of suspended sediment transport by proglacial sediment sinks, particularly lakes.
	Downstream initial increase followed by decrease in suspended sediment concentration, in the absence of major lateral inputs of sediment.

Source: Gurnell (1987).

will also be delivered by transport along supraglacial, englacial, subglacial and ice-marginal routes. Proglacial rivers away from the ice margin will have abundant sources of sediment in outwash materials. Supraglacial streams carry very variable concentrations of suspended sediment. Hammer and Smith (1983) considered that concentrations were low in supraglacial channels compared with proglacial streams. There is little information on suspended sediment transport by englacial streams but more for subglacial streams (see, for example, Vivian, 1970; Vivian and Zumstein, 1973; Hagen *et al.*, 1983). Concentrations vary from between 15–200 mg 1^{-1} to 785 mg 1^{-1}, the latter level occurring during flash floods. Streams fed by snow melt and groundwater generally have low suspended sediment concentrations (Gurnell, 1982; 1983).

Studies of downstream variations in suspended sediment concentration in proglacial streams have produced conflicting results. Increases in a downstream direction have been noted by many workers (for example, Mills, 1979; Fahnestock, 1963; Maizels, 1978) but the opposite trend has also been reported (Gurnell, 1982). The differences are probably related to the spatial distribution of sediment sources and sinks along the stream channel. Lakes are particularly important sediment traps and reference can be made to a number of quantitative studies (Mathews, 1956; Gilbert, 1975; Gilbert and Shaw, 1981; Smith *et al.*, 1982).

Suspended sediment concentration · in mountain streams is clearly related to a number of environmental factors of which the most important are the discharge regime of the rivers and the nature and availability of sediment. Glacierised basins and basins that have been recently glaciated will possess a considerable store of material to be reworked gradually as erosion penetrates the basin. Some general trends in suspended sediment concentrations in glacierised catchments are shown in Table 4.4. In other basins deforestation, land-use pressures and trampling will all increase the movement of sand and silt particles to the rivers.

Solute transport

Transport of dissolved load in mountain rivers appears to be similar to that of lowland rivers but there is likely to be considerable regional variation because of differences in chemical weathering rates and water sources. Unlike suspended sediment, solute load concentration decreases with increasing discharge because of the dilution effects of quick flows. Thus dissolved load is more uniformly distributed over time. The dissolved load in rivers draining tropical mountains and mountains with considerable soil and vegetation cover is generally higher than in rivers draining high alpine regions. But, as noted in Chapter 3, chemical denudation must not be underestimated in such regions. The annual dissolved load measured at Dainya Bridge on the Hunza River is equivalent to a gross yield of 90 t km^{-2} yr^{-1} (Ferguson, 1984), about three times the world average (Meybeck, 1976). But even so, this was only 2 per cent of the total river load.

Values of solute loads obtained from glacierised basins also indicate quite high values. A cationic denudation rate of 930 μequiv m^{-2} yr^{-1} was obtained from the South Cascade glacier basin in the Rocky Mountains (Reynolds and Johnson, 1972) and a rate of 947 μequiv m^{-2} yr^{-1} for the Berendon glacier basin in British Columbia (Eyles *et al.*, 1982).

Rates appear to be related to rainfall amounts. Chemical weathering, including ion exchange, seems to be very effective in Alpine basins despite the near absence of normal biologic and pedogenic processes (Souchez and Lemmens, 1987).

BED FORMS

Bed forms in steep channels are usually extremely complicated, depending on a combination of channel steepness, nature of bed material and flow characteristics. Bed material can be arranged in a variety of forms ranging from sand sheets and gravel clusters to transverse ribs, step-pools and bars and braids. Sand sheets are migrating accumulations of bed load one or two grain diameters thick that alternate between fine and coarse particles (Whiting and Dietrich, 1985; Lisle, 1987). Where bed load is a mixture of sand and gravel, the material may be sorted into alternating mobile zones of high grain roughness (gravel) and low grain roughness (sand). The gravel component travels rapidly across the smooth sand zones and congregates downstream where other gravel particles produce higher grain-to-grain friction. Observations and experiments have shown that the formation of such features and sand and gravel transport is discontinuous and unpredictable. Also it is difficult to relate such transport to the standard hydraulic variables.

The great variation in size of bed load in mountain channels can create clusters of particles where small particles cluster around one or more large particles (Brayshaw *et al.*, 1983). Clusters have a significant role in sediment transport because particles in the cluster do not move until the key particles do. The size of clusters tends to be small in comparison with the size of the channel. A variant of clusters, known as 'transverse ribs', also occurs where lines of large clasts, usually one or two diameters wide, stretch across the channel (Koster, 1978). Transverse ribs in gravel rivers may correspond to step-pool sequences in steep cobble channels.

Step-pool sequences are common bedforms in channels with relatively low throughput rates of sediment and slopes greater than 7 per cent (Whittaker, 1987). Steps are thought to form during low rates of transport of large clasts or during the final stages of a debris flow or torrent (Sawada *et al.*, 1983). Three types of steps are possible (Hayward, 1980; Whittaker, 1987):

(a) Boulder steps, which consists of a group of boulders arranged in a straight or curved line across the channel.
(b) Riffle steps, which are a collection of larger than average size sediments that steepen the channel. Riffle steps may incorporate boulder steps.
(c) Rock steps, which may occur where the channel is cut into bedrock.

In forested areas some steps may be associated with log jams and tree obstructions (Heede 1972a; 1972b; Keller and Swanson, 1979).

Sediment transport in step-pool systems has distinctive characteristics. Sediment for transport is derived from limited sites within the catchment and once in the stream system the sediment is stored in the pools where transport ceases. The time of travel of sediment in a series of pools and steps can be conceptualised as the time of travel for one pool multiplied by the number of pools (Beltaos, 1982). Thus bed-load transport rates vary spatially and temporally. Whittaker and Jaeggi (1982) had made the comparison between step-pool systems in boulder channels and riffle-pool sequences in channels dominated by gravel and cobble material. Riffle and pool rivers generally possess gentler slopes with bar formation and meandering. Riffle-pool sequences are generally spaced five to seven times the channel width. Step-pool systems possess shorter wavelengths, and there appears to be no correlation with channel width. The formation of steps and pools is associated with high-intensity, low-frequency floods and heterogeneity of bed material size — both characteristics of mountain areas.

Bars are characteristic of gravel rivers. The simplest morphology is of alternating diagonal bars where each bar is a tongue of gravel extending from a deep pool along one bank at the upstream end to a diagonal front at the downstream end. It is thought that bars form in gravel channels to transport bed load where entrainment thresholds are, on average, exceeded by only a small factor. Bars are also associated with braiding which is characteristic of glacial outwash plains, semi-arid regions of low relief which receive discharge from mountain areas, as well as highland areas in a variety of climates. Braided channels consist of two or more channels, divided by bars or islands, but with one channel usually dominant. Such rivers also possess high width–depth ratios, steep slopes and large bed loads. Braids are caused by the deposition of coarse material which the river is locally incompetent to transport and which acts as a focus for further deposition. When the bar is large enough the main flow becomes concentrated in the flanking channels which erode the banks and scour the beds until a central bar emerges as an island (Davoran and Mosley, 1986).

Knighton (1984) has summarised the conditions conducive to braiding:

(a) An abundant bed load which should also contain size fractions which the stream is locally incompetent to transport.
(b) Erodible banks which are an important source of material as well as being necessary for the channel widening characteristic of braided reaches. Without erodible banks incipient bars would be destroyed.
(c) A highly variable discharge which, in combination with high rates of sediment supply, contributes to bank erosion and irregular bed-load movement, all conducive to bar formation.
(d) Steep slopes are important because there are indications that braiding develops when channel slope is above a threshold value. Also, in

theory, the degree of braiding seems to increase as slope steepens. The critical factor may be stream power, which is defined as the product of discharge and channel slope.

None of these conditions is exclusive to mountain regions but it can be argued that it is in mountains that they are best developed.

BASIN SEDIMENT BUDGETS

The sediment budget of a drainage basin is a quantitative statement of the relations between sediment mobilisation and discharge and of associated changes in storage. It is the most complete analysis of a drainage basin possible because it forces attention on the linkages between sediment transport, including slope processes, channel morphology and discharge. Drainage basins, including channel systems, contain various stores, such as hillslopes, swales, hollows or zero-order channels, alluvial fans, active channel and bed bars, terraces characterised by different levels of activity. The transit times and fate of particles entering a particular store are sensitive to interchanges with stores having different turnover times.

To produce a basin sediment budget the volume of stores and the age distribution of the sediment stores must be measured. This is often very difficult to achieve, especially in mountain drainage basins with complex interactions of geomorphological processes and sediment stores of a variety of ages and origins. Even when sediment reaches the streams, its downstream transfer is discontinuous due to concentrations of input and tendencies for storage of varying lengths of time. Swanson *et al.* (1982) have noted that the volume of material temporarily stored in bars and along channels is often more than ten times the average annual yield of total particulate sediment. Also most of the sediment in channels can only be moved at high flows. Because of these problems it is not surprising that only a few small mountain drainage basins have been analysed in this way. However, the results of such analyses do provide a valuable insight into the behaviour of mountain drainage basins and allow a few general principles to be established.

One of the most intensive studies has been conducted in the basin of Lone Tree Creek, 14 km north-west of San Francisco on the western slope of Mount Tamalpais (Lehre, 1981). The small (1.74 km^2), relatively undisturbed basin, underlain by greywacke melange, possesses narrow steep-walled canyons (20–40°), separated by gently rounded upland ridge crests (2–17°). The basin has a drainage density of 8.98 km km^{-2} and a relief ratio of 190 m km^{-1}. The model of the linkages between storage sites and erosional processes proposed by Lehre (1981) is shown in Figure 4.10. Swales are a major component of the model and emphasise the importance of zero-order channels mentioned earlier. Swales are created by debris slides, producing depressions in which sheet erosion and rilling operate. A gully may develop which joins the scar to the channel system, and with time the swale may become filled, producing a smooth, U-shaped, colluvium-filled depression. The process is repeated when the colluvium fails in heavy rainstorms. Many other works (for example Tsukamoto and Minematsu, 1987; Alger and Ellen, 1987; Reneau and Dietrich, 1987) have noted the association between swales and debris flows. Material on hillslopes outside of swales is transported by creep, surface wash and occasionally landslides, and eventually enters the channel system through gully, bank or headcut erosion. Material reaching the channel bed may be moved from the catchment as suspended load and bed load, stored temporarily in bars and channel deposits or stored more permanently in terraces or behind debris jams.

The three main components of the basin are sediment yield, the sediment mobilised (moved any distance) and sediment production (reaching or giving access to a channel). The redistribution of material on slopes is the difference between the total amount of sediment mobilised and that produced by slope process. It is the amount of material moved from one hillslope store to another farther downslope. Negative values indicate that the transfer of sediment to bed/bank storage exceeds new mobilisation on hillslopes. The difference between production to channels and sediment yield gives the change in bed/bank storage. Negative values indicate that removal of sediment from bed/bank storage exceeds resupply from hillslope processes. Positive values represent the combination of new accretions to storage from slopes above and redistribution of material in bed/bank storage.

Lehre (1981) found that in dry years and in wet years without extreme flow events most of the sediment mobilised returns to storage. Large net removal of sediment from storage occurs in flow events with recurrence intervals greater than ten or fifteen years. Table 4.5 summarises mobilisation and production of sediment for the years 1971–4. In an average or dry year gully wall and slide scar erosion are chief mobilisers of sediment, whereas in years of extreme rainfall and flow events debris slides and flows are responsible for over 80 per cent of all sediment reaching channels. As nearly all the material is contributed directly to the channel during conditions of high flow, most of it is carried out of the basin.

The balance of mobilisation, production and storage for the three years is shown in Table 4.6. Three years is only a short time in the overall long-term behaviour of the basin but it gives a very good insight into the linkages between sediment and channel processes.

Debris slides and debris flows were the most important erosional agent, accounting for at least 53 per cent of the estimated long-term particulate yield. This seems to be true of most mountainous areas, whether in the alpine zone with appreciable bare areas or in tropical zones with a dense vegetation cover. The extreme role of debris flows in

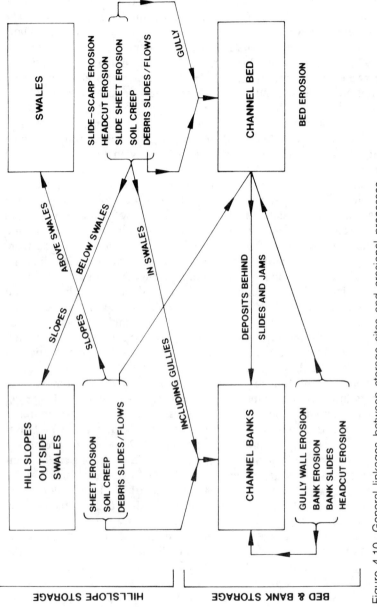

Figure 4.10. General linkages between storage sites and erosional processes. *Source:* Lehre (1981). **Note:** boxes indicate storage elements; listed below each box are erosional processes mainly responsible for mobilising sediment in that element. Arrows show transfers between elements; labels on arrows qualify or restrict location of transfers.

Table 4.5. *Mobilisation and production of material in Lone Tree Creek, California, 1971–4 average (tonnes per square kilometre per year) (after Lehre, 1981).*

Source	Mobilisation	Production
Landslides	948	594
Slide scar erosion	45	28
Slide scar sheet erosion	68	53
Headcut erosion	18	17
Soil creep	6	
Hillslope sheet erosion	4	159
Gully scar		
Bank erosion		16
Bed erosion		37
Total	1089	904

Table 4.6. *Summary of sediment movement and storage in Lone Tree Creek, California, 1971–1974 (after Lehre, 1981).*

Year	1971-2	1972-3	1973-4	1971-4
Rainfall (mm)	640	1215	1090	2945
Recurrence interval of peak flow (yrs)	< 1.5	15–20	3–5	
Mobilisation on slopes (t km^{-2})	86	1219	1960	3265
Production to channels (t km^{-2})	148	985	1575	2708
Redistribution on slopes (t km^{-2})	−71	317	389	635
Yield (t km^{-2})	24	1420	630	2074
Bed and bank storage (t km^{-2})	124	−435	945	634

basin sediment dynamics has been noted by Kang Zicheng and Li Jing (1987) in Tibet. In one basin studied (26 km²) debris flows contributed 5.2 × 10⁶ m³ and 7.8 × 10⁶ m³ of sediment in 1964 and 1965 respectively. Pickup *et al.* (1981) have emphasised the contribution of landslides in the tropical rainforest mountains of the Fly River basin, Papua New Guinea. Many of the slides are small and occur in weathered regolith and reworked alluvium and colluvium. Such sliding is almost continuous and adds essentially fine material to the rivers because the regolith is highly weathered. Major landslides, with a larger recurrence interval, cause debris flows' and more substantial channel changes. Slope wash is also an important contributor of sediment, with saturated soils and high rainfall intensities producing overland flow. There is often a well-developed network of rills and gullies beneath the rainforest. Loffler (1977), also in Papua New Guinea, has noted that the intensity of overland flow decreases significantly with altitude where throughflow operating below the dense root mat is more important. Widespread saturation overland flow has also been reported on kaolin-rich soils on valleyside slopes in mountainous rainforest in Queensland, Australia (Bonell and Gilmour, 1978).

The role of debris flows in sediment routing over longer time-scales has been noted by Benda and Dunne (1987) in the Oregon Coast Ranges of the USA. They were able to date forty-six debris flows in a fifth-order basin by dating buried charcoal. This suggested that debris flows occurred in first- and second-order channels with recurrence intervals of 1500 and 750 years, respectively. These time intervals were used to construct sediment budgets for the basins. The first- and second-order basins stored the majority of sediments supplied to them and release it periodically by debris flows. This shows a strong stochastic character to the routing of sediments that includes alluvial channels alternating between aggradation, degradation and armouring.

Similar results were obtained by Dietrich and Dunne (1978) in the Rock Creek basin in the Coast Range of Oregon. Using a slightly more complicated model, they were able to establish the linkages, transport rates and time in storage. Debris flows were again found to be important and the role of alluvial fans as long-term stores was stressed. Debris fans at tributary mouths store the equivalent of 800 years of bed-load discharge and 87 years of suspended load discharge. The role of alluvial fans in storing sediments has also been stressed by Loughran et al. (1981) in the mountains of eastern Australia. Material, in moving from one store to another, can either remain unsorted or unaltered, or it can experience breakdown or sorting during transport.

The average residence time of soil on hillslopes was estimated to be 20 000 years and rates of soil creep and sediment discharge from the basin are in balance. Approximately half of the soil transported to channels is carried away as suspended load, with the remainder being stored in debris fans, channels and the floodplain. Large streamside obstructions and bedrock beds can also trap sediment temporarily (Lisle, 1986). Residence times in these stores increase downvalley from decades to about 10 000 years.

The discontinuous nature of sediment transport in mountain rivers has already been stressed. Sediment budget analysis allows estimates to be made of rate of transport. Dietrich and Dunne (1978) estimate transport rate through gravel bars in the Rock Creek basin as 236 m yr^{-1}. But they also point out that if the velocity of all particles in a mobile bed is considered, including those which are under the armour layer but still move occasionally, the average distance travelled per year will be lower.

Similar results have been obtained in mountain rivers in Hokkaido, Japan (Nakamura et al., 1987). Average transport rate of bed load, although relatively constant for individual rivers, varied in the range 30–155 m yr^{-1} from river to river. Storage times also varied from 26.1 years to 14.9 years, and from 17.5 to 12.1 years for sinuous and straight sections.

The characteristic hydrology of mountain streams also means that movement of bed material occurs during very limited time periods. In the East Fork River on the Pacific slope of the Wind River Mountains, Wyoming, most of the bed load is moved during the few weeks of spring snowmelt runoff (Meade *et al.*, 1981). Despite the annual scour and fill the configuration of the bed is similar from one low-water season to the next. During low-water periods, movable bed load is stored in distinct areas separated by areas mostly free of movable material. These storage areas are 500–600 m apart and contain an average of 2500–3000 tonnes of movable material, an amount equivalent to the annual bed-load discharge. This implies that during a typical runoff season, the bed material is moved about 500 m from one storage area to the next.

Long term changes

The construction of sediment budgets, such as those just described, although indicating short-term fluctuations, has to assume that the basin is in approximate steady-state. Over longer time periods, with changes of climate and human impact, this will probably not be true. Thus, it is important to consider the longer-term changes that have occurred in mountain drainage basins. In the California Coast Range drainage basin, discussed earlier, Lehre (1981) has estimated that landslide frequency has increased tenfold in the past 50–150 years. This increase seems to coincide with the introduction of cattle to the basin and the conversion of grasslands form longer-rooted native bunch grasses to shallower-rooted introduced annual grasses as well as trampling and heavy grazing. Most other mountain areas are also experiencing increased erosion because of land-use changes and human activity and are far from a steady-state situation. However, not all long-term changes are due to human activity. In the mountains of North Island, New Zealand, Grant (1981) has identified five major erosion and sedimentation periods since the thirteenth century. Each appears to have been caused by a sustained increase in the frequency of major rainstorms related to long-term climatic fluctuations.

Chapter 5

SLOPE FORM AND EVOLUTION

INTRODUCTION

Mountains are characterised by a great variety of slope types and it is difficult to generalise. However, there have been a number of attempts to describe distinctive elements in particular areas or in particular environments. Barsch and Caine (1984) attempt to make the distinction between slope form in high mountains, with a relative relief of more than 1000 m, and that in lesser mountains. However, as they point out, apart from the work of Evans (1972b) and that of Ford *et al.* (1981) there is little quantitative data on which to base these generalisations. High mountain areas are characterised by a steep topography, dominated by rock walls of more than 60° and steep slopes between 35° and 60°. Flat valley bottoms and high plateaus may also occur. In lesser mountains slopes steeper than 35° are relatively infrequent.

Barsch and Caine (1984) also make a distinction between 'Alp type' relief and 'Rocky Mountain type' relief. In the Alps rock walls and steep slopes are usually the result of glacial erosion. Other typical forms of glacial erosion, such as horns, arêtes and cirques, also occur. The lower slopes are dominated by post-glacial talus and fan slopes. Areas of low relief exist only on the shoulders of glacial troughs, as remnants of Tertiary surfaces, and on the floors on infilled valleys (Figure 5.1a). As a contrast, in the southern Rocky Mountains sharp bedrock ridges and horns occur alongside wide, rounded interfluves and flat-topped mountains. These flat surfaces are also usually the remnants of Tertiary erosion surfaces (Figure 5.1b).

There are other mountain areas, which include most mountains in arid and semi-arid areas, that do not quite fit these two models. The main difference is lack of intense glaciation. Nevertheless, slopes are steep and relative relief great due to a combination of tectonic activity and intense

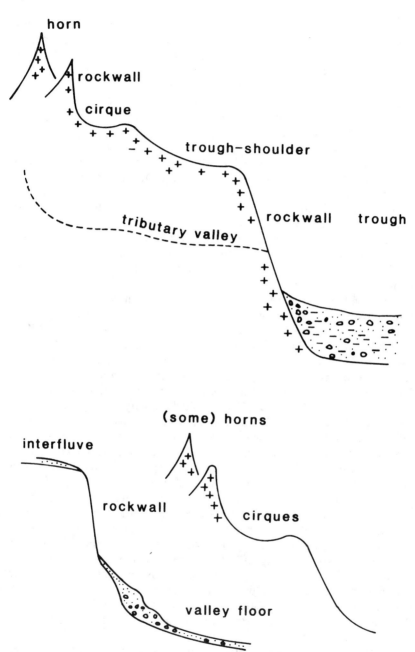

Figure 5.1. Schematic interpretation of mountain slopes (after Barsch and Caine, 1984).

dissection. Many of these areas are dominated by Richter denudation slopes at angles of 30–35° (Hollermann, 1973a). Richter slopes are examined in greater detail in a later section. Thus the distinctions emphasised by Barsch and Caine (1984) are the intensity of glacial erosion and the amount and type of low relief.

Specific slope forms in mountains can be related to the operation of a specific set of processes. Chapter 3 has demonstrated the importance of periglacial processes in alpine mountains and three of the slope forms suggested by French (1976) to be typical of periglacial areas are extremely common in mountains. These slope forms are the free-face/talus slope, smooth debris mantled slopes and stepped profiles (Figure 5.2). The free-face/talus slope profile is one of the most conspicuous in mountains (Figure 5.2a). It is composed of a vertical or near-vertical rockwall below which is a talus or scree slope followed by a footslope complex dominated by slope wash, debris flow and solifluction activity. This is very similar to the standard four-element slope model comprising a waxing slope or crest, free-face, constant or debris slope and waning, pediment or footslope.

Jahn (1960) has described this type of profile in rather greater detail from Spitsbergen. The scarp section is composed of weathering rock walls (above 40°) and rock slopes covered by a thin talus layer inclined at 25–40°. This is followed by a zone of dry talus cones with an inclination of 30–40° and of humid slopes (15–25°) created by avalanches at the outlets of gullies. The third component is a zone of solifluction terraces consisting of high talus terraces and stone garlands of short length inclined at angles of 15–25°, medium-length terraces and rock streams at angles of 10–15° and long and low terraces and rock streams at angles of 3–10°. The lowest zone is one of sedimentation by slope water and of declining solifluction influence with slopes inclined at angles of 1–5°. Although this is a fairly standard slope form, areas differ in the presence and extent of the lower slope elements.

The second type of slope form (Figure 5.2b) is common in many areas. It is characterised by relatively smooth profiles with no abrupt breaks of slope and with a cover of frost-shattered and soliflucted debris. Occasional rock outcrops or tors may interrupt the smooth slope form. Slope angles range from 10° to 25–30°. Many such slopes appear to be dominated by mass movement although it is possible that some are Richter slopes only covered by a thin veneer of superficial material. The stepped profile (Figure 5.2c) is one dominated by erosional terraces which are usually cryoplanation terraces (see Chapter 3). Tors often dominate such slopes.

It is clear from this brief introduction that, although a great variety of slope forms exist in mountains, there are a number of common elements in many of them. It is usual to differentiate weathering-limited slopes from transport-limited slopes. Weathering-limited slopes exist where the production of material by weathering cannot keep pace with

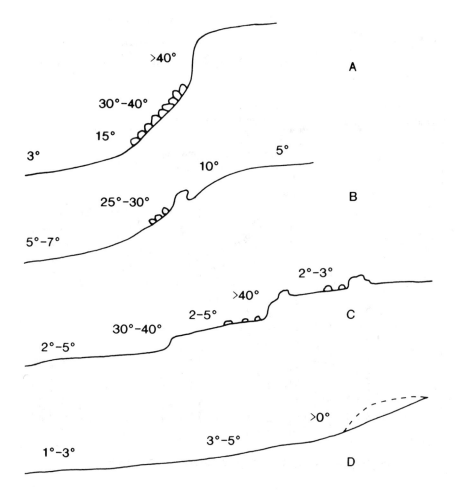

Figure 5.2. Slope forms characteristics of periglacial areas (after French, 1976).

the rate of removal, whereas on transport-limited slopes downslope transport is unable to remove all the material produced by weathering. On transport-limited slopes downslope movement of material is effected by rapid mass movements such as slides and flows, slow mass movement such as creep and heave, erosion by surface and subsurface water runoff and solution processes. It is also clear that in many instances there are relationships between the slope forms and certain process assemblages.

Such interactions have been incorporated into the nine-unit landsurface model of Dalrymple *et al.* (1968), refined in Conacher and Dalrymple (1977). It is an attempt to integrate slope form, geomorphological and pedological processes (Figure 5.3). It subdivides slope profiles and yet

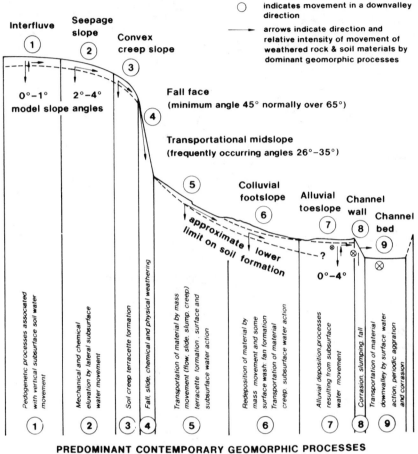

Figure 5.3. Hypothetical nine-unit landsurface model.

Source: Dalrymple *et al.* (1968).

integrates the components by considering water and material movement. On units 1 and 2, pedogenetic processes and vertical water movement dominate. The convex creep slope (unit 3) is characterised by both pedogenetic and geomorphological processes. The free-face and transportational midslope, units 4 and 5, are controlled essentially by the processes of weathering and mass movement. The colluvial footslope (unit 6) contains both geomorphological and pedogenetic processes. The alluvial footslope (unit 7) is controlled by subsurface water movement and periodic incursions by the river in flood. Units 8 and 9 are fluvially controlled.

The important point to remember is that the nine-unit landsurface is a model and it is not expected that all the components will occur on all slopes. Mountains characterised by intense river incision will possess few slopes with units 6 and 7 whereas valleys with well-developed river terraces may possess all the units. The model draws attention to such features and emphasises the integration between form and process and should be applicable to all slopes, and especially slopes in mountain regions. The slopes described by Jahn (1960) from Spitsbergen are encompassed in units 4, 5, 6 and 7. The free-face/talus model would comprise basically units 4 and 5. Even the general slope forms described by Barsch and Caine (1984) and discussed earlier can be fitted into this scheme. The level terrace remnants and trough shoulders would be largely composed of units 1, 2 and 3, perhaps also comprising units 6 and 7. If applied in a rigorous and systematic way, the nine-unit model would enable comparisons to be made between contrasting mountain regions.

The model is used here as a means of identifying and selecting the major slope components present in mountains yet at the same remembering that slope profiles are integrated systems. The slope systems examined in greater detail are:

(a) rock slopes, including Richter slopes;
(b) the free-face/talus slope combination;
(c) slopes dominated by colluvium or weathered regolith, sometimes called 'threshold slopes'.

ROCK SLOPES

Rock slopes dominate most high mountains, especially those that have been heavily glaciated. As noted in Chapter 3, rockfalls and rockslides are important mass movement processes. Thus the evolution of rock slopes is determined by the strength and ultimate failure of rock masses. A rock mass may become unstable because of disintegration of the rock material, inadequate strength along joints or other discontinuities and undermining of the slope by erosional agencies. It is unusual for exposed rock to be unjointed or unbedded; thus the behaviour of rock masses is related to these characteristics.

The characteristics of joints of greatest importance to rock stability are dip and orientation, spacing, the nature of the joint surfaces, joint thickness, joint continuity and joint infill material. Joint orientation and dip will control the orientation of any applied load and stresses and will determine the mode of failure. Failure by sliding or toppling is more likely to occur on a joint plane which dips towards a slope rather than on one which dips away from a slope. Selby (1980) has devised a strength classification for joint orientations. Joints dipping into the slope possess greater strength than those dipping out of the slope. Random orientation of joints will be favourable to slope stability in hard rocks with rough joints, whereas it will be unfavourable in weak or shattered rock. Rocks with strong planar joints dipping steeply into a slope will fail in cross joints.

Joint spacing is probably the most important factor in governing rock stability. The smaller the joint spacing the more likely a rock is to fail. Joint spacing will also control the size of boulders involved in rockfalls or rockslides and indirectly the nature and inclination of the boulder slope that may exist at the slope base. Allochthonous boulder slopes or blockfields will also be governed by the joint spacing in the original rock. The nature of the joint surface will dictate the resistance to movement of blocks with increased roughness creating greater resistance. Joint surfaces may be described as smooth, rough or wavy; rough and way surfaces will provide greater strength. Roughness will also determine how much of the rock is in contact. This is known as the *joint contact* factor, and where blocks are balanced on only a few protuberances this value will be extremely low.

Joint thickness is defined as the closure which must occur for the joint surface to be perfectly in contact and is the equivalent of the maximum cavity depth within the joint zones. Joint thickness will determine the ease with which water can penetrate the joints and will be a major factor in governing frost wedging. Joint thickness also provides an indication of the extent to which the rock mass may deform under an applied load without the main rock material being affected.

Joint continuity or persistence is important because continuous joints create extensive zones of weakness. Terzaghi (1960; 1962a; 1962b) has used this principle to illustrate the progressive failure of jointed rock. The total cohesion or resistance of a jointed rock will be equal to the combined strength of the intact blocks plus the friction between the joint planes. The larger the bridges of intact rock between the joints the greater is the effective cohesion. Slow yield will cause these bridges to fail one by one increasing the stress on the remaining intact rock and initiating a chain reaction or progressive failure until the slope eventually fails. Terzaghi (1960) has suggested that the Turtle Mountain slide of 1903, near Frank, Alberta, was a result of this process. Joint infill material will affect the behaviour of the joints and will, in general, lead to less strength especially if the infill is weathered clay and water. Cleft or joint water pressures are important factors in determining the timing of many rockfalls.

It is important that these factors are integrated in such a way that they can be used to explain not only the behaviour of rock slopes but also their form and mode of evolution. Thus for geomorphic purposes a classification is required which will provide a basis for understanding the features of rock masses which give them resistance to processes of weathering and erosion and which should be universally applicable so that it can provide a common basis of measurement (Selby, 1987). This was achieved by Selby (1980) in his rock mass strength classification and

Table 5.1. Rock mass strength classification.

Parameter	1 Very strong	2 Strong	3 Moderate	4 Weak	5 Very weak
Intact rock strength (N-type Schmidt Hammer 'R')	100-60 $r = 20$	60-50 $r = 18$	50-40 $r = 14$	40-35 $r = 10$	35-10 $r = 5$
Weathering	Unweathered $r = 10$	Slightly weathered $r = 9$	Moderately weathered $r = 7$	Highly weathered $r = 5$	Completely weathered $r = 3$
Spacing of joints	> 3 m $r = 30$	3-1 m $r = 28$	1-0.3 m $r = 21$	300-50 mm $r = 15$	< 50 mm $r = 8$
Joint orientations	Very favourable. Steep dips into slope, cross joints interlock.	Favourable. Moderate dips into slope.	Fair. Horizontal dips, or nearly vertical (hard rocks only).	Unfavourable. Moderate dips out of slope.	Very unfavourable. Steep dips out of slope.
Width of joints	< 0.1 mm $r = 7$	0.1-1 mm $r = 6$	1-5 mm $r = 5$	5-20 mm $r = 4$	> 20 mm $r = 2$
Continuity of joints	None continuous $r = 7$	Few continuous $r = 6$	Continuous, no infill $r = 5$	Continuous, thin infill $r = 4$	Continuous thick infill $r = 1$
Outflow of groundwater	None	Trace	Slight < 2.5 l min^{-1} m^{-2}	Moderate 2.5-12.5 l min^{-1} m^{-2}	Great > 12.5 l min^{-1} m^{-2}
			$r = 5$	$r = 4$	$r = 3$
Total rating	$r = 6$ 100-91	$r = 5$ 90-71	$r = 4$ 70-51	$r = 3$ 50-26	$r = 1$ < 26

Source: Selby (1982).

rating, which has since been used to explain the form of many rock slopes. The rock mass strength classification incorporates intact strength, degree of weathering, spacing, orientation, width and continuity of joints and the outflow of groundwater (Table 5.1). Each of these is given an overall relative rating subdivided into five classes. Class 1 indicates very strong and class 5 very weak rock. Intact rock strength is given a 20 per cent rating, separation of joints 30 per cent, joint orientations 20 per cent, and joint width, continuity and water flow are collectively given 20 per cent. This classification can be used to examine the nature of bare rock slopes by means of a concept known as strength equilibrium slopes.

STRENGTH EQUILIBRIUM SLOPES

On many slopes there appears to be an approximate condition of limiting equilibrium between the inclination of the slope and the resistance of the rock mass as measured by application of the rock mass strength classification (Selby, 1982). Such slopes have been called 'strength equilibrium slopes' and application of the rock mass strength classification will demonstrate whether a slope is in a state of strength equilibrium. Hillslopes with inclinations which have evolved so that their form is in equilibrium with the mass strength of their rocks are those which are not controlled by tectonic, structural, erosional or depositional processes. Equilibrium slopes usually require considerable periods of time, usually more than 10 000 years, for this adjustment to occur. If the data for slopes which are in equilibrium are plotted, with rock mass strength rating against average slope angle supported by each rock unit, these data points will fall within a strength equilibrium envelope (Figure 5.4). Using data points from a variety of slopes, Moon (1984) has specified the use of 90% confidence limits derived from the standard error of the regression of slope angle against rock mass strength. Data points for slopes which are out of equilibrium will fall above or below this envelope. This relationship will therefore determine whether a particular slope is governed by its rock mass strength and will provide an approximate indicator of the maximum angle of slope a stable rock mass can support.

There are five main situations where rock slopes may be out of equilibrium:

(a) Where structural controls are dominant a slope may be able to maintain very steep slopes. Newly exposed rock devoid of open joints will be in this category, as will slopes which are buttressed and protected by lower slopes.
(b) Basal undercutting by rivers may keep cliffs at angles too steep for the mass strength of the rocks.
(c) Rocks which are governed largely by solutional processes may have

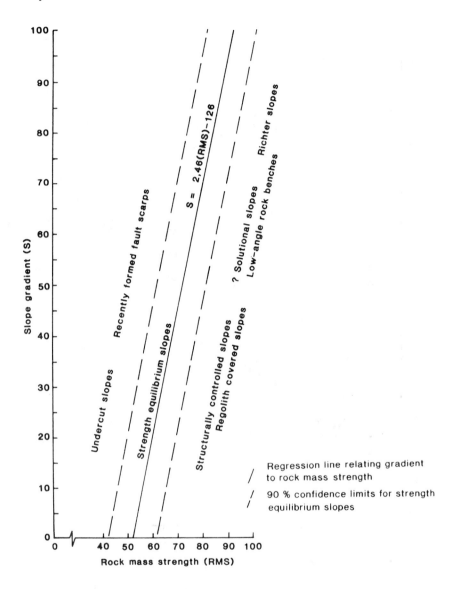

Figure 5.4. Rock mass strength equilibrium envelope.

Source: Moon (1984).

slope profiles which are out of equilibrium.

(d) Rock slopes may possess forms which are relict from previous conditions and processes. Long rectilinear slopes in many mountain areas were apparently formed as Richter denudation slopes (Bakker and Le Heux, 1952) and have since lost their mantle of talus by *in-situ*

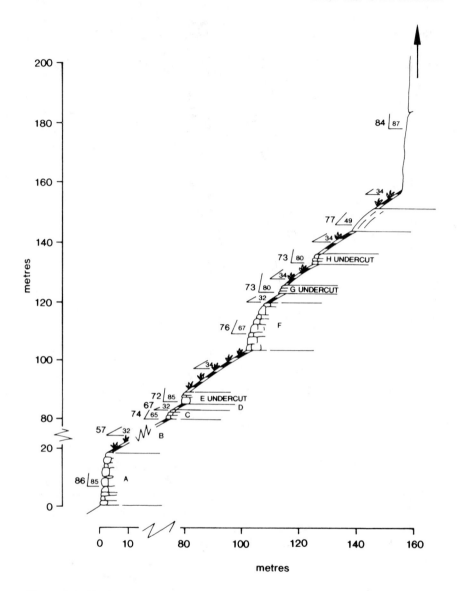

Figure 5.5. Rock mass strength criteria on the basalt slopes of the Drakensberg, South Africa.

Source: Moon and Selby (1983).

comminution and removal by wind or water.

(e) Strength equilibrium slopes are confined to weathering-limited situations where only a very thin veneer of superficial material exists. Transport-limited slopes will be governed by the nature and strength of the regolith cover.

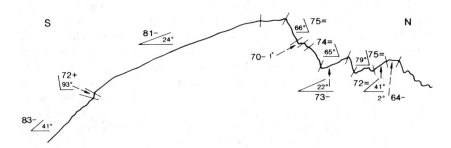

Figure 5.6. Rock mass strength values on the limb of a fold in Nepal.
Source: Selby (1987).

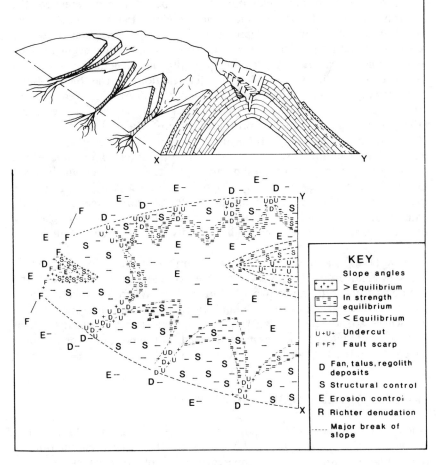

Figure 5.7. Strength equilibrium relationships on a eroded anticline.
Source: Selby (1987).

Analysis of some specific strength equilibrium slopes will demonstrate the applicability of the concept. One of the areas analysed is the Drakensberg Mountains of South Africa (Moon and Selby, 1983). Most of the outcrops possess widely spaced joints and have high mass strength ratings and steep inclinations (Figure 5.5). All the units except E have data points which lie within the strength equilibrium envelope but there are also a number of anomalies. Unit I is controlled by unloading joints, which indicates a structural control on the slope. Some of the lower outcrops have high slope angles for their mass strength because they are being undercut by the headwards extension of talus- and soil-covered slopes below. There are also some small soil covered slopes with inclinations of 32–34° which are governed by shallow landslides and appear to be threshold slopes.

The rock mass strength classification can be used to differentiate the controls on slope form and inclination (Selby, 1987). Over-steepened slopes are usually short-term features of a landscape. They start to evolve towards strength equilibrium angles as soon as the joints begin to open and the blocks can fall to alter the slope profile. Structural control will be most evident in an area of active deformation such as on the flank of a growing anticline and is most commonly observed where erosion has stripped cover beds to leave a resistant rock body at the surface. Selby (1987) uses as an example the profile of a limb of a fold in gneiss about 10 km south of Cho Oyu in the Khumbu Himal, Nepal (Figure 5.6). The mass strength numbers show that all dip slope units, except one which is the crown slope of a planar slide and over-steepened, have inclinations below those which could be supported by the rock strength. Therefore these slopes are structurally controlled. All the scarp slope units are in equilibrium. This and many other examples demonstrate that glacially oversteepened slopes are changing rapidly into strength equilibrium forms.

Selby (1987) has also shown how the procedure can be extended to maps of landforms (Figure 5.7). A letter system has been added to the symbols to illustrate the controls on slope form. Where the rock has been stripped to reveal a clear joint plane, then the symbol S is appropriate, but where erosion has left an irregular surface then the erosional form will be dominant and the symbol E is used. All depositional or regolith-covered slopes have been classified together.

FREE-FACE/TALUS TRANSFORMATION

One of the most conspicuous slope assemblages in high mountains is that comprising a bare rock face followed by a substantial accumulation of talus (Figure 5.8). Although apparently a simple type of slope form, there is still considerable speculation concerning the relationships between the developing free-face and talus form and evolution. Differences of

Figure 5.8. Systems of scree slopes on basalt ledges on a mountain slope in Berufjordur, eastern Iceland. (Photo: author.)

opinion regarding the forms and attributes of scree slopes and the processes acting on them arise from several sources (Thornes, 1971). They may represent differences of time scale in the phenomena investigated. They may also arise from a lack of integrated studies of the scree environment as a whole. Thus 'there is a need to examine screes in terms of the sub-systems in which they operate and in terms of multi-dimensional space, rather than as pairs or triplicates of attributes' (Thornes, 1971). They may represent differences of time scale in the phenomena investigated. They may also arise from a lack of integrated studies of the scree environment as a whole. Thus 'there is a need to examine screes in terms of the sub-systems in which they operate and in terms of multi-dimensional space, rather than as pairs or triplicates of attributes' (Thornes, 1971, p. 49).

The slope form, texture and fabric of talus have been analysed most commonly (see, for example, Bones, 1973; Carniel and Scheidegger, 1974; Albjar *et al.*, 1979; Church *et al.*, 1979) but debris movement has been less extensively studied. Debris movement can be measured directly (Church *et al.*, 1979) or indirectly (Gray, 1972; Luckman, 1978b). Rapp (1960a) monitored the accumulation of debris on snow patches, vegetation, sack carpets and wire netting at the upper apex of scree cones; Gardner (1972) used fabric netting placed downslope from the debris cone apex; and Luckman (1972) used polyethylene squares placed at various points. Rapp estimated aggradation to be 20–30 mm yr^{-1} in Karkevagge, Scandinavia. Luckman (1978b) observed preferential downslope accumulation, usually in midslope positions due to snow avalanche effects, while Gardner (1972) identified multiple processes including snow avalanches and rockfalls.

Gardner (1983b) monitored six slopes in the Mount Rae area of the Canadian Rocky Mountains between 1975 and 1982. Annual rates of accumulation were extremely variable between points on the same slope, between slopes and between years. Annual accretion rates of 30–40 mm yr^{-1} were similar to those recorded by Rapp (1960a) in Karkevagge. Extremely high rates were associated with extreme events such as debris flows.

Rates of accumulation appear to reflect local-scale climatic control in conjunction with the nature of the rock composing the free face. Freeze-thaw activity is one of the most important processes generating talus, but, as seen in Chapter 3, the timing and frequency of rockfalls can be controlled by a variety of factors. Thus Hyers (1982) has shown how the free-face fracture network can be modified by insolation. Talus accumulation may increase with decreasing topoclimatic severity (Madole, 1972) and only when a threshold of minimum topoclimatic severity is exceeded does jointing become an important determinant (Morris, 1981).

It is necessary at the outset to differentiate between the various types of talus and the processes involved in their formation because there is no doubt that there has been some past confusion. The simplest talus type is the rockfall talus where accumulations of rock debris occur by the dominant process of discrete rockfall. Within this category Rapp (1960b) has shown how rockfall talus varies from a simple talus sheet below a simple rock wall to talus cones where a rockfall occurs from a wedge, funnel or chute. If such chutes are close together talus cones may intersect to form compound talus slopes.

It is probably unusual for talus sheets to be created simply by rockfall, with processes such as snow and slush avalanches and debris flows being also involved. Brunsden *et al.* (1984) have distinguished between patterned talus, talus fans and fans. A similar classification has been used by White (1981). Patterned talus sheets usually dominate slopes in mountain valleys. They occur below steep shallow ravines or chutes that act as small catchments for water, debris and avalanches. Such talus accumulations are generally larger than those produced by simple rockfall processes alone. The formative processes are dominated by rockfall but there is substantial redistribution of material by mass movement processes. The result is a variety of surface patterns including stripes of alternating coarse and fine material, shallow scallops, festoons and lobes of coarse material separated by arcuate patches of fine material and a complex of lobes and shallow levee-flanked channels (Brunsden *et al.*, 1984).

Where there is sufficient water to generate true runoff, rockfall debris, valley-confined mudflows and overland flow combine to produce complex debris slopes. Such slopes are usually intermediate in inclination between true screes with angles of between 30–36° and fans produced by running water or mudflows with angles of 5–10°. Fans can be subdivided into alluvial fans and mudflow fans. Alluvial fans tend to possess smooth, slightly concave surfaces sloping at angles less than about 7°. Mudflow fans are gently concave with angles up to 20°, with a micro-topography characterised by shallow, levee-flanked channels and lobes of coarse material. Figure 5.9 shows a mudflow fan in Austerdaler, north Iceland in the foreground, with discrete debris channels and talus cones in the background.

CLIFF-TO-TALUS TRANSFORMATION

A number of classic models have been proposed to explain the cliff to talus transformation by rockfall alone. The main features were deduced by Fisher (1866), although Lawson (1915) seems to have developed similar ideas independently. The models were modified by Lehmann (1933; 1934) who added the effects of cliff angle, scree angle and rock/scree volume ratio. The Fisher–Lehmann model was further refined by Bakker and Le Heux (1946; 1947; 1950; 1952) and Scheidegger (1961a; 1961b). Similar ideas have been expressed by Wood (1942). In the Fisher–Lehmann model it is assumed that the whole free face is

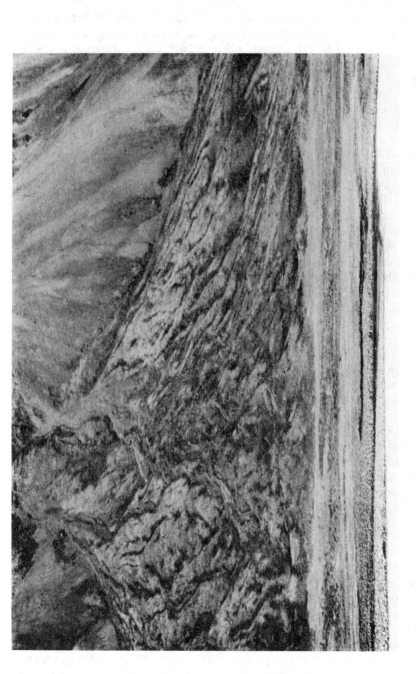

Figure 5.9. Composite debris flow fan with characteristic levees contrasting with steeper and simpler scree cones, north Iceland. (Photo: author.)

equally exposed to weathering and that in unit time a uniformly thick layer of rock falls away from it. This rock accumulates at the base as a scree at the angle of repose of the material. As the rock face retreats the talus slope extends upslope until it completely obliterates the free face. The problem that has concerned most workers is the nature of the bedrock slope beneath the scree. It is argued that this buried rock slope will be essentially convex, its precise form being determined by the ratio of the amount of material produced to that accumulating. One interesting conclusion of this work is that in situations where all the scree is removed the residual rock surface is a rectilinear slope at the angle at which scree would accumulate if present, irrespective of whether cliff recession is parallel, central rectilinear or otherwise (Young, 1972). This result was stated as a geomorphological law by Bakker and Le Heux (1952). This rectilinear slope is termed a *Richter denudation slope* and is one of the rock slopes mentioned in the previous section that does not satisfy the mass strength requirements.

The only realistic way that the transformation of the slope form over time can be assessed is with a space-for-time substitution. This is what Graf (1971) has attempted by dating slopes relatively with respect to the position of sequential terminal moraines in the Beartooth Mountains of Montana and Wyoming. Slopes within the youngest moraines he interpreted as having only evolved since the glacier retreated from that moraine. He measured the relative percentage of the valley cross-section profile occupied by free face, scree slope and low alluvial slope (Table 5.2). There appears to be a general decrease in the size of the free face and an increase in the size of the scree slope as the slope evolves, which is in accord with the general theories.

Similar results were obtained by Caine (1983) in the mountains of north-eastern Tasmania. Mountain slopes in this area consist basically of three slope components, a cliff, talus and footslope. Regression models demonstrated that all three components are reduced consistently with mountain size but that relative importance in the landscape varies. As the mountain summit area is reduced talus height is reduced much more rapidly than summit area, footslope height is reduced at a slightly lower rate than summit area, and cliff height reduced at a much slower rate. A suggested developmental sequence is shown in Figure 5.10, and agrees well with Graf (1971) and the general theories. The sequence is based on an average cliff retreat rate of 0.03 mm yr^{-1}.

Olyphant (1983) has examined cliff-to-talus transformation on twenty-seven cirque walls in the Sangre de Crosto Mountains of southern Colorado. He found that differences in potential insolation, altitude, windward/leeward aspect and bedrock fracturing explain 66 per cent of the observed variation in the degree of postglacial cliff-to-talus transformation. Free-face fracture density possessed the strongest statistical effect. The average rockfall weathering rates ranged between 0.05 and 0.82 mm yr^{-1}. Also the degree of cliff-to-talus transformation was

Table 5.2. *Relative percentage of valley cross section occupied by different slope types on slopes of different ages.*

	Free face	Scree slope	Alluvial slope
Neoglacial	43.0	23.5	0.0
Pinedale IV	24.8	36.4	0.0
Pinedale III	43.3	37.5	2.6
Pinedale II	43.3	46.1	6.5
Pinedale I	23.5	58.2	13.3
Bull Lake II	3.8	76.3	18.9

Source: Graf (1971).

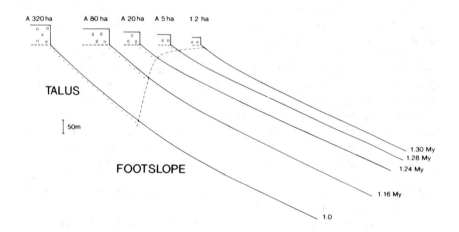

Figure 5.10. The development of mountain slopes in north-eastern Tasmania.
Source: Caine (1983).

about 11 per cent greater on windward than on leeward aspects when other effects were controlled.

TALUS ACCUMULATION AND REDISTRIBUTION

Three important models of talus accumulation and redistribution have been proposed:

1. Talus creep model: This suggests that talus behaves as a conveyor belt with material moving down as more material is added to the top. This is analogous to the queuing theory developed by Thornes (1971) but which is difficult to apply in practice. There is little evidence to suggest that screes develop in this orderly manner.

2. Rockfall model: This has been developed especially by Statham (1973; 1976) and Kirkby and Statham (1975) and considers the energy involved in moving individual particles down a talus slope. Screes developed under rockfall are subject to the impact energy of the particles hitting the scree surface. This impact may move other particles downslope and these surface movements will be responsible for regrading the slope. The mathematical model that has been developed involves stones travelling a specific distance over the talus according to the Poisson distribution. Thus most stones do not travel far. The mean travel distance decreases as the headwall height decreases and the talus builds up by linear translation of material parallel to the basal surface. The final slope should be at the angle of response of the material. However, Andrews (1961) found that many screes occurred with angles well below the maximum stable angle and frequently possessed concave profiles. Screes possessing high rates of basal removal are generally steeper than adjacent screes and angles of repose may be attained where there are very high rates of deposition (Chandler, 1973). Concave profiles have been noted by many workers which is difficult to explain simply by a rockfall model. Observations (see Gardner, 1969c; Caine, 1969) have shown that most particles do not travel far down the talus initially. Kirkby and Statham (1975) further argue that the model is only valid if the influx of new material is slow enough to give one-at-a-time movement without dry avalanching.

3. Slush avalanching model: As noted in chapter 3, slush avalanching is an important process on talus slopes. Caine (1969) observed that the amount of material carried by rockfall in the Two Thumb Range in the Southern Alps of New Zealand decreased with distance downslope. Redistribution by slush avalanching reversed this trend to produce a depositional layer which increased with distance downslope. Caine (1969) combined these observations to produce a simulation model which indicated that concave form could be produced by the redistribution of material by slush avalanches. In practice screes dominated by slush advances tend to possess very irregular profiles with an extensive lower concavity. In some instances these lower concavities may develop into avalanche bounder tongues. Howarth and Bones (1972) have suggested that rockfall scree slopes are steeper than scree slopes dominated by meltwater runoff. In their study on Devon Island, Canada, they found that 85 per cent of rockfall slopes were completely or dominantly concave and 95 per cent of meltwater dominated slopes were concave.

However, these models do not explain all the situations observed in mountain regions. Many workers have noted an upslope decline in particle size on screes which would not be expected if avalanching were dominant. Also, preferred particle orientation parallel to the slope has been observed (see, for example, McSaveney, 1972; Andrle and Abrahams, 1989). Rapid mass movement would destroy any sorting. Gray (1973) has also noted that such avalanches only rarely (i.e. at intervals greater than 65–85 years) extended to the base of taluses in central Yukon. Observations by Whitehouse and McSaveney (1983) in the Craigieburn Range, New Zealand, do not substantiate any of the models discussed above. Surface colour and weathering rind thickness on boulders indicated that talus surface age increased downslope, so that there is a decreased probability of disturbance at lower levels. Most fresh material is deposited on upper slopes and material is redistributed by dry avalanching.

These observations do not support talus accumulation models that presume the addition of wedges or sheets that cover sequentially all the surface. The presence of an older basal fringe of large boulders developed by rockfall and snow avalanching conflicts with models predicting the greatest accumulation at the talus base. The age distributions on debris flow taluses indicate an irregular episodic accumulation of lobate debris flows with intervening smoothing by snow glide and snow avalanching.

Attention has so far focused on movements of coarser components of the scree by sporadic and often catastrophic processes. Much less attention has been given to movement of particles by overland flow, even though Jahn (1960) and Soons and Rayner (1968) have noted its occurrence. Dingwall (1972) has shown that in the high alpine area of Alberta overland flow is capable of eroding measurable quantities of sediment from bare surfaces during summer months. Overland flow must be regarded as an important process effecting the translocation of debris on alpine slopes.

Caine (1983) has noted a number of unusual features of talus slopes in the mountains of north-eastern Tasmania. Most of the talus sheets are of the simple type as the homogeneity of the dolerite bedrock minimises couloir and gully development. Slope angles are generally much less than 38° and frequency analysis suggests peaks at 34° and 26°. This is a much broader range of slope angles than has been generally reported and reflects a lack of a simple rectilinear or concave profile form. Profiles are locally complex and frequently convex with a bulging lower section and relatively gentle gradients near the top. Bulging and terracing suggest a measure of mechanical failure in the talus. This is not in accord with the simple model for talus development by rockfall accumulation. The talus generally consists of large clasts which are moderately well sorted and there is no evidence of the downslope increase in clast size reported by other workers (see, for example, Bones, 1973; Washburn, 1970). This is probably the result of the uniformity of particle size. There is also no base fringe of large blocks and no preferred orientation in talus samples which suggests little movement of the surface debris. Block size, block form and block fabric appear to be unrelated to factors such as cliff height, slope length or slope angle.

Caine (1983) has attempted to explain the lower than usual angle of the talus sheets in the Tasmanian Mountains by shallow sliding, a process which would explain the irregularities in the profiles. Sliding at shallow depths within fully drained talus has been analysed by Scheidegger (1961b), but the mechanism proposed seems unable to explain the specific features. The possibility of sliding is enhanced if the talus were not fully drained, which might occur with ground freezing and the creation of high pore pressures. For the Tasmanian taluses with a limiting angle ϕ of 42° and an angle of 34° the ratio of pore water pressure to total vertical stress (r_u) is 0.17. With a slope angle of 25°, r_u = 0.40. Impeded drainage due to ground freezing seems capable of producing such pressures.

This analysis of talus slope form and development has demonstrated that a number of conflicting issues still need clarification. It is unlikely that the evolution of any scree slope is dominated by the operation of a single process, such as rockfall-induced movement and redistribution. Most slopes are affected, as least temporarily, by slush avalanching or debris flows (Figure 5.11). The important questions that need answering relate to the timing and frequency of such events. Is it feasible to assume that given sufficient time these mass movement processes gradually work backwards and forwards over the scree surface producing a slope form in 'steady state'? Answers to these questions require not only accurate monitoring of current rates of movement and accumulation but the use of techniques such as lichenometry and weathering rind thicknesses. Also, once screes become inactive and vegetated they begin to evolve under the influence of a markedly different set of processes and behave in a similar fashion to other slopes with weathered mantles. The evolution of such slopes is now examined.

SLOPES WITH RESIDUAL SOIL MANTLES

Many mountain areas are dominated by slopes covered with residual soil mantles and existing at angles ranging from about 20° to 35°. In the nine-unit model discussed earlier these are the transportational midslopes (unit 5) grading into the colluvial layers at the slope foot. They may extend away from a free face, but in many areas, especially in the lower areas of tropical and subtropical mountains, they dominate a landscape without free faces. The foothills and lower slopes of the Himalayas are classic examples where such slopes extend for several thousand metres from valley floor to ridge crest.

Detailed quantitative information is lacking because of the nature of the terrain and problems of accessibility. Such slopes appear to have evolved either from *in-situ* weathering of the rock composing the slope or by transformation of a scree slope into a sandy soil. Observations of these slopes have indicated that many are subject to frequent shallow

Figure 5.11. Erosion and reactivation of stable screes demonstrates the cyclic nature of many mountain slope processes. (Photo: author.)

landsliding and it appears that landsliding is the main process involved in their development. Such slopes have been called threshold slopes in the sense that there is a single threshold angle above which rapid mass movement will occur from time to time and below which the slope is stable. Two categories of threshold slope angle are frequent. The first is known as the *frictional* type, and refers to slopes composed of dry rock material. The second, known as the *semi-frictional* type, relates to situations where water flow occurs through the slope material parallel to the ground surface with the water table at the surface. Variability in regolith properties means that there is no single threshold angle for a given location but a range of such angles. These threshold angles will change with time as weathering alters the nature of the slope materials.

It has been demonstrated that weathering affects the shear strength, bulk and permeability properties of rock and its derived soil cover (Francis, 1987). Thus if weathering modifies those factors which control landsliding, then weathering must control slope evolution. Some of the changes brought about by weathering are summarised in Table 5.3. The transformations that are thought to occur have been summarised by Carson and Kirkby (1972). The initial state of many of the slopes will have been a scree slope standing approximately at the angle of repose, but continuous weathering will progressively change the character of the mantle. The coarse scree material will be broken down into soil and the mantle will pass through a sequence of talus, mixed talus and colluvium and soil. This will produce a cohesionless sandy soil with an angle of internal friction of about 35°. This is very similar to that of the original scree but the voids will be smaller and it is possible for positive pore pressures to develop which reduce the strength of the material. Small landslides may be initiated. The stable slope for this type of material would be

$$\tan\Theta = \frac{1}{2}\tan\phi'$$

where Θ is the angle of slope and ϕ' is the effective angle of internal friction. A ϕ'-value of 35° gives a slope angle (Θ) of about 20°, an angle which appears relatively common on such slopes.

This synthesis may have to be modified to take account of the fact that mixtures of rocky rubble and sandy particles, due to greater interlocking among particles, often possess higher ϕ' values than pure scree or soil. Thus ϕ'-values of 43–45° are quite common in slope mantles that are in the transition from scree to soil. Substituting a ϕ'-value of 45° in the above equation produces a slope angle of 26.6°. Angles of this order are extremely common in a wide range of environments.

Carson and Kirkby (1972) have provided a summary of the way such slopes develop. As the slope evolves, steeper slopes are replaced by those with a gentler slope. On deep, clay-based regoliths there may be only one change in slope from a temporary angle of stability to an ultimate angle.

Table 5.3. Rock and soil properties relevant to slope evolution (from Francis, 1987).

	Rock and soil particle assemblage properties relevant to slope evolution				
	Of most relevance to rock slope/scree evolution		Of most relevant to soil-covered slope evolution		
Material characteristic, controlled by weathering state	Unweathered rock, totally interlocked particles, directionally dependent maximum density granular soil, 'rockfill'	Slightly weathered rock, open discontinuities, typical granular soil/'rockfill'	Transitional granular/cohesive soil	Highly weathered normally consolidated cohesive soil	Completely weathered, normally consolidated residual clay cohesive soil
Main control of shear strength behaviour	Void ratio and discontinuity orientations	Void ratio and particle mineralogy	Clay/granular particle proportions	Clay content, sensitivity, and mineralogy	
Determination and magnitude of shear strength properties	In-situ model testing, ϕ' between 70° and ϕ' of 35°, directionally dependent	In-situ/model testing, large-scale laboratory testing, ϕ' between 50° and ϕ'_{cv} of 30°	Laboratory testing, ϕ' between 45° and ϕ'_f of 5°	Laboratory testing, in-situ testing, ϕ' between 30° and ϕ'_f of 5°	Laboratory testing, in-situ, ϕ' between 30° and ϕ'_f of 5°

However, a strong well-jointed rock mass is likely to pass through several phases. There will be the initial phase when the cliff face is replaced by a scree slope, a second phase with a change from the scree slope to a taluvial slope, and a third phase when a soil mantled slope develops. It needs to be stressed that these slope changes are the result of shallow landslides initiated mostly by high water pressures in the slope materials. The specific angles that are created will depend on the characteristics of the mantle and the pore pressure patterns which occur within it. The complexity of water flow in mountain slope material has been stressed by Okunishi and Okimura (1987). Thus a wide variety of threshold slopes must exist in mountain landscapes, although certain angles appear to be especially common. These are:

(i) 43–45° slopes on fractured and jointed rocks where, because of the high degree of packing, the ϕ' angle is about 43–45°. Such slopes are common in mountain regions.
(ii) 33–38° slopes mantled by the same type of material as in the 43–45° slopes but with looser packing. These are essentially scree slopes existing at the angle of repose of loose cohesionless material.
(iii) 25–28° slopes which are taluvial slopes developed from scree or regolith as described above.
(iv) 19–21° sandy slopes developed as described above. This slope may be regarded as the ultimate stable slope for sandy material.

Other changes may be initiated by processes such as soil creep and slope wash to produce additional slope components superimposed on the above sequence.

SPECIFIC EXAMPLES

It is important to assess the manner in which the individual 'models' of slope type, discussed above, combine to form the landscape of mountain regions. Slopes in recently deglaciated areas will possess different slope forms to those areas that have been developing for longer periods. Rock slopes might not have reached the condition of rock mass strength equilibrium and scree slopes may be in their infancy. Most alpine regions will be dominated by the free-face/talus slope form with modifications depending on the interaction between material input at the top of the slope and material removal at the slope base by river activity and perhaps glacial action. The role of basal erosion in determining slope evolution has been amply demonstrated by Richards and Lorriman (1987).

It is not possible to provide a complete synthesis of slope form and evolution in mountain areas but the principles can be examined in two examples, the first from the author's unpublished work on some Icelandic slopes and the second from published work on slopes in the Nepal Himalaya.

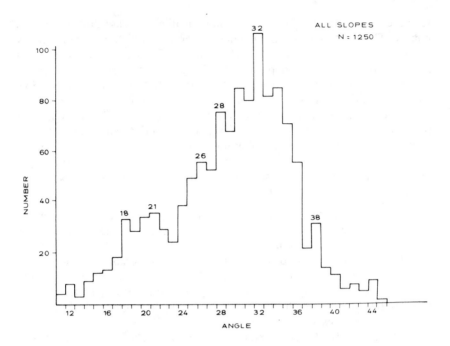

Figure 5.12. Frequency histogram of slope angles on terrace slopes in Vestur-daler, Iceland.

The slopes examined in Iceland were on a series of river terraces in Vesturdaler, north Iceland. Four main terraces exist, and if the assumption is made that the slopes on the terrace front have developed by normal subaerial processes once they are protected by a lower terrace, then the slopes may be placed in an evolutionary sequence. Unfortunately it is not possible to date the terraces accurately and no specific rates of change can be measured. However, the sequence of slope profiles enables changes in slope type to be deciphered (Figure 5.12). The terrace slope currently being undercut is maintained at the angle of repose (32–33°) of the coarse cobbles, up to 50 cm long, which dominate the terrace material. Apart from being composed of rounded material, this is equivalent to the scree slopes discussed earlier. Movement of material is continuous across the whole slope by a combination of debris sliding and individual block movement and there is no soil or vegetation cover. The slope form is a simple, single rectilinear component at the angle of repose.

As soon as the slope is protected by a lower terrace fragment a number of changes takes place. A lower accumulation slope (25–28°) develops by a combination of mass movement and slope wash and the crest of the slope becomes more rounded, and an additional component, also 25–28°, develops. These changes are accompanied by the development of a rudimentary soil cover and the growth of vegetation from both the slope base and slope crest. The central 32° component is much diminished in size. The next major change occurs when the slope is almost completely vegetated and a lower 18–21° component develops by processes which include localised slumping, accelerated soil creep and the occasional alluvial fan. Mass movement also continues on the other slope components.

The form of the oldest slopes has been simplified by the elimination of the central 32° component, the creation of a single main component at 25–28° and an upper and lower component of 18–20°. Mass movements are now extremely localised, and accelerated soil creep appears to be dominant. It is interesting that the angle of the slope component matches almost exactly the sequence of slope changes suggested by Carson and Kirkby (1972), discussed above. Analysis of other slopes in Iceland also indicates that the sequence is very common.

Some of the most complex slope forms exist on mountains that do not penetrate the alpine altitudinal zone. A good example is the Low Himalaya zone of India and Nepal, which has been described in some detail by Brunsden *et al.* (1981). They examined an area in eastern Nepal close to the Sikkim border which included the first two hill ranges which form the Low and part of the Middle Himalaya. The climate of the area is humid sub tropical, with a winter dry season. Mean annual precipitation varies from 1000 mm to over 2000 mm, with 80 per cent concentrated in the period from May to September. Recent uplift and the high and intense precipitation have combined to produce an intensely dissected landscape with a relative relief up to 1500 m. Each phase of uplift has resulted in rapid incision of the main drainage ways and a polycyclic landscape whereby slope components are repeated.

The general landscape that has been produced is characterised by deep valleys with long side-slopes (20–50°), river terraces and alluvial fans. The major slopes are convex-concave in form with three main slope facets (Figure 5.13):

(i) Gentle (less than 35°) convex-concave upper slopes with reasonably level (less than 10°) crestal areas.

(ii) Midslope units with slopes generally at or above 30° but locally increasing to 50° on old landslide scars and rock outcrops.

(iii) Extremely steep (40–90°) unstable lower slopes immediately above the incising rivers.

The relative importance of these components varies with the intensity of the slope processes and the rapidity of river incision. Where drainage density is high, knife-edge ridges have been created by the meeting of lower and middle hillslope units.

In some areas the model described above is modified by a number of relict features which include successive river terraces, elevated fans, tilted fan remnants and boulder fields backed by major scars which may reflect

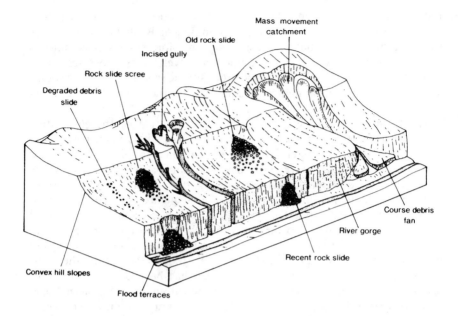

Figure ·5.13. Schematic diagram of the three major slope elements and associated processes in the Low Himalaya zone of India and Nepal.

Source: Brunsden *et al.* (1981).

earthquake-generated failures. Alluvial fans are conspicuous features of most mountain landscapes (see, for example, Hoppe and Ekman, 1964; McPherson and Hurst, 1972). Extensive boulder fields and screes may also be relict features. Screes, at angles of 30–37°, below steep faces developed in phyllites, schists and quartzites, appear to be relict forms related to colder climate periods when the subtropical forest cover did not exist and frost weathering was an active process.

An interesting feature of the slopes is that regolith is well developed, especially on the upper slopes, and development appears to be controlled by factors discussed in the section on threshold slopes. Weathering penetrates rapidly into the foliated, jointed and crushed rocks. This is evident on some of the gneissic slopes, where the regolith is up to 20 m thick. This means that if the vegetation cover is destroyed, accelerated soil erosion and landsliding are initiated. The general effect of deforestation on land degradation in the Himalayas is examined in greater detail in Chapter 9. The nature of the slopes and the mantle of regolith determine that mass movements are the processes controlling hillslope form.

Rockfalls occur mainly on the steep slopes developed on quartzite and schist. Individual falls are usually small and include wedge, planar and toppling failures. Rockslides occur on shales, phyllites and schist, are usually shallow, and result in debris cones below arcuate scars. Such rockfalls and rockslides are of great significance in slope evolution. The arcuate scars develop into wide concave hollows between sharp ridges creating a characteristic scalloped slope form. This appears to be the dominant form of slope evolution on the steep slopes (40–90°).

Debris slides are the commonest form of mass movement. They create shallow (1–3 m) scars elongate (100–1000 m) downslope which are accentuated by gully erosion. The slides sometimes develop into debris flows which travel downslope along narrow parallel tracks. They are usually related to heavy rainstorms and are widely distributed on regolith covered slopes over 30°. Mudslides are relatively uncommon and are associated with deeply weathered phyllites and schists. Rotational slides are also uncommon and confined to areas where deeply weathered shales and schists have been undercut by rivers or deep gullies.

This analysis of part of the Low Himalaya has demonstrated that slope evolution is largely governed by mass movement processes, especially small rockslides and debris slides. But other processes, such as surface and subsurface water erosion, may be locally important. The operation of such processes is largely governed by the nature of the soil mantle and the density of vegetation cover. Overland flow is particularly noticeable on weathered gneiss slopes, whereas on other lithologies a denser vegetation cover and coarser regolith leads to greater infiltration and the development of percolines, spring seepage and perennial gullies. There may be a threshold erosion condition where the increase of bare unstable ground becomes so extensive that there is insufficient vegetation cover and stable slope area to dissipate runoff and prevent continuing instability. Thus an 'autocatalytic condition then exists where bare ground promotes runoff, gullying, new landsliding, increase in bare area and a further increase in runoff' (Brunsden *et al.*, 1981, p. 54).

CONCLUSION

Slope form in mountain areas is generally complex, but it is possible to distinguish a number of frequently occurring assemblages. These assemblages can be explained by the factors and concepts discussed in this chapter. Once clear 'models' have been established, rapid reconnaissance of less accessible areas will allow judgements to be made concerning the processes operating on the slopes. This will then enable applied or management problems to be placed on a firmer scientific footing. Such judgement is essential to projects involving hazard mapping. This procedure is examined in greater detail in Chapter 8.

Chapter 6

GLACIATION OF MOUNTAINS

The distribution of glaciers is a function of the distribution and amount of accumulating snow and of the nature and amount of incoming energy and its utilisation at the glacier surface during the summer ablation period. An ice mass exists, in simple terms, where, on average, the amount of water accumulation equals or exceeds the amount of summer ablation. Some glaciers, such as those of the Colorado Front Range, USA, exist primarily because of extremely high winter accumulation which is able to counteract high ablation rates in summer. In other areas a balance is achieved with lower accumulation and ablation rates. As a result glaciers exist in a variety of environments and topographic positions.

It is estimated that the current aggregate area of the world's glaciers is about 14.9×10^6 km^2 (Flint, 1971) although larger figures are given by Shumskii et al. (1964) totalling 16.2×10^6 km^2. Some of this discrepancy is due to the fact that many ice-covered regions — for example, in South America and Central Asia (Grosval'd and Kotlyakov, 1969) — are still imperfectly mapped. The Antarctic ice sheet accounts for 12.5×10^6 km^2 and Greenland possesses 1.7×10^6 km^2. This leaves 700 000 km^2 of glacier ice scattered around the world, mostly in major mountain ranges. It is this last group that is mostly of interest in this chapter, although parts of Greenland exhibit characteristics of mountain glaciation. Table 6.1 provides a more detailed analysis of the distribution of glaciers, though some of the values are only approximations because many of the mountain glaciers, especially in South America, are imperfectly mapped, as just noted. Types of glacier vary considerably from region to region and from mountain range to mountain range, so before examining the distribution of glaciers in more detail it is useful to consider some of the more important classifications of glacier type that have been proposed.

Table 6.1. *Size and distribution of ice masses.*

Region	Area (km²)
South Polar	
Antarctic ice sheet	12 535 000
Other Antarctic glaciers	50 000
Subantarctic islands	3 000
North	Polar
Greenland	1 802 600
Canadian Arctic	153 169
Iceland	12 173
Spitsbergen and Nordaustlandet	58 016
Other Arctic islands	55 658
North American continent	
Alaska	51 476
Other	25 404
South American Cordillerra	26 500
European continent	
Scandinavia	3 810
Alps	3 600
Caucasus	1 805
Other	61
Asian continent	
Himalaya	33 200
K'un Lun Chains	16 700
Karakorams, etc.	16 000
Other	49 121
African continent	12
Pacific region	1 015

Source: Flint (1971).

TYPES OF GLACIER

A number of classifications of glacier type exist (see, for example, Avsiuk, 1955; Court, 1957). The scheme proposed by Miller (1973) identifies four thermal types — polar, sub-polar, sub-temperate, and temperate — but is difficult to use in practice. One of the most comprehensive schemes is that of Ahlmann (1948). He actually suggested three different classifications: a thermal classification; a classification based on degree of activity; and a morphological classification based on the size and form of ice.

Thermal classification

Ahlmann (1935) distinguished between temperate and cold (polar) glaciers. Temperate glaciers are approximately at pressure melting point throughout their thickness, meltwater is present at the glacier base and substantial rates of basal sliding may occur. Cold glaciers are generally frozen to the bedrock on which they rest, meltwater at depth is absent, and all movement must occur by deformation in the ice. However, many glaciers do not fall into these simple categories. In a high mountain range it is possible for a glacier to be completely cold in the high-altitude accumulation zone but to be at the pressure melting point in its lower, more temperate zone. This has important geomorphological implications when considering glacial erosion and deposition.

Dynamic classification

The dynamic activity of a glacier is associated with its mass balance as well as with its thermal regime. Three types of glacier activity have been proposed: active, passive or inactive, and dead. Active glaciers are usually fed by a continuous ice stream from an accumulation zone. The glaciers on the west side of the Southern Alps of New Zealand are a good example (Embleton and King, 1975). Precipitation and ablation are high, slopes are steep and the glaciers are very active. Suggate (1950) recorded a rate of 525 cm per day in 1894. Current rates are less than this and the glaciers are generally thinning and retreating. A dynamically active glacier is one which is flowing fast, irrespective of whether it is retreating or advancing. Embleton and King (1975) note a subgroup of active glaciers which has been termed 'regenerated' glaciers. These are fed entirely by ice avalanches, move only short distances from their avalanche fan head, but their activity is shown by large terminal moraines. The Glacier de Nantes in the French Alps is one such regenerated glacier.

  Passive glaciers occur where the supply of snow to the accumulation zone is low, such as in the lee of a mountain range, or where the slopes are gentle. Dead ice is that which no longer receives supply from an accumulation zone and its movement is entirely dependent on the slope. Compared to active glaciers, the number of passive and dead glaciers is small, which makes the classification of limited use in regional synthesis. However, it must be recognised that glaciers may change from one type to another if major climatic changes occur. The most useful classification is that based on morphology, since it also reflects geomorphological activity and resultant landforms and deposits.

Morphological classification

One of the simplest classifications reflects the interactions between glacier ice and topography and differentiates ice sheets or ice caps which are superimposed on the underlying topography from glaciers which are constrained by topography and influenced in their form and direction of flow by the shape of the ground (Ostrem, 1974). This basic distinction allowed Linton (1957) to evaluate the effects of former glacial activity in Scotland, Norway and New Zealand. However, he also recognised that there was an intermediate situation in which the central parts of ice bodies may be wholly submerging the uplands but whose marginal portions were constrained to some degree by pre-existing topography. Thus 'the tendency for free radial outflow exists powerfully but it is canalised by the relief of the land along favourably disposed preglacial valleys, or along new troughs produced by the integration of pre-existing valley elements by breaching of watersheds'. (Linton, 1957, p. 311).

To take account of such problems the following, more complete, classification has been proposed:

(a) niche, wall-sided, cliff or apron glacier;
(b) cirque glacier;
(c) valley glacier — alpine type;
(d) valley glacier — outlet type;
(e) transection glacier;
(f) piedmont glacier;
(g) floating glacier tongues and ice shelves;
(h) mountain ice cap;
(i) glacier cap or ice cap;
(j) continental ice sheet.

Space does not permit a detailed examination of all these types, emphasis being given to those most significant in mountain areas. However, at the height of the Quaternary glaciations many mountain areas were affected by very different ice masses from those present today and many of the landforms have probably been inherited from these conditions. Thus it is important to consider all types, however briefly. A more comprehensive account has been provided by Embleton and King (1975).

Niche glaciers consist of small wedges of ice lying in shallow hollows on the upper slopes of mountains. They are often associated with rock benches, are mostly isolated features and may represent early stages in cirque glacier development. Groom (1959) has described such glaciers from Spitzbergen and Galibert (1965) and Rothlisberger (1974) have discussed what they called 'ice aprons' in the European Alps.

Cirque glaciers are small ice masses, wide in relation to their length, which exist in small armchair-shaped hollows. Although they often possess large termainal moraines, they lie in a true rock basin. They may be separate ice masses or simply the head of a valley glacier.

Valley glaciers (alpine type) flow from cirques or icefields, with the valleys often radiating from the main massif. The altitude range of valley glaciers is high in relation to their size and the bedrock slope is steep. Tributary glaciers may joint the main valley glacier with surfaces which are approximately conformable. Glacier lengths are usually 20–30 km, although the Hubbard Glacier in North America is 120 km long. As with river basins, a hierarchy of valley glaciers exists with good correlations between glacier widths and number of tributaries or cirque-collecting grounds. Ahlmann (1948) has suggested four sub-types of valley glacier. The first type has a large proportion of its area just above its median height with little area at highest and lowest levels; the Rhône Glacier or Hintereisferner is of this type. The second type has a greater area in the upper part, an example being the Grosser Aletschgletscher. The third type has most of its area at lower levels and is fed largely by ice avalanches, it is exemplified by many of the glaciers of Central Asia. The fourth group has large lateral tributaries and has the greatest proportion of its area a little below median height; the glaciers of north-west Spitsbergen are an example.

Valley glaciers (outlet type) are fed from an ice cap and not a series of cirque basins but in other respects are very similar. The glaciers draining the Norwegian and Icelandic ice caps (Figure 6.1) are of this type.

Transection glaciers occupy much of a mountain group from which glaciers flow in several, often radiating, directions. They are smaller and often too dissected to be classed as mountain ice caps. Glaciers of this type produce glacial breaches and cols.

Piedmont glaciers form when valley glaciers emerge from mountains and flow onto the surrounding lowland. Skeidararjokull in Iceland and the Malaspina in Alaska are examples.

Floating ice tongues are restricted to high latitudes where glaciers reach sea level.

Mountain ice caps are accumulation areas, usually upland plateau surfaces, from which glaciers, usually of the outlet type, flow in many directions (Figure 6.2). Vatnajökull in Iceland, which is 600–700 m thick and rests on a plateau of about 1200 m elevation, is an excellent example.

Glacier caps are small ice masses that form at relatively low levels on flat terrain in the high Arctic; the Barnes ice cap on Baffin Island is an example. Such ice masses are probably inactive geomorphologically.

Continental ice sheets are restricted to the ice sheets of Antarctica and Greenland. During the Quaternary glaciations the North American Laurentide and the European ice sheets also existed.

REGIONAL ANALYSIS

This analysis concentrates on providing a general summary of the essential characteristics of the glacier systems of the main mountain areas of the world. Reference can be made to Table 6.1 to assess the relative size of the systems discussed. Thus the Antarctic ice sheet, although rising to a height of several thousand metres, is largely ignored, as are the low-

Figure 6.1. Outlet glacier of Eyjfjallasjokull, southern Iceland. (Photo: author.)

Figure 6.2. Myrdalsjokull ice cap, southern Iceland. (Photo: author.)

lying Arctic areas. Greenland can be included in an account of mountain regions because of a combination of altitude and the relative relief of its glaciers. Considerable areas of mountain topography are also exposed. About 7 per cent of the ice mass lies between 3050 and 3390 m and the firn line is at about 1400 m (Bauer, 1955). The main glacier areas in the USSR are the Atlantic-Arctic, the Atlantic-Eurasian, East Siberian and Pacific-Asian areas which can be divided into nineteen glacier areas, fourteen of which are of mountain type. Cirque glaciers are dominant in the northern Urals and east Siberia, valley glaciers in the Caucasus and the Tien Shan, and transection glaciers in the central Asian mountains.

Scandinavia has a variety of glacier types, from mountain ice caps to valley glaciers. The glaciers have exhibited phases of both advance and retreat in recent years, but mass balance calculations are complicated by diverse topographic conditions and local weather variations. Mass balances of many of the Icelandic glaciers also vary considerably from year to year. These variations are often related to size of catchment area in relation to local fluctuations in climate, with some glaciers advancing and others retreating. This is also a characteristic exhibited by many Alaskan glaciers, some of which may show periods of very rapid movement or surges.

There are a number of small ice caps and glaciers in South America. The Southern Patagonian ice cap is probably the most impressive, but small ice caps exist in various locations in the Andes. Mass balance measurements have been conducted on the Quelccaya ice cap of Peru (Hastenrath, 1978).

The glaciers of the European Alps, mainly of the valley glacier or cirque type, have generally been retreating over the last thirty years or so. Thus the Hintereisferner glacier has been retreating at an average rate of 25 m yr^{-1}. The Southern Alps of New Zealand possess a number of glaciers. There are major differences between glaciers on the wetter western slopes and those on the drier eastern slopes. The Tasman is the longest glacier and flows south-east on the eastern side at a considerably slower rate than the Franz Joseph on the west side. The activity of the Franz Joseph Glacier has been intensively studied by Suggate (1950). The glacier retreated slowly from 1910 to 1922, then advanced a few metres until 1933 when a rapid retreat started and continued until 1946. A renewed advance then occurred until 1951 followed by a retreat until the early 1960s and then another advance. These changes appear to be related to changes in precipitation amounts.

The small glaciers that exist on the equatorial mountains of Africa have been extensively researched by Hastenrath (1984). It must be remembered that the former glacier extent was much larger and six glacial episodes have been recognised on Mount Kilimanjaro (Downie, 1964). The regimes of these glaciers are related to different seasonal changes to those already discussed. Penck glacier on Mount Kilimanjaro is the longest and one of the steepest, falling from 5800 m to 4600 m in

2.4 km. The main accumulation period is from March to June, and there may be two major ablation seasons from July to September and during January and February. Another unusual feature is that the main accumulation zone appears to be in the lower part of the glacier, with melting taking place at the top of the mountain. There are twelve small glaciers on Mount Kenya; the largest, Lewis glacier, has an area of 0.36 km^2 and descends to 4480 m. Accumulation occurs from March to May and from November to December. Glaciers in such environments can only survive where protected from direct sunlight and exposed to precipitation.

THE DYNAMICS OF GLACIERS

Mass balance

The dynamics of glaciers will be considered with specific reference to valley glaciers, which are often the most important types in mountain areas. It is important to understand the basic dynamics because erosion, debris transport and deposition are related to glacier dynamics. Knowledge of such relationships also enables landscapes shaped by past glacial action to be explained. The most important feature of any glacier is its budget or mass balance, which is the relative magnitude of mass gained from precipitation, usually snow, and mass lost through melt processes (Anon, 1966; Mayo *et al.*, 1972). Mass balance varies throughout the year, depending on a number of factors, some of which are assessed in Table 6.2. These considerations are important because the amount of potential mechanical work a glacier can undertake is related to the amount of snow and ice added each year in the accumulation basins at high altitude and transferred downslope. Glaciers with high mass transfer, where large inputs of mass are balanced by large exports downvalley, are strong erosive agent (Andrews, 1972). Such glaciers are found especially in west coastal mountains such as British Columbia, Norway and Chile. Rockfalls from valleysides may blanket the ablation zone, reduce icemelt and contribute to a more positive mass balance (Bull and Marangunic, 1968).

The winter mass balance (b_w) is usually expressed as cubic metres of water equivalent to enable comparison with summer meltwater balance (b_s). The changing net balance can be examined by plotting balance data (b_w/b_s) against the altitude zones of the glacier (Figure 6.3). The point where the net balance (b_n) is zero is known as the *equilibrium line*. This line divides the glacier into two unequal areas: an upper area of net gain, and a lower ablation area of net mass loss. Glaciers characterised by high mass transfer have a high gradient of net balance, expressed in millimetres of water equivalent per metre change in elevation, at the equilibrium line. This value is sometimes referred to as the

Table 6.2. *Mass balance relationships of a medium-sized glacier.*

Season	Spatial variation	Mass balance characteristics
Autumn	Snow accumulation at higher altitudes. Ablation of ice continues at low altitudes	Snow mass increasing Ice mass decreasing Total mass constant
Winter	Snow accumulation over whole glacier. Little ablation	Snow mass increasing Ice mass constant Total mass increasing
Spring	Snow accumulation at higher altitudes. Ablation of winter snow at low altitudes	Snow mass constant Ice mass constant Total mass constant
Summer	Little snow accumulation except at high altitudes. Ablation over much of glacier.	Snow mass decreasing Ice mass decreasing Total mass decreasing

Source: Mayo *et al.* (1972); Sugden and John (1976).

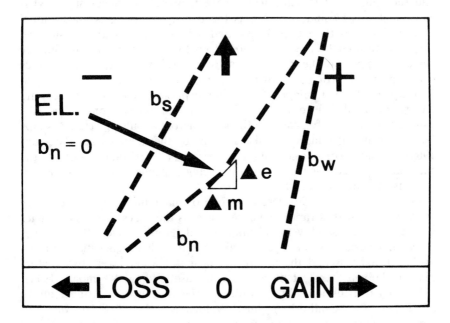

Figure 6.3. Mass balance relationships of a valley glacier.

'energy of glaciation' (Shumskii, 1950) or the 'activity index' (Meier, 1960; 1961). Temperate glaciers have high activity values. Also the net balance is steeper in maritime climates than in more continental areas and the temperature at the equilibrium line is higher. Thus the South Cascade Glacier, Washington, has a net balance gradient of 17 mm m^{-1}, whereas the Gulkana Glacier, in the interior of Alaska, has a gradient of 6 mm m^{-1} (Meier et al., 1971).

Ice movement

Ice flow in valley glaciers is strongly affected by frictional retardation against the valleysides and irregularities on the valley floor. Glaciers move in response to gravity and movement occurs by internal strain deformation and sliding where ice is at pressure melting point at its base. Stress and strain are related by a power law (Glen, 1953; 1955) where strain rate, ξ, is equal to $\beta\sigma\eta$ where β and η are constants and σ is the value of applied shear stress. η commonly ranges in value from 2 to 4 and the value of β is dependent on ice temperature. Shear stress (τ) is equivalent to $\varrho gh \sin\alpha$, where ϱ is the density of ice, g is gravitational acceleration, h is the ice thickness, and α is the slope of the ice surface. This equation enables vertical velocity profiles to be calculated. Surface velocities appear to be proportional to the third power of the surface slope and the fourth power of the ice thickness (when $\eta = 3$). This relationship has been developed for a simple parallel sided slab of ice and has to be modified to account for frictional retardation against the valleysides and also for variable thickness across a valley glacier. A shape factor can be introduced to take account of the lateral drag (Paterson, 1969). Basal debris also enhances frictional resistance against the substrate, with coefficients of friction being dependent on debris concentrations (Boulton et al., 1979).

In a transverse section there is a zone of maximum flow in the centre at the surface with a rapid reduction towards the edges. Velocity changes with depth are less easy to calculate and determine (Nye, 1965a). Lines of equal velocity appear to be parabolic but with irregularities imposed by bed roughness and the nature of the basal water layer, two factors that are known to affect the process of basal sliding (Raymond, 1971).

Velocity distribution along a valley glacier is related to distance above or below the equilibrium line. Velocities of 200–800 m yr^{-1} on temperate glaciers are not uncommon in the vicinity of the equilibrium line with velocity declining up- and downglacier (Eyles, 1983). Ice accelerating towards the equilibrium line experiences extending flow which changes to compressive flow below the line. Above the equilibrium line there is a downward velocity component, whereas below the equilibrium line there is an upward component. Local patterns of flow are created by variations in the long profile of the valley floor. Compressive flow will occur in hollows and depressions and extending flow occurs as ice moves over substantial high points. Slip lines or trajectories of maximum shear stress are directed upward in compressive flow and directed downward in extending flow (Nye, 1952; Nye and Martin, 1968). Ice velocity is also related to the area of the base actually in contact with the bed with decoupling occurring in the lee of bedrock highs. The size of subglacial cavities exhibits seasonal variation.

Periodic changes in the volume of snow and ice in the accumulation zone may lead to bulges being moved downglacier as kinematic waves. Such a bulge arriving at the snout of the Nisqually glacier in Washington raised the ice surface level by more than 30 m (Meier and Johnson, 1962). Ice velocities of about four times normal can be experienced (Nye, 1960; 1965b). Kinematic waves usually originate in the vicinity of the equilibrium line.

A number of glaciers are affected by periodic surges in which ice is transmitted downglacier at greatly increased speeds (Harrison, 1964; Robin and Barnes, 1969). Meier and Post (1969), after detailed study of 204 surging glaciers in western North America, observed that part of the glacier acts as a reservoir which fills with ice before being moved downglacier at high speeds. Velocities may be increased by as much as 100 times. A surge may have a dramatic effect on glacial transport. Lateral moraines may be left in high elevations and tributary glaciers shorn off. Periodic surging may produce unusual morainic loops.

Surges tend to have a periodicity of 15–100 years with the periodicity for any one glacier being relatively constant. Some glaciers may be permanently surging. The exact mechanism of surge initiation is still unknown, with the high velocities and trigger mechanisms difficult to explain. High velocities may be related to a greater than average amount of basal water and the surges themselves may be self-induced.

GLACIAL EROSION

An excellent summary of the processes involved in glacial erosion has been provided by Fenn (1987). The main erosive processes are plucking or quarrying, crushing, shearing and abrasion. Plucking involves the removal of blocks of rock when adhesion between ice and the block exceeds the adhesion between the block and parent material. Thus most plucking involves jointed or fissured rock. The mechanisms involve plucking out when the block is excavated by ice deforming around it, wedging out when the rock gives under pressure from overriding rock particles, and freezing on when the block is lifted by freezing to the base of the ice. There are several conditions that would favour such erosion. Highly jointed rock where high pore water and joint water pressures can develop are especially susceptible (Moran *et al.*, 1980; Addison, 1981). Rock is also weakened in zones of frost penetration (Eyles and Menzies, 1983) and in the lee of bedrock protrusions where stress unloading may be present. The freezing-on process is enhanced in situations where reduction of pressure occurs, such as in subglacial cavities (Rothlisberger and Iken, 1981), and where the glacier margin freezes to its bed during cold seasons (Anderson *et al.*, 1982). Plucking will also occur where basal stresses are large enough to shear off parts of the bedrock (Hutter and Olunloyo, 1981) and where subglacial sediments are so cemented into the ice that they behave as bedrock elements (Boulton, 1979).

Crushing and shearing refer to the process of erosion by the force created by the ice-debris mass. This involves failure due to overburden pressure and due to shear failure by pressure on the upstream face of an obstacle. Such stresses are likely to be generated wherever heavily jointed rock occurs, where cold patches (Robin, 1976) and entrained debris enhance basal stresses (Hagen *et al.*, 1983).

Abrasion is the wearing down of the bedrock surface by particles trapped in the basal layers of the ice. The mechanisms involve abrasive wear and brittle fracture. The extent of the glacial abrasion depends very much on the speed, thickness and amount of basal sliding of the glacier. Abrasion also relies on an abundance of debris in the basal layers of the ice and preferably debris harder than the underlying bedrock. Thus the rock formations over which the glacier has moved will be important, as will the production of debris by weathering and mass movement from the valleysides above the glacier which, once incorporated into the ice body, may gradually work itself to the basal layers.

Two theoretical models for rate of abrasion have been developed. Boulton, (1974; 1979) has argued that rate of abrasion is directly dependent on the effective normal pressure and on the concentration, size and shape of particles in traction with the bed and inversely dependent on bed material hardness. Hallet (1979; 1981) believes that rate of abrasion is dependent on the flux of particles in contact with the bed, the effective force with which the particles press against the bed and the relative hardness of the bed and the abrading particle. The significance of many of these factors is seen in the landscapes produced by glacial erosion and the amount and nature of debris available for transport and deposition.

GLACIAL TRANSPORT AND DEPOSITION

Sediment is transported by glaciers along supraglacial, englacial and subglacial routes (Figure 6.4). The debris-rich basal layer is thin (less than 1 m thick) in most mountain glaciers, compared with polar glaciers which have numerous freezings-on of basal meltwater. The greater proportion of the englacial load of mountain glaciers is derived from supraglacial sources. Large quantities of debris will fall onto the surface of valley glaciers from adjacent rock faces. This may produce supraglacial lateral moraines which are well developed below the equilibrium line and which may be directly fed by rockfalls. Supraglacial

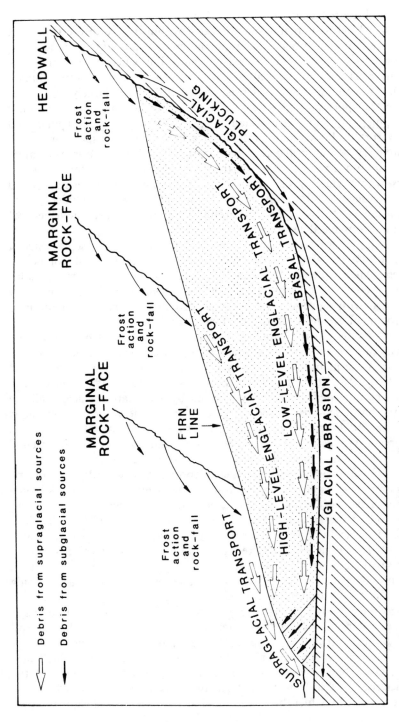

Figure 6.4. Debris transport paths within a valley glacier.

debris is little affected by the glacial processes.

Englacial debris is mainly incorporated within the ice in the firn zone, and since it is of supraglacial origin the greatest amount is found along the glacier margins, along the glacier centre line where two or more ice streams exist and in the lee of nunataks and rock exposures (Small, 1987). As with supraglacial debris, englacial debris is largely unaffected by the transport processes and retains its original characteristics. Thus both types of debris are usually coarse, heterogeneous, angular and poorly sorted. Basal debris will be affected by the transport processes. Boulton (1978) has noted that rock particles in basal debris were more spherical and rounder than particles in supraglacial material. Debris layers within glaciers may be formed as sedimentary layering derived from the accumulation zone or by sliding into crevasses. Under certain circumstances debris layers may result from the upward movement of material from basal layers.

Englacial concentrations occur as longitudinal debris septa (Sharp, 1949) at the margins of glaciers and along centre lines of compound glaciers. Medial debris septa are produced when two or more valley glaciers coalesce. Transverse debris concentrations are probably associated with ogive structures found on temperate glaciers at the base of prominent ice falls (King and Lewis, 1961; Small, 1987).

Debris content in the basal layer may reach 10 per cent by volume and consists of fine sand with pebbles and the occasional boulder. Bottom-freezing is the main mechanism by which rock particles are incorporated into the ice (Boulton, 1970). Observations in natural subglacial cavities and of natural surfaces recently exposed by a retreating glacier indicate that the basal sediment layer is irregular in thickness and discontinuous (Souchez and Lorrain, 1987). The grain size distribution of the particles is bimodal, reflecting the clast-size group and the mineral fragments, respectively. This is markedly different to englacial and supraglacial debris. These differences allow glacial deposits to be identified with some accuracy. The sediment concentration in meltwaters provides some indication of the erosion processes beneath the glacier (Collins, 1979).

Meltout processes occur at the surface and at the base of the ice. The fabric characteristics of till formed by melting processes will resemble those of the parent ice unless reorganised by later processes. Till formed on the surface is usually called 'ablation till'. Melting will be gradual as heat has to penetrate the blanket of ablation till. The basal melting process is more complicated. A gradual layer-by-layer accumulation of till, retaining the fabric characteristics of the dirty ice, will only occur if the meltwater can escape. If water is not able to escape the till will become saturated and flowage may occur. Also, the accumulated water may cause the ice to lift off its bed, creating cavities into which new material will be dropped. Thus many meltout tills possess some of the characteristics of lodgement, deformation and flow tills.

The process by which till is deposited at the base of a glacier is known as 'lodgement'. There appear to be three mechanisms involved. Particles may collect, one by one, through frictional interference, and lodgement will also occur when the force imposed by the moving ice on particles or debris rich ice in traction over the glacier bed is not able to overcome the frictional drag between these and the bed. The third mechanism occurs when basal melting brings slow-moving debris bands close to the floor and allows deposition as the ice melts from beneath.

Sliding and flowing of ablation till has been known for a long time. Boulton (1978) has distinguished two elements in flow tills. These are an upper, allochthonous, far-travelled, component of subaerial derivation, and a lower parautochthonous, little travelled, component which has accumulated basally. The allochthonous element has undergone frequent failure and remoulding and surface water has transported away fines, leaving a till depleted of fines. The parautochthonous element has not been subjected to surface processes and retains a massive character.

Lawson (1982), working on the Matanuska Glacier in Alaska, recognised four main types of flow based on water content. Plug flow occurs at the lowest water content, then there is a graduation to a very fluid flow at high water contents. All flow types have distinct morphological and sedimentary characteristics and are from tends to hundreds of metres long. They occur preferentially on the margins of glaciers with strong upward compressive flow and on the Matanuska Glacier account for 95 per cent of the till.

The result of these processes is to produced a glaciated mountain valley sediment system composed of a number of major components (Figure 6.5). Some of these components comprise the types of glacial deposits discussed above, together with fluvio-glacial landforms such as eskers, kames and sandur (outwash plains). Material is also provided by processes acting on the valley slopes above the glacial limit, possibly producing a sequence such as is illustrated in Figure 6.6. These deposits will have been superimposed onto the 'classic' landscapes of glacial erosion.

GLACIATED MOUNTAIN LANDSCAPES

Matthes (1930), in his introduction to his classic study of the Yosemite Valley in the Sierra Nevada of North America, has argued that there are three kinds of evidence of glacial action in mountain regions: grooved and polished rock surfaces; characteristic topographic forms; and deposits of ice-borne debris. In the Sierra Nevada there are extensive polished and grooved surfaces as well as U-shaped, trough-like canyons having spurless parallel walls and stairwise-descending floors that often hold shallow lake basins on their treads. Also characteristic are hanging side-valleys. Matthes might also have mentioned the armchair-like hollows called 'cirques'. However, he did point out that not all glaciated mountains possessed these forms in abundance.

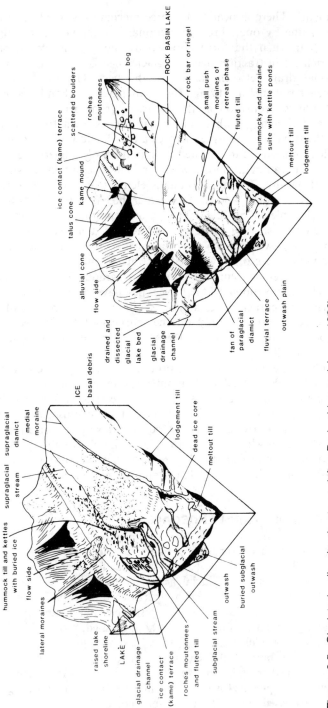

Figure 6.5. Glaciated valley landsystems (after Derbyshire and Love, 1986).

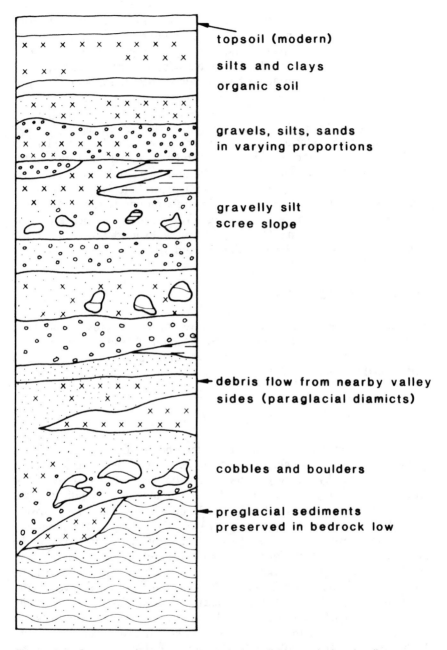

Figure 6.6. Some possible sequences of deposits in a glaciated valley.

The development of such features depends on a number of factors, such as nature and intensity of glaciation, number of glacial epochs and the nature of the rock type. Some rock types appear to inhibit glacial erosion. In large areas of the Sierra Nevada, on granitic rocks, many canyons and valleys appear to have retained their pre-glacial V-shapes. Many other valleys lack some of the features discussed above, such as valley steps and rock basins. Also side-valleys can be left hanging by processes other than glacial action. Thus, Matthes thought the most reliable record of glacial activity was the debris left behind. Gardner (1972) has also stressed that the most distinctive glacial landforms in the Canadian Rocky Mountains are terminal and lateral moraines. This may be generally true but in tectonically and geomorphologically active high-mountain areas the evidence of previous glaciations is sometimes more obvious from erosional landforms than from sediments which are easily removed by denudation. This was the view expressed by Clapperton (1987) in his study of the Ecuadorian Andes.

FORMS OF GLACIAL EROSION

A comprehensive classification of features of glacial erosion has been provided by Sugden and John (1976). All the features may occur in mountain regions but the most distinctive landscapes are those associated with troughs and cirques.

Glacial troughs

Four types of glacial trough have been described by Linton (1963): alpine troughs, Icelandic troughs, composite and intrusive troughs. Alpine troughs are cut by valley glaciers, originating in a cirque or series of cirques and occupying a pre-glacial drainage system. Icelandic troughs are closed at one end by a trough head and are associated with erosion beneath an ice cap with ice descending from the ice cap and spilling over the trough head as outlet glaciers. The outlet glaciers from the Josteldalsbre ice cap in Norway are good examples. Other examples are the Ventisquero (glacier) Dickson and Ventisquero Grey flowing off the Hielo Sur, the Southern Patagonian ice cap. The most distinctive feature of landscapes produced by outlet glaciers is the contrast between heavily glaciated troughs and unmodified intervening slopes and plateau summits. Such landscapes have been described in east Greenland (Bretz, 1936; Sugden, 1974), Scotland (Linton and Moisley, 1960), and Norway (Dahl, 1965). The plateau surfaces may be covered with thick regolith with fragile landforms of pre-glacial age such as tors.

Composite troughs are often open at both ends, forming through valleys, and are the result of breaching of watersheds by glacial diffluence and transfluence. New valleys are created and drainage changes initiated. Five categories have been suggested: simple diffluent troughs, multiple diffluent troughs; troughs formed by simple transfluence; multiple transfluence troughs; and radiative dispersal systems. Radiative dispersal systems are spectacularly developed in the Fiordland District of South Island, New Zealand, and in southern Norway. In these areas Linton (1957) has described a series of radiating glacial troughs apparently unrelated to pre-glacial topography. The Fiordland District is a mountain block structurally distinct from the Southern Alps and separated by the tectonic depression of the Waian valley and Lake Te Anau on the east. Summits rise from 1200 m in the south to 2800 m in the north. The system of troughs in southern Norway radiate from the upland in all directions except north, north-east and east. Linton (1957) argues that the radial pattern is the result of an ice body, large enough to cover the whole area, which descends *en masse* in a simple radial outward movement unimpeded by topography. The fourth type of trough is the intrusive or inverse trough, whereby ice pushes up valley against the direction of the pre-glacial drainage. Such troughs tend to occur in the lower hill areas.

Much interest has centred on the cross and long profiles of glacial troughs. Glacial troughs are traditionally described as being U-shaped, although Davis (1916) described them in Mission Range, Montana, as having a catenary pattern. But many workers have been found that a parabola of the form $y = ax^b$ is a better approximation for the shape. Values of b obtained by Svensson (1959) in Sweden were close to 2 but Graf (1970), working in the Beartooth Mountains of Montana and Wyoming obtained values in the range 1.5–1.8, and Drewry (1972) in Antarctica found values of 1.6 and 2.3 for two sides of a subglacial valley. Aniya and Welch (1981) obtained b-values of less than 1 to over 5 for valleys in the Victoria Valley of Antarctica, but those developed on granite had values between 1.5 and 2.0. They argued that higher values were the result of intense lateral erosion rather than deepening. Aniya and Naruse (1985) obtained b-values ranging from about 1 to over 3 for a glaciated valley in Patagonia.

Kanasewich (1963) and Corbato (1965) used a parabolic model for the bedrock cross-profile when deducing the thickness of a glacier from gravimetric surveys. Doornkamp and King (1971) found that an equation of the form $y^2 = a + bx^2$ fitted best profiles in the Front Range of Colorado. More recently Wheeler (1984) has challenged the simple parabolic formula noted above, noting several drawbacks for its use. Such curves cannot be used with negative horizontal distances, thus they have always had to be fitted to the two sides of the valley separately. Secondly, the curve has no turning point and is constrained by the zero-datum of the ordinate. Wheeler (1984) suggests that the quadratic equation $y = a + bx + cx^2$, a form of parabola, is a better one to use. It also allows both sides of the valley to be modelled at the same time.

The formation of such a parabolic shape can be achieved by the simple widening of a V-shaped pre-glacial valley with or without over-deepening. The widening process is usually accompanied by a reduction in the sinuosity of the valley floor and the truncation of spurs. It may also lead to the production of hanging valleys, although these are more often the result of over-deepening. Ice-eroded spurs projecting into the main valley at a junction with a tributary glacier have been termed 'bastions'. Bastions may be created when tributary ice pushes the major ice stream away from the valleyside which consequently suffers less than normal erosion (Davies, 1969). The comparatively smooth parabolic shape may be interrupted by structural benches, as described by Cotton (1942) for some glacial troughs in New Zealand.

Johnson (1970) attempted to explain the shape of a glaciated valley by analogy to a channel eroded by debris flow in which no flow occurred at the midpoint of the valley bottom and upper parts of each valley side. Reynaud (1973) attributed the shape to the characteristic distribution of shear stresses and Boulton (1974) to the distribution of abrasion rates along the cross-profile which were determined by the effective normal pressure and glacier flow velocity. The most compelling argument is that erosion will produce a cross-profile in which ice is transported most effi-ciently. Using this principle, Hirano and Aniya (1988) have shown that the resultant cross-profile takes a catenary form. But examination of actual data suggests that there are two types of cross-profile develop-ment. One type, represented by data from the Rockies, is development from a shallow, wide V-shaped valley to a deep U-shaped valley; and the other type, represented by examples from Patagonia and Antarctica, is from a steep and narrow V-shaped to a wide, broad U-shaped valley. The Patagonia and Antarctic model can be construed as representing a widening rather than a deepening process under continental ice, whereas the rocky Mountain model represents overdeepening by a valley glacier.

The formation of the trough will be aided by the process known as 'unloading' or 'dilatation' (see Chapter 3). There is no doubt that the retreat of glaciers from deeply incised valleys has permitted the fracturing and uparching of rock on valleysides and floors (Lewis, 1954; Harland, 1957; Gage, 1966). Sheeting joints have frequently been observed parallel to glacial facets (see, for example, Battey, 1960) and there is the possibility that dilatation can occur beneath a valley glacier thus enhanc-ing the erosive effect of the ice. Dilatation will occur as rock is gradually eroded by the glacier.

The parabolic shape of troughs is often abruptly terminated in an upper shoulder. In Icelandic-type troughs this break of slope is approx-imately rectangular and the ground above the trough edge is usually part of the plateau surface. In alpine-type troughs there is usually an intermediate slope, the trough shoulder, between the steeper trough wall below and a steeper slope above. Several explanations have been put forward to explain trough shoulders. The simplest explanation is that they represent part of the original valleyside which has not been over-steepened by glacial action. The contrast in slope angle between trough wall and trough shoulder may be accentuated by the trough shoulder being reduced by frost action and mass movement. Some trough shoulders may reflect a previous valley-in-valley form or perhaps multiple glaciation. A more intriguing possibility, summarised by Thompson (1962), is that shoulders are produced by periglacial mass movement in interglacial times. Many workers have pointed to the coincidence between the interglacial tree line and the trough edge.

The sides of the troughs may exhibit marked asymmetry with the steepest slopes being associated with the most active glacierisation. In north temperate latitudes the steepest slopes tend to face north and north-east, while in south temperate areas they tend to be south-facing (Evans, 1972a; Sugden and John, 1976). In equatorial zones asymmetry is weak and in many Antarctic mountains the steepest slopes face towards the northern sectors.

The long profiles of glacial troughs are often characterised by irregular convex breaks of slope or steps producing a glacial stairway Reverse slopes and basins are also common. The steps often show signs of glacial abrasion and plucking and the basins frequently contain lakes or have been infilled with glacial drift and alluvium. Valley steps may represent pre-existing irregularities, possibly nickpoints, which have been exaggerated by glacial action. Other steps appear to be controlled by sharp discontinuities in the bedrock. The exact mechanisms whereby these irregularities become enhanced by glacial action is still unknown but variations in the nature of glacial flow downvalley appear to be extremely important. The idea of extending and compressing flow, discussed earlier, seems to be crucial in this respect. Observations by Nye (1959), on Austerdalsbreen in Norway, have shown that irregularities in the bed influence glacier flow in a systematic way and different types of flow can influence erosion on the bed. The process becomes self-regulating, thus once irregularities have been created they become accentuated.

Cirques

The glacial cirque is one of the most characteristic forms of glacial erosion. Thus 'the close association between cirques and areas of present or former glacierisation, observations of cirques containing cirque glaciers, and the virtual impossibility of devising suitable non-glacial hypotheses in all cases, gradually forced universal acceptance of the view that most cirques are features of glacial erosion, though even now, the processes responsible are not fully understood' (Embleton and King, 1975, p. 207). A cirque has been defined as a hollow, open downstream but bounded upstream by the crest of a steep slope (headwall) which is arcuate in plan around a more gently sloping floor (Evans and Cox, 1974). Cirques vary in size between small depressions a few score metres across to extremely large features several kilometres in diameter found in areas such as the Himalayas.

There have been numerous studies of cirque shape, size and orientation. The general shape of cirques has been examined by Galibert (1965) and Derbyshire (1968); length–height ratios have been measured by Manley (1959) and Andrews (1965). Other morphometric indices have been devised by Evans (1969) and Andrews and Dugdale (1971), and the form of cirque long profiles has been studied by Lewis (1960), Haynes (1968) and Sugden (1969). Most work, though, has concentrated on cirque orientation and elevation in an attempt to relate cirque development to climatological factors (see, for example, Derbyshire, 1964; Temple, 1965; Lewis, 1938; King and Gage, 1961; Schytt, 1959; Seddon, 1957; Andrews, 1965; Evans, 1969; Unwin, 1973). In the northern hemisphere the majority of cirques face directions between north and east, whereas in the southern hemisphere a southerly or south-easterly aspect is dominant. The shade factor will be more important in middle and high latitudes because in tropical areas the greater angle of the sun's rays reduces the shading effect of mountains. Many studies have also shown that cirques have been preferentially developed on the lee side of mountains with respect to the prevailing snow-bearing winds. Thus, in the Front Range of the Rocky Mountains in Colorado, cirques are best developed in the eastern flanks.

It has been shown that cirque aspect plays a major role in determining seasonal change in the spatial variation of potential direct insolation (Isard, 1983). In the middle latitudes of the northern hemisphere low sun elevation in winter means that in south-facing cirques, relatively unobstructed steep backwall slopes receive the largest amount of insolation. During summer, high solar elevation decreases the spatial variation of potential direct insolation to south-facing cirques. North-facing cirques receive very little direct solar radiation in winter but the spatial variation in summer is large. Maximum incidence angles occur during the summer solstice to slopes of east or west aspect.

A most significant relationship has been established between the elevation of cirques and the elevation of the local snow line or firn line. The firn line on a cirque glacier usually occurs about three-fifths of the way between the snout and the upper limit of ice. This relationship enables estimates to be made on the firn line or the snow line during the formation of relic cirques. It is thought that the climatic snow line will not usually lie at more than a few hundred metres above the elevations of the cirque floors. Thus, in Rocky Mountain National Park in Colorado, the mean level of cirque floors would indicate an average altitude of the former snow line at 3000 m rather than the present 4200 m. Similar calculations have been made in many other areas such as the Andes (Hastenrath, 1971) and Japan (Hoshiai and Kobayashi, 1957). This type

Figure 6.3. *Grouping of cirques in the Gongga Mountains of south-west China.*

Slope orientation	1st group Elevation (m)	No. of cirques	2nd group Elevation (m)	No. of cirques	3rd group Elevation (m)	No. of cirques	4th group Elevation (m)	No. of cirques
N	4830	6	4680	46	4450	42	4200	7
W	4920	6	4700	43	4540	40	–	–
S	4880	6	4640	22	4470	36	–	–
E	4830	6	4610	2	4410	7	4150	8

Source: Liu Shuzheng and Zhong Xianhao (1987).

of analysis is complicated in areas such as Tasmania, where multiple cirque levels occur. In the Gongga Mountains of south-west china, four groups of cirques occur at different elevations, the latter varying with aspect (Table 6.3). In some cases the multiple levels may be due to changes in the position of the snow line so that the cirques are not contemporary but of different generations. Where cirques appear to be contemporary, 'reconstructed glaciers' may have formed in lower hollows from avalanche ice descending from higher cirques (Cotton, 1942).

As mentioned earlier, the exact mechanism of corrie formation is still unknown. Two sets of processes have to be considered; the development and retreat of the headwall and the development and lowering of the floor. Cirque enlargement takes place predominantly by recession of the headwall by frost shattering, basal sapping, dilatation jointing and removal of basal debris by glacial action. It is also possible that steep niche glaciers or ice aprons in couloirs may shape the headwall. The lowering of the floor is probably produced by the strong rotational component of glacier flow. Cirque glaciers are short but have a high net balance gradient, producing strong vertical components. Maximum erosion occurs in the vicinity of the equilibrium line.

A number of attempts have been made to integrate the glacial land-forms discussed above into models of landscape evolution. Willard (1904) talks about sapping at low levels and retrogressive undercutting by individual ice streams consuming the pre-glacial topography leaving sinking ridges, meandering dulled divides, low cols or passes and passage-ways of transection pointing to piracy and wide shiftings of glacial drainage. In addition, thin arêtes, needle-pointed Matterhorn pyramids with incurving slopes and subdued spires with radiating spurs enclosing basin lakes also exist.

The general sequence of glacial erosion appears to be an increase in the number and size of cirques with an extension of glacial troughs and the production of hanging valleys, arêtes and horns. Davis (1900; 1906) has described a model whereby a normally eroded, round and smooth mountain pass (Figure 6.7a) is sharpened by frost action and glacial erosion

Figure 6.7. Transformation of a landscape by glacial erosion (after Davis, 1900).

(Figure 6.7b) into a landscape dominated by cirques, glacial troughs and arêtes with little pre-glacial topography remaining (Figure 6.7c). The landscape portrayed in Figure 6.7b may be an equilibrium form. It is certainly common to many mountain areas. Arêtes and horns appear to be self-preserving and the overall altitude of a mountain chain may be maintained by isostatic uplift. Linton (1963; 1964) has suggested a model involving divide elimination. The widening of glacial troughs would truncate and shorten pre-existing divides, with widening as far as the equilibrium line. Eventually some type of equilibrium would be achieved.

All the major landforms of glaciated mountains discussed above are not solely the result of glacial action, although glacial action may be dominant. The pre-glacial topography and degree of weathering have been major factors in determining the nature and location of the features, periglacial activity on slopes above valley glaciers has modified the slope forms, and in deglacierised areas postglacial modification has been intense. Some mountains exhibit abundant evidence of glacial action, whereas in other areas glacial action has been slight. This emphasises the complexity of mountain landscapes, a point which has been stressed many times in this book.

Examples of glaciated mountains

It is instructive to examine specific examples of glaciated mountain landscapes to assess the manner in which the individual landform types discussed above combine to provide distinctive assemblages and also enable the glacial history of the areas to be understood. The examples — the Ecuadorian Andes (Clapperton, 1987), Japanese Mountains (Yoshikawa *et al.*, 1981) and Papua New Guinea (Loffler, 1977) — have been chosen to provide a wide spectrum of landscape types. The Cordillerra de los Andes in Ecuador forms a north–south barrier 100–120 km wide rising to over 4000 m above the lowlands of the Guayas and Amazon basins. The landscape is dominated by two parallel mountain ranges forming dissected plateau-like surfaces at 3500–4500 m altitude surmounted by strato-volcanoes such as Chimborazo (6310 m). The permanent glaciers occur on the twelve highest central volcanoes, all of which are over 4800 m in elevation (Hastenrath, 1984). The glaciation limits are consistently lower on the eastern sides of individual mountains and in the eastern cordillerra reflecting the dominant easterly source of precipitation.

The most conspicuous landforms of glacial erosion are those of areal scouring, glacial troughs and cirques. Areal scouring is widespread above 3800 m, consisting of abraded rock knolls and craggy 'stoss-and-lee' forms interspersed with bedrock basins filled with peat or water. The extent of the ice-scoured terrain west of Cuenca indicates the presence of a former ice cap almost 2500 km^2 in area, 3°S of the Equator and

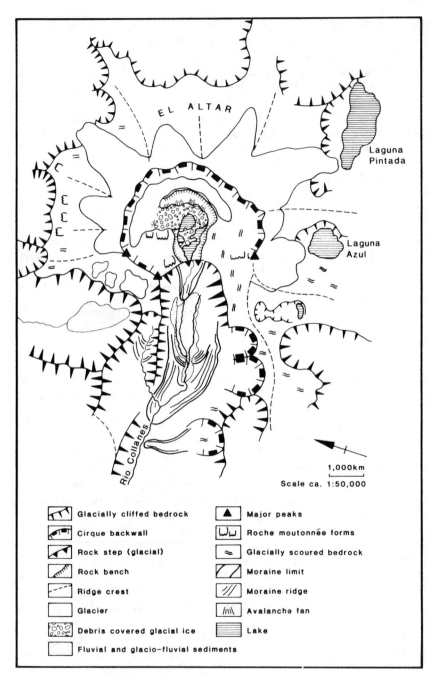

Figure 6.8. Glacial landforms of El Altar, Ecuador.
Source: Clapperton (1987).

35–70 km from the Pacific Ocean. Ice-scoured features also occur on the flanks of the larger central volcanoes and indicate that the former ice covers of Cayambe, El Altar and the Chimborazo-Carihuairazo Massif merged with surrounding ice caps. Glacial troughs 8–20 km long and 200 m deep are prominent in the eastern cordillerra and on the flanks of some central volcanoes. Troughs are absent from active volcanoes such as Cotopaxi, Tungurahua and Sangay, are poorly developed on infrequently active or dormant volcanoes such as Pichincha, Antisana, Cayambe and Chimborazo, but are well developed on extinct volcanoes such as Carihuairazo, El Altar and Sincholagua.

Cirques occur in various stages of development. Well-developed forms are confined to the same areas as well-developed troughs, namely the eastern cordillerra and the higher, extinct volcanoes. There is an interesting relationship between glaciation and volcanic activity in that the best cirques have resulted from the enlargement of hollows created by explosive volcanism. Many are 3, km in diameter and over 1000 m deep. The suite of features discussed above is well developed on El Altar (Figure 6.8).

Landforms of glacial deposition are also conspicuous in the Ecuadorian Andes, especially moraine systems. On the south-east flanks of Chimborazo, 270 m high moraines occur. There are three morphologically fresh moraine stages which enabled Clapperton (1987) to produce a glacial sequence for the area. The three moraine systems have been conceptualised into Neoglacial, Late-Glacial and Full- Glacial stages. Glaciation limits and Equilibrium Line Altitudes (ELAs) have been estimated for the Western and Eastern Cordillerras (Figure 6.9).

It is interesting to compare the glacial geomorphology of the Ecuadorian Andes with mountains that no longer possess ice masses but which were formerly glaciated. The higher Japanese mountains provide such an example, although it was only in 1902 that the existence of glacial landforms was established (Yamasaki, 1902). Since then studies have established that glaciers extended down as far as a height of about 1000 m in central Japan. Glacial landforms are easily recognisable above 2000 m in the Hida, Kiso and Akaishi ranges in central Japan, and above 1400 m on the Hidaka Range in Hokkaido. However, as in the Andes, some Holocene volcanoes higher than these levels have not been glaciated. Cirques in Japanese mountains are small, with steep floors and glacial troughs are rare and short with a number of atypical features that gives rise to speculation on their separation from fluvial valleys. Cirques occur at levels higher than 2400 m in central Japan and higher than 1500 m in Hokkaido and are found on east-facing slopes, the leeward slopes to the prevailing westerlies. End moraines are small but have been used to construct a two stage glacial sequence.

The mountains of Papua New Guinea have been free of ice for 10 000 years but the glacial landforms are remarkably well preserved, attesting to the slow rate of subsequent denudation. The Pleistocene glaciation of

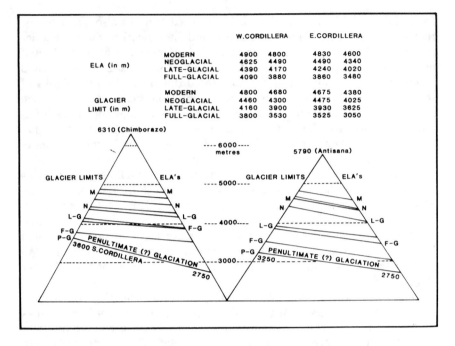

Figure 6.9. Glaciation limits and Equilibrium Line Altitudes for Western and Eastern Cordillerras of Ecuador.

Source: Clapperton (1987).

Papua New Guinea has been admirably summarised by Loffler (1972). Traces of glaciation are visible on about twenty mountains that exceeded the Pleistocene snowline, which was at approximately 3550 m. Glaciation varied from ice cap glaciation on plateaux to valley glaciation on the incised mountain landscapes. Examples of the first type occur on Mount Giluwe, the Saruwaged Range, Mount Albert Edward and Mount Scratchley. Mount Wilhelm, the Kubor Range and the Star Mountains are typical of the second type.

Cirques with steep back walls up to 700 m high and broad flat valley floors are grouped around the eastern peaks of Mount Giluwe. Their form and location suggest that they are merely glacially modified amphitheatre valley heads and not the result of intense glacial erosion. The landforms on Mount Wilhelm (4510 m) are very different, with deep troughs with stepped valley profiles and over-deepened cirques. Most cirques are situated west and south-west of the north-south trending ridge, and have been attributed by Reiner (1960) to the pre-glacial asymmetry of the Mount Wilhelm massif, favouring extension of the ice in a westerly and south-westerly direction. Most glacial troughs on Mount Wilhelm have lateral moraines but, in contrast with Mount Giluwe, recessional moraines are rare. Glacial landforms on the other peaks that possessed a Pleistocene ice cover are essentially similar to either the Mount Giluwe-type ice cap or the Mount Wilhelm-type valley glaciation. The evidence suggests that the last glaciation was the most extensive. This is different to the situation in higher latitudes but is typical of smaller mountain glaciers in the middle and lower latitudes (Galloway *et al.*, 1973).

Chapter 7

VOLCANOES AS MOUNTAINS

INTRODUCTION

Some justification is needed for examining volcanoes as separate examples of mountains. It is certainly true that volcanic mountains are subject to the same processes once they are established as mountains. However, it can be argued that volcanic mountains are distinctive because of their shape, the nature of their materials and the often continuing relationship between volcanic activity and the more 'normal' processes characteristic of all mountains. Volcanoes are also subjected to some processes — such as pyroclastic flows, ash falls and volcanic mudflows or lahars — which other mountains do not experience. There is further justification in treating volcanic mountains separately in terms of the impact they have had on human activity. Volcanoes, because of the inherent fertility of soils developed on many volcanic materials, have always been favoured areas for settlement, but this has to be balanced by the continuing threat posed by the volcanoes themselves. Thus volcanic mountains possess all the hazards and problems of mountains world-wide but they also possess the additional hazards posed by actual and potential volcanic activity.

Volcanic mountains are also interesting in that they may form completely isolated individual mountain systems, for example, Mount Cameroon in West Africa or Mount Ararat in Asia Minor, as well as forming the major components of the cordilleran mountain systems found in North and South America. The Western Cordillerra of Canada is a good example. There are no active volcanoes today but in the Quaternary Period there were 150 active volcanoes in British Columbia and the Yukon, twenty of which are high composite volcanoes of mountain status. Mount Edziza, 50 km south-east of Telegraph Creek in north British Columbia, is a typical example. It began to form about 40 million years ago and has erupted three times in the last 1800 years, the last eruption being 1000 years ago. Since then it has undergone an intense period of dissection. It is hoped that the following account justifies the separate treatment of volcanic mountains. It is impossible to discuss in detail all aspects of volcanology; attention is therefore focused on the distinctive elements of volcanic mountains. More substantial treatments can be found in the many excellent books on volcanoes (such as Ollier, 1969; Green and Short, 1971; Macdonald, 1972; Francis, 1976; Williams and McBirney, 1979; Bullard, 1984).

NATURE OF VOLCANIC MATERIALS

Knowledge of the nature of volcanic materials is essential to an understanding of the way in which individual volcanoes have been constructed as well as the ways in which subsequent weathering and erosion modify the features. Volcanic rocks are igneous rocks that form at the Earth's surface either by the gradual outpouring of magma or by the ejection of material by explosive activity. The latter group of rocks are known collectively as *pyroclastic* rocks, the fine-grained products forming ash, tephra or tuff with the coarser products solidifying as volcanic breccia. The rapidity with which magma cools on the surface produces essentially fine-grained rocks, or, if cooling is especially rapid, rocks with no crystalline texture are formed.

Igneous rocks are classified according to their chemical and mineralogical composition and texture. The proportion of silica, the dominant oxide, is particularly important, as it is used to form a four-way classification. Those igneous rocks which contain more than 66 per cent silica are called acid igneous rocks; rocks with between 52–66 per cent are called intermediate igneous rocks; basic igneous rocks possess 45–52 per cent silica, and ultrabasic rocks less than 45 per cent. Acid rocks tend to be light in colour because of the preponderance of light-coloured minerals such as quartz and feldspars, whereas intermediate and basic rocks are darker because of a greater proportion of dark-coloured ferromagnesian minerals such as augite, olivine and hornblende. The nature of the feldspar minerals also helps to sub-divide the igneous rocks. In acid rocks, alkali feldspars such as orthoclase, microcline and albite-rich plagioclase occur, whereas in basic rocks, calc-alkali feldspars such as plagioclase are dominant. The most important volcanic rocks are basalt, a basic rock, and andesite, an intermediate rock.

There are a number of ways in which an original uniform magma can be altered into rocks of differing composition. *Differentiation* is a process whereby gravity separation occurs, with the heavier minerals sinking into the magma body as crystallisation occurs. Magma can react with the walls and roof of the magma chamber, some of which may be *assimilated*. Assimilation of sandstone will add silica to the magma, making it more acid. Complete melting of formerly solid rocks to form a new magma is called *anatexis* and will generally produce acid magmas. Although there can be a great variation in lava types produced by neighbouring volcanoes, many volcanic areas are characterised by similar suites of rocks. Suites reflect deep-seated petrographic provinces and the relationship of volcanoes to the different plate boundaries discussed earlier. There appear to be three main suites. In the *calc-alkaline suite* rocks are relatively rich in calcium and they tend to be acid in nature. This suite is found in the volcanoes of island arcs and the Pacific borders of North and South America. It is also associated with mountain-building and is sometimes called the 'orogenic' suite, which is better than its old name of 'Pacific' suite. The *alkaline* suite is composed of basic rocks rich in alkalis, especially sodium. It used to be called the 'Atlantic' suite, which was somewhat confusing since this suite contains the dominant rocks of the central Pacific; 'non-orogenic' suite is a much better designation. The third type is the *potassic* suite, with basic to intermediate rocks relatively rich in potassium. It is sometimes known as the 'Mediterranean' suite since such rocks are characteristic of Mediterranean volcanoes.

The alkaline suite is often derived by differentiation of primary basalt magma, the potassic suite by assimilation of limestones, marls and other sedimentary rocks by differentiates of primary basalt magma, and the calc-alkaline suite by the melting of crustal rocks to form a magma by anatexis followed by assimilation and differentiation (Ollier, 1969). The nature of the materials and the form of the eruption process determines, to a large extent, the size and configuration of the volcano.

TYPES OF VOLCANO

Volcanoes can be classified by the degree of violence of the eruption. The names are derived from type volcanoes or type areas. In the *Icelandic* type, fissure eruption is dominant, with great outpourings of basaltic lava; whereas in *Hawaiian* eruptions activity from a central vent is more pronounced. *Strombolian* eruptions are more explosive with a higher proportion of pyroclastic material. Eruptions dominated by pyroclastic rocks can be similarly classified. *Vulcanian* eruptions are violent, producing large quantities of ash as well as relatively viscous lava. *Vesuvian* is a more violent form of Strombolian or Vulcanian which scatters ash over wide areas; and *Plinian* is even more violent, with large quantities of ash. *Pelean* is the name given to very violent eruptions and the explosion of viscous magma. The magma is intermediate to acid, large quantities of pumice are erupted, and *nuées ardentes* (see below) are characteristic.

The landforms produced are related to these eruption characteristics. Large volcanic mountains tend to be *lava shields*, *strato-volcanoes* or a mixture sometimes called *composite*. Lava shields have gentle slopes (less than 7°) and convex outlines, the best known examples being on the Hawaiian islands. Strato-volcanoes are the large, classic cone-shaped volcanoes such as Fujiyama, which rises 3700 m above its floor. The Mounts Rainier and Shasta in the Cascade range of North America and Popocatepetl and Orizaba in Mexico are other examples. They tend to be composed of alternating sequences of lava flows and pyroclastics. Such volcanoes approximate a conical form only if built from a central vent. Elongated mountains, such as Hekla in Iceland, are created by eruptions from elongate fissures. Their nature and formation are examined in greater detail below. The remainder of volcanic landforms tend individually to be smaller in scale but as they occur frequently on the larger volcanic mountains they are worth considering briefly. *Lava domes* are smaller and more convex in shape than lava shields. Smaller-scale central eruptions may produce straight-sided *lava cones*. Slopes are generally low, but on some volcanoes may be up to 45°. *Lava mounds* are ancient basaltic volcanoes with no craters, one example being Mount Cotterill, Victoria, Australia.

Because they are more viscous, acid lavas produce different landforms. *Cumulo-domes* are convex domes, often formed of rhyolite. Examples occur in North Island, New Zealand; Lassen Peak, California; and in the Auvergne, France. The term 'mammelon' has often been used instead of cumulo-dome but Cotton (1944) suggests this term should be reserved for successive flows of trachyte. *Tholoids* are cumulo-domes that form in the craters of larger volcanoes and *plug domes* are created by the most viscous magma, which moves up the volcanic vent as a solid mass. *Spines* are smaller types of plug domes.

The shape of volcanoes composed of pyroclastic materials is determined by the nature of those materials. Coarse particles tend to produce steeper slopes than fine particles: *Scoria cones* or *cinder cones* are single, steep or gently concave features. They range up to 700 m or more in height but most are 30–300 m high. Most pyroclastic cones are essentially symmetrical with smooth slopes of 25–40°. They often overlap along a common fissure. Eruptions of constant strength tend to produce uniform slopes corresponding to the maximum angle of repose of the ejecta (Williams and McBirney, 1979). Most ejecta are coarsest and thickest near the vent, but ejecta from violent eruptions diminish outward in thickness and size less rapidly than do ejecta from weak eruptions. If the eruptions are weak the slopes tend to be concave, and convex slopes may develop when eruptions are more violent. Almost all youthful cones less than 1000 m high have straight or even convex slopes. Larger cones, especially eroded ones, possess concave slopes. Asymmetrical cones result from the greater accumulation of ejecta on leeward sides or from multiple conduits. *Scoria mounds* are scoria cones with no apparent crater and *nested scoria cones* are late-phase features which form in the centre of a large crater.

Most large volcanoes possess craters on their summits but some possess much larger depressions, or calderas, created by volcanic collapse. Eruption may continue to the point where the upper part of the magma chamber is emptied and no longer supports the top of the cone, which collapses. It is sometimes difficult to distinguish an explosive crater from a caldera. But a frequency analysis of the dimensions of calderas show a minimum diameter of about 1 km. Thus craters more than a kilometre across may be calderas. Several types of caldera have been identified. Small calderas, 1–6 km in diameter, and situated on top of strato-volcanoes, are called *Haruna*-type calderas. The *Krakatoan* type is formed by the foundering of the tops of large composite volcanoes following explosive eruptions; it is much larger in size, varying in diameters in the range 10–20 km. In the *Katmai* type, collapse results from drainage of the central magma chamber to feed new volcanoes, and, in *Valles*-type calderas, foundering takes place along arcuate fractures independent of pre-existing volcanoes, but as a result of the discharge of large volumes of siliceous pumice. *Hawaiian*-type calderas are formed by the collapse of the tops of shield volcanoes during the later stages of their growth. The *Galapagos* type is similar except that collapse results from injection of magma and eruptions of lava from circumferential fissures near the summit. Two less common types are the *Masaya* type, formed by piecemeal subsidence of a broad shallow depression occupying the centre of a low shield; and the *Atitlan* type, formed by subsidence associated with eruptions from volcanoes near the rim.

VOLCANIC LANDSCAPES

The way in which many of the features discussed above combine to produce distinctive volcanic mountain landscapes can be assessed by examining a number of specific examples. Nyamlagira, in Zaire, is a basalt dome rising 1500 m above its base to a maximum height of 3056 m. The sides of the volcano slope gently at about 7° to a steep-walled sunken crater 2 km across and 75 m deep. It has been built up by a succession of lava flows. Muhavura (4129 m), a simple composite cone, is the easternmost volcano of Virunga Range in south-west Uganda on the Rwanda frontier. It is an almost perfectly symmetrical cone, rising nearly 3000 m above its base, composed of successive layers of lava and ash. Kilimanjaro (5895 m) is a complex cone rising 4000 m above the surrounding plains. It is in essence three separate volcanoes with, in order of formation, Shira in the west, Mawenzi in the east, and Kibo in the centre. Shira and Mawenzi were originally two separate volcanoes 25 km apart. Little remains of Shira except the western rim of a caldera known as Shira ridge. Mawenzi is also severely dissected. There are more than 250 small cinder and lava cones on the flanks of Kilimanjaro.

In Papua New Guinea there are two irregular volcanic clusters on the mainland. The largest is centrally located in the southern part of the highland region, with a smaller area in the south-east in the Cape Vogel Basin. The highland and southern fold mountain volcanoes consist of fifteen major centres including Mounts Bosavi, Sisa, Giluwe, Hagen and Italibu. The great majority are steep-sided conical volcanoes but Mount Giluwe is a lava-shield volcano with gently sloping convex sides with numerous eruptive sites on its lower flanks marked by scoria cones with craters and scoria mounds without craters as well as lava domes and maars (Blake and Loffler, 1971). Maars are craters that extend below ground level with rims of pyroclastic material. None of the highland volcanoes are active. Mount Giluwe last erupted 200 000 years ago.

The Mount Lamington group is the most famous group of volcanoes in the Cape Vogel Basin. Mount Lamington is an active strato-volcano and the Hydrographers Sesara volcanoes are extinct and deeply dissected. Mount Lamington erupted violently in 1951 with catastrophic *nuées ardentes*. Three thousand people were killed and within a few minutes all living matter in an area of approximately 225 km^2 was destroyed. Mount Victory is also an active strato-volcano with lava domes and flows in a rugged summit area flanked by gently sloping footslopes with volcano-alluvial fans. Several young adventive cones occur on its flanks. It last erupted in the 1890s. Balbi volcano, in the Northern Bismarck Sea Volcanic Arc, exemplifies the complex nature of large volcanoes. It is a strato-volcano with a number of coalescing cones and a summit area that contains a variety of lava forms plus ash cones, craters, domes, spines and solfatara fields.

Japan possesses about 200 Quaternary volcanoes, of which sixty have erupted in historic times (Yoshikawa *et al.*, 1981). They are mostly strato-volcanoes and large caldera volcanoes. The largest strato-volcano is Mount Fuji, 3776 m high and 35 km across at the base. The major large caldera volcanoes are Kutcharo, Shikotsu, Aso and Aira. Analysis of the strato-volcanoes has established that there are usually two phases in their development (Aramaki and Yamasaki, 1963; Moriya, 1979). In the early stage, mafic to intermediate magmas are ejected repeatedly as pyroclastic falls and lava flows producing the 'classic' cone-shaped volcano. At this stage the amount of pyroclastic material is greater than that of lava, but gradually lava comes to dominate. There is also a change to material richer in silica. Pyroclastic flows are uncommon and small in size in the early stage. Early-stage volcanoes in Japan include Fuji, Yotei, Iwaki and Iwate, all exhibiting a symmetrical cone shape with no noticeable pyroclastic flows and volcanic fans at their bases. In the late stage, magmas are ejected with violent explosions as pyroclastic falls, as well as viscous lava flows. On the summits, lava domes or Haruna-type calderas with or without central cones are present, while at the volcanoes' base volcanic fans, mudflows and pyroclastic flows develop. The shape can then be separated into two components: the main part and the outer skirt. The main features of this developmental sequence are shown in Figure 7.1. Examples of late-stage volcanoes are

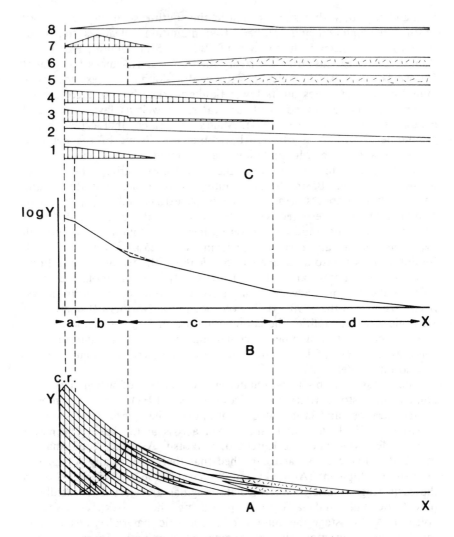

Figure 7.1. Schematic profile and structure of strato-volcanies.

Source: simplified from Suzuki (1975).
Note: A: Structure proposed by Suzuki for andesitic strato-volcanies.
     B: Semi-logarithmic expression of A's profile. The profile is divided generally
       into four straight segments with different slope decrements. a, summit
       part slightly gentler than b; b, main part of the cone; c, skirt part; d,
       peripheral part.
     C: Schematic diagram showing the variation of total thickness of various
       materials constituting the volcano with distance from the source. 1, explo-
       sion breccia deposits distributed ballistically; 2, pyroclastic air–fall
       deposits transported by wind; 3, lava flows; 4, poorly vesiculated
       pyroclastic flow deposits; 5, well-vesiculated pyroclastic flow deposits; 6,
       mudflow deposits; 7, talus deposits; 8, fluvial deposits.

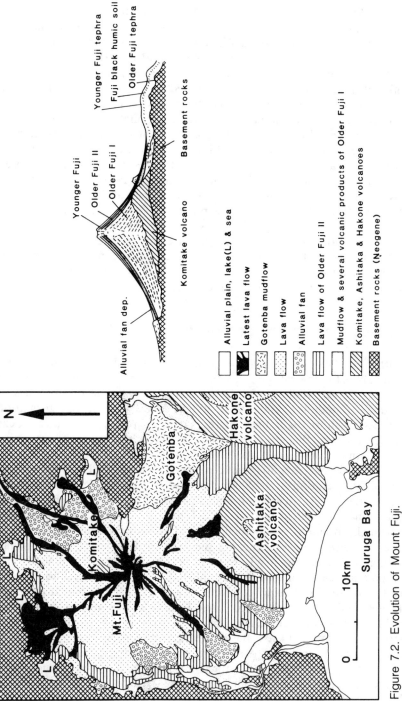

Figure 7.2. Evolution of Mount Fuji.

Source: Yoshikawa *et al.* (1981).

Asama (Aramaki, 1963) and Akagi (Moriya, 1970).

As mentioned above, Mount Fuji is an early-stage strato-volcano, it is superimposed on Komitake, a Pleistocene strato-volcano (Tsuya, 1940). Fuji exhibits slight elongation on a NNW–SSE axis (Figure 7.2). There is a 100 m deep crater, 700 m across on its summit, and it possesses sixty flank volcanoes. Fuji is actually composed of two volcanoes, Older Fuji and Younger Fuji. The former was formed over a period from 80 000 to 10 000 years ago and the latter commenced formation about 5000 years ago. Older Fuji also exhibits two stages: an early longlasting stage with repeated explosive eruptions resulting in a pyroclastics-enriched strato-volcano; and a later shorter stage when effusive eruptions superimposed a lava-enriched volcano on the former (Machida, 1967). Ash layers from stage 1 occur at the eastern foot as hundreds of beds 100 m thick. Mudflows also occurred in the later phases of stage 1. The quiescent time between Older and Younger Fuji is shown by a buried soil. During this period the mountain was eroded by streams, and alluvial fans were created at its foot. Eruptions of Younger Fuji built up the present features by an accumulation of lava flows and pyroclastic falls, and the last eruption occurred in 1707 (Tsuya, 1955).

WEATHERING AND SOIL FORMATION

Weathering and erosion commence as soon as volcanic landforms are created. Weathering rates on volcanic materials can be extremely high. In St Vincent, West Indies, a fertile soil had formed within twenty years of eruption of andesitic ash (Ollier, 1969). The speed of soil formation underlies much of the human interest in volcanic mountains. Basic rocks weather quicker than acidic rocks. Basaltic ash erupted from some Central American volcanoes is often planted with corn the year after it falls. But siliceous rhyolitic pumice in Guatemala and El Salvador shows only slight alteration after 2000 years and it probably takes up to 5000 years to develop sufficient soil and weathered material to support crops. Weathering and soil formation are affected markedly by climate and on individual mountains by microclimatic differences. Weathering rates may be ten to twenty times greater on the windward slopes of the Hawaiian mountains than on the dry leeward slopes. Ferromagnesian minerals and feldspars are easily altered to clay minerals and ion oxides, the ultimate weathering product being a brown, base-rich heavy soil. In tropical climates and in very leached sites kaolin is a common weathering product, whereas in temperate climates and in badly drained sites montmorillonite is more likely. Basalt weathers do produce very productive soils. The more acid the rock the lower is its subsequent fertility.

Weathering may produce distinctive landform features. Pseudo-karst features are present on kaolinised pyroxene andesites of the Caliman Massif, eastern Carpathians, and near Huaron, Peru, there are stone towers similar to some tropical karst but produced by frost action on Tertiary ignimbrite (Tricart *et al.*, 1962).

There have been a number of studies that have attempted to quantify the rate at which weathering takes place on volcanic material. Hay (1960) found that a 400-year-old ash on St Vincent had weathered to form a clay soil 2 m thick and volcanic glass had decomposed at a rate of 15 mg cm^{-2} yr^{-1}. Ruxton (1968), working on ash falls from Mount Lamington, Papua New Guinea, analysed the loss of mobile elements, principally silica, and estimated that the average rate of weathering is halved about every 5000 years. The rate of clay formation was up to 14 mg cm^{-2} yr^{-1}. Ruxton (1968) also determined rates of weathering on the 650 000-year-old Hydrographers strato-volcano, Papua New Guinea. Weathering profiles at sites with little erosion possess an upper zone of 1.5–7.5 m of silty clay above a lower 15–30 m zone of clayey silt. The average loss of silica was calculated as 4–6 mg cm^{-2} yr^{-1} very similar to values of 3.8 mg cm^{-2} yr^{-1} for Oahu, Hawaii, and 4 mg cm^{-2} yr^{-1} in St Vincent.

The relationships between weathering, volcanic materials, landforms and soil formation can be assessed by examining soils developed in volcanic areas. The volcanic soils of New Zealand have been extensively studied (Gibbs, 1980; Gibbs and Wells, 1966). There appears to be a developmental sequence from recent soils through yellow-brown pumice soils and yellow-brown loams to brown granular loams. Recent soils tend to be less than 900 years old, yellow-brown pumice soils vary in age from 900 to 5000 years and yellow-brown soils are up to 15 000 years old. The rate of formation will be affected by the climate, altitude and topography. The deepest and best-developed soils occur in the areas protected from erosion. Soils on the steep sides of volcanic mountains are kept in a perpetual state of youth by erosion and by periodic inundations of fresh volcanic material. Accumulation of ash in thin layers usually allows the existing vegetation to survive and soil processes to continue but at a modified rate. Accumulation of thick ash layers requires recolonisation of the surface and a slow development of the vegetation and soil processes. The erupted materials may consist of particles of markedly different size such as ash lapilli or blocks; the larger the particles the slower the rate of soil formation.

The youngest volcanic soils in New Zealand are represented by the Ngauruhoe, Tarawera, Rotomahana, Burrell and Rangitoto soils. Ngauruhoe soils are derived from andesitic sands erupted intermittently from Mount Ngauruhoe and Ruapehu over the last 400 years. The soils show limited horizon development with some C horizon material at the surface and carbon in the subsoil due to periodic burial of organic material. They are extremely susceptible to wind erosion. Tarawera soils are formed on basaltic lapilli and ash erupted from Mount Tarawera in 1886. Profiles exhibit 2–3 cm of dark greyish-brown gravelly sand over greyish-brown and black gravel and sand. Rapid weathering of the surface layers has produced high values of cations and available phosphorus. Rotomahana soils have developed on rhyolitic sand, silt and gravel erupted from Lake Rotomahana in 1886 and possess a number of distinctive characteristics. They possess an A horizon of average thickness of 7 cm, with a moderately developed structure and nutrient content. This more rapid rate of soil formation has been attributed to hydrothermal pre-weathering of the material under the lake before the eruption. It has been estimated that this pre-weathering produced between 10 per cent and 15 per cent of the clay content and made cations available (Gibbs, 1980).

Burrell soils have formed on andesitic lapilli and ash erupted from Mount Egmont between 250 and 400 years ago. They are gravelly and sand soils with thick AC horizons due to organic matter coating the mineral particles to depths of up to 50 cm. The deep penetration of organic matter is thought to be due to the extremely high rainfall of over 500 mm per year and intense leaching of organic compounds through the coarse-textured materials. The soils possess high C/N ratios, moderate to strong acidity, and low cation content. Rangitoto soils are formed on basaltic ash erupted from Mount Rangitoto between 500 and 800 years ago. The soils are weakly leached, weakly acid and high in nutrient cations. This is reflected in highly productive grassland. Profiles possess 20 cm of black friable sandy loam A horizon over dark brown to greyish-brown sand AC and C horizons.

Yellow-brown pumice soils have formed on pumiceous rhyolite tephra erupted over large areas of North Island, New Zealand, between about 900 and 5000 years ago. Most soils possess well developed A, B and C horizons with moderate levels of carbon, low levels of available nitrogen, moderate C/N ratios and low-percentage base saturation. The formation of three separate horizons on fine sandy pumice can occur in about 100 years, whereas in gravelly pumice or pumice flows up to 500 years may be needed to develop a distinct B horizon. Development to the yellow-brown loam stage seems to require between 3000 and 5000 years, depending on moisture conditions and particle size. Pumice soils may lose their structure if the land is converted to pasture. They are less able to retain water and runoff and erosion increases producing gullies up to 25 m deep and 150 m long (Selby, 1966). Gullies cut in pumice are discontinuous and characterised by vertical headwalls and sidewalls (Blong, 1966). The erodibility of volcanic soils has also been noted in other areas (Jungerius, 1975).

Areas near volcanoes are likely to exhibit extremely complex layered soils as a result of periodic ash accumulation (Ottersberg and Nielsen, 1977; Gerrard, 1985; Limbird, 1985). If the ash falls are thick enough former soils will be completely buried and preserved as fossil soils or palaeosols. Sections through slope deposits on Mount Egmont show at least four palaeosols below the presently forming soil. Soils buried for less than 200 years show very similar characteristics to the same soils that have escaped burial. But after being buried for more than 200 years changes begin to take place. The thickness of the A horizon becomes reduced as a result of lack of fresh organic material and the depletion of the buried organic matter. Gibbs (1980) has noted that the buried A horizons of yellow-brown earths and recent soils show a paler colour and loss of structure. The old dark grey A horizon may become mottled pale grey and brown, possibly as a result of an accumulation of soil water at the junction of the old and new materials. However, buried A horizons of many yellow-brown pumice soils and yellow-brown loams show little change in colour or structure after burial for periods of up to 20 000 years. This is thought to be due to the stability of the humus-allophane colloid complex under moist conditions. This brief account of certain types of volcanic soil demonstrates a further facet of volcanic mountains that set them aside from other mountain types. Their modes of erosion and dissection are also somewhat different.

EROSION AND DISSECTION OF VOLCANIC MOUNTAINS

Dissection of volcanic mountains begins as soon as the eruption takes place by a combination of fluvial action and mass movement. Large strato-volcanoes have a long active life and there will have been many sequences of volcanic activity, erosion and gullying, with gullies being repeatedly filled with lava and pyroclastic material and new gullies created. It is possible that over a long period of time some sort of steady state is achieved. There are interesting possibilities here for studies of magnitude–frequency relationships and it is surprising that there have been so few detailed studies of the erosional development of volcanoes. On volcanoes with short repose times there will be little chance for erosive processes to establish a general pattern, whereas with longer repose times erosional processes will be able to shape the volcanic form. As will be seen later, volcanoes vary considerably in the pattern and length of their repose times. Some exhibit essentially random repose times, whereas others are decidedly non-random. Much will depend on the nature of the eruption — for example, whether it affects the whole volcano or only part of it. This means that some parts of the mountain may be dominated by erosional processes, while other parts are determined by repeated but small-scale eruptive activity. Much will also depend on the nature of the material being erupted. Lava, pyroclastic flows and ash falls will have different effects on the long-term development of the volcano.

The erosion of volcanoes is largely determined by their size, shape and the nature of the materials. Large volcanoes will possess different climates at different altitudes and on different aspects. The large volcanic mountains of South America possess environments ranging from hot humid to cold periglacial and even glacial climates. El Misti, Peru, a composite cone of lava and ash, has larger and more persistent snowfields on its western side. This side is, therefore, in a more advanced state of dissection (Bullard, 1962). Steeper volcanoes will suffer greater dissection than those with gentler slopes.

It is probably the nature of the materials which has the greatest influence on the erosional development of volcanoes. Pyroclastic material is more easily eroded than lava, and, within the variety of pyroclastic materials that can occur, welded pyroclastics are more resistant to erosion than non-welded parts. Welded zones stand out as prominent landscape features. Unwelded materials may weather to produce silicic solutions that may be reprecipitated near the ground surface, causing case hardening. The porosity and permeability of the material will determine the ease with which water infiltrates and seeps through the rock rather than running down the slope across the surface. Intercalcated lava may lead to perched water tables. All volcanic materials except the most massive basalt flows are permeable, therefore, if erosion is to occur, high water tables, weathering to produce clay or rainfall so intense that infiltration cannot occur fast enough, are required. However, it is possible for erosion to occur despite high porosity. Vulcan, Papua New Guinea, is only thirty years old yet is already well gullied (Ollier and Mackenzie, 1974). Pyroclastic fall deposits are especially permeable. On some of the younger volcanoes of Hawaii no well-defined channels exist even when rainfall exceeds 5000 mm. There is often a circular pattern of decreasing grain size spreading away from a large central vent. Thus the coarsest scoria is found close to vents, followed by moderately coarse well-sorted lapilli and scoria and further still well-sorted fine ash. Therefore on a large pyroclastic cone the upper and middle slopes are porous and the lower slopes more impermeable, thus springs and erosion commence on the lower slopes.

Many volcanoes contain lakes in craters or calderas which not only pose a risk to life and property (see below) but may provide a source of water for the erosion of the lower slopes. In general scoria cones are usually too permeable to hold water, but there are exceptions, such as Muhavura in Uganda which rises to about 3000 m with a small lake in its crater. Volcanic activity may create temporary lakes by blocking valleys. The merging footslopes of Mounts Soaru and Karimui, Papua New Guinea, blocked temporarily the Tua River to form a lake which was quickly drained.

Volcanoes generally possess a radial drainage pattern and even actively growing volcanoes are dissected by numerous ravines. Craters and calderas may possess centripetal patterns. Breaching of crater rims will impose a certain asymmetry on the drainage pattern and may lead to greater dissection on one portion of the volcano. In theory, when drainage is first initiated on a volcanic cone there is little erosion near the rim because of a lack of waterflow. Erosion will be at a maximum on the mid-slopes and will decline towards the lower slopes because of declining gradient and deposition of material. The larger valleys produce a regular pattern known as 'parasol ribbing'. Cotton (1944) viewed ribbing as a result of the avalanching of hot ash but most of the gullies seem to be due to fluvial action.

There is no doubt that freshly deposited, but cold, ash is prone to avalanching that will erode grooves similar to parasol ribbing. Each groove may be scoured many times with the ash coming to rest as fans at the foot of the volcano, such as has occurred on Barcena, Mexico (Richards, 1965). Such grooves have edges which commonly intersect and are remarkably uniform in size and length. In fluvial ribbing there are often shorter valleys which originate between longer valleys on the middle and lower slopes.

Erosion in the early stages can be extremely rapid (Kadomura *et al.*, 1983). Richards (1965) has charted the early phases of erosion on Barcena, Mexico. Eruption started on 1 August 1952, when the island was covered in ash. There was no evidence of erosion on 12 September, but by 20 September a number of gullies had developed and dry waterfalls up to 10 m high had been created. By May 1955 gullies were well established, with fairly smooth long profiles. A similar set of events occurred on the scoria cone Vulcan, Papua New Guinea (Ollier and Brown, 1971). The volcano erupted in 1937 and the survey was undertaken in 1967. Erosion was greatest in mid-slope positions, nearly 33 mm yr^{-1} at a distance of 60 m from the crater rim. The average amount of erosion was 18 mm yr^{-1}. Studies following the 1980 eruption of Mount St Helens have been able to establish the scale of post-eruption erosion (see, for example, Collins *et al.*, 1983; Lehre *et al.*, 1983).

Dissection produces two types of eroded cone; the 'parasol' effect of radial valleys discussed above, and a 'planeze' landscape where a small number of main valleys are responsible for most of the erosion. The distinction between the two may be a function of scale and nature of material. Parasol ribbing is more characteristic of small cones composed of easily erodible material, whereas planezes are more characteristic of large-strato-volcanoes. Planezes are triangular remnants of the original volcano which are gradually consumed as erosion progresses. Eventually the planezes are eroded away, leaving a residual volcano, and, finally in the skeleton stage, only a few dykes, necks and sills remain. Kear (1957) has classified these different stages for some New Zealand volcanoes. The planeze stage is generally middle Pleistocene to Holocene in age, the residual stage is of Pliocene–Pleistocene age, and the skeleton stage has only been reached on volcanoes that have been continuously eroded since Miocene times. The large Tertiary volcano, Napak, in Uganda, still possesses planeze remnants even though 97 per cent has been removed by erosion. On Mount Rainier, USA, planezes that extend up to 3000 m are known as 'wedges'. On higher slopes wedges are replaced by long walls of rock arranged radially from the summit known as 'cleavers' because they split the descending ice mass into separate lobes.

Ruxton and McDougall (1967) have estimated the rate of erosion of the Hydrographers Volcano by reconstructing the original surface from the planeze remnants. Estimated denudation rates ranged from 80 mm yr^{-1} at a height of 60 m to 750 mm yr^{-1} at a height of 760 m. There was a linear correlation between the rate of denudation and height, the average maximum slope angle and the average slope length.

The large volcanic mountains of Papua New Guinea demonstrate these effects admirably. Mount Karimui (2750 m), in the southern foothills of Kubor Range in the Central Highlands, possesses extensive planeze remnants on three sides. The symmetry has been affected on the northern side by large amphitheatre valleys, the largest of which has developed along the groove formed where individual cones intersect. Also, four major valleys radiate from an erosion caldera. Mount Bosavi (2600 m) has a simple conical shape with numerous deep valleys with planezes converging to produce a topography of angular ridges and V-shaped valleys. A 600 m deep valley has breached the crater on the SSE side. The valleys are undergoing rapid erosion by landslides and rockfalls. All the volcanoes except Mount Hagen are deeply dissected on their southern and, to a lesser degree, their northern flanks, while the east and west flanks are generally less dissected, forming planeze surfaces. Mount Hagen, the northernmost volcano, is deeply dissected towards the north and west. This asymmetry in dissection may reflect the drainage and slope of the pre-volcanic landscape. After the eruptions the lowest base levels were to the south, except for Mount Hagen which was to the north. Higher rainfall on the southern slopes associated with the prevailing south-east trades may have contributed to the strong dissection of the south slopes on the southernmost volcanoes such as Mounts Favenc, Duau, Murray and possibly Bosavi (Ollier and Mackenzie, 1974). There appears to be a general relationship in the humid tropics between degree of dissection and direction of the dominant trade winds, with a tendency for breached craters to face the rain-bearing winds.

Many of the valleys, discussed in general terms above, possess distinctive, broad, amphitheatre-like head portions. Such features may be distinctive to volcanoes. Alternate resistant and non-resistant beds dipping downslope seem especially favourable for their development. Non-resistant beds are undercut to produce waterfalls which eventually coalesce into one large fall. Streams soon reach the stage where they possess a 'fall point' above which the stream continues to cut down but below which lateral erosion is dominant (Stearns, 1966). River capture of the upper slopes, where streams are converging upwards, helps to enlarge the valley. The steep headwalls and sides of the valleys are dissected into many spurs by close-set ravines, chutes or flutes. An alternative process has been suggested by Wentworth (1943). He argues that the valleys have been eroded by shallow landsliding, and when the steep edges of neighbouring valleys retreat they will intersect to form steep ridges which will eventually be consumed. Occasionally one amphitheatre valley may grow to a large size and may breach a crater. Caldera of La Palma, Canaries, may be of this type. Valleys on volcanic mountains also reflect the climatic environment. On the summits of Fuji and some other volcanoes in Hokkaido, valleys are not V-shaped but spoon-shaped as a result of periglacial processes.

PYROCLASTIC FLOWS, SURGES, LAHARS AND DEBRIS AVALANCHES

Some of the most devastating events associated with volcanic mountains are related to pyroclastic flows, mudflows and large-scale slope failures. The eruptions at Mont Pelee and the Soufrière of St Vincent in the West Indies and in the Valley of Ten Thousand Smokes, Alaska, in the early years of the twentieth century alerted earth scientists to the effects of such flows. The pyroclastic flows erupted in the West Indies in 1902 were called *nuées ardentes* or 'glowing avalanches'. Many other types of pyroclastic flow occur, varying considerably in temperature and material content They range from gas-rich to gas-poor and the ejecta vary from almost lithic to almost wholly pumiceous. The largest, most gas-rich flows usually take place during the initial volcanic phases and their deposits consist almost entirely of fresh effervescing magma. Hot flows are much more mobile than cold flows and avalanches. Cold avalanches generally travel distances three to five times the distance they descend, whereas hot flows frequently travel ten to twenty times the height of the fall.

Major slope failures of one form or another are significant degradational processes on volcanic mountains (Siebert *et al.*, 1987). Such failures and the associated explosive eruptions have resulted in more than 20 000 deaths in the past 400 years. Slope failures, including those with a magmatic component (Bezymianny type) and those solely phreatic (Bandai type), create huge horseshoe-shaped depressions. The paroxysmal phase of a Bezymianny eruption may include powerful lateral explosions and pumiceous pyroclastic flows. Massive volcanic landslides can also occur without related explosive eruptions, as happened on Unzen volcano, Japan, in 1792. Large-scale deformation often precedes slope failure, but Bandai-type eruptions are difficult to anticipate because they often climax suddenly without precursory eruptions and may be preceded by only short periods of seismicity.

Since the eruptions at Mont Pelee and Soufriere pyroclastic flows of one form or another have been extensively studied but there is still considerable confusion as to their nature and mode of movement. The terminology is also confusing, but it is possible to make the fundamental distinction between pyroclastic surges and flows, volcanic mudflows (lahars) and volcanic debris avalanches, although they often occur on the same mountain at the same time.

Pyroclastic surges and flows

It has long been known that so called 'pyroclastic' flows appear to have
a turbulent regime in their upper gas-rich parts and a laminar regime
closer to the surface where the density and viscosity are higher. Their
high mobility is due to their expanded fluidised state. A small amount
of gas at high temperature and pressure is probably sufficient to create
the fluidisation. Local velocities can be as high as 150 m s^{-1}. Most
pyroclastic flows are poorly sorted and unstratified and in general the
greater the volume of flow the less is the viscosity. Thus pumice flows
can be extremely large (up to 100 km^3), scoria flows are usually
intermediate in size (0.1 km^3) and lithic flows generally small
(0.001 km^3).

Detailed analysis of pyroclastic deposits has enabled several different
components to be identified. The term 'pyroclastic surge' or 'base surge'
was used by Richards (1959) following his study of Barcena, Mexico. The
terms were also used to describe some of the events following the eruption of Taal volcano (Moore, 1967; Moore *et al.*, 1966). Base surges have
been described as density currents with deposits containing characteristic
dune-like structures (Fisher and Waters, 1970; Waters and Fisher, 1971).
Pyroclastic surges are essentially low-concentration pyroclastic flows and
are normal accompaniments of pyroclastic flows (Sparks, 1976; Sparks
and Walker, 1973). Fisher (1979) has identified three types of surge.
Ground surges occur immediately below a pyroclastic flow and may
originate from a collapsing eruption cloud or directly from a crater. A
base surge originates from a collapsing phreatomagmatic eruption and is
not associated with hot pyroclastic flows. Ash cloud deposits are
mechanical segregates from the top of pyroclastic flows. *Nuées ardentes*
is the general term for turbulent pyroclastic density currents (Fisher and
Heiken, 1982; Fisher, 1983). Localised hot rock avalanches also occur,
such as those that occurred during the growth of a composite dome
within the crater of Mount St Helens between 1980 and 1987 (Mellors *et
al.*, 1988).

The distinction between the types of movement is crucial to an analysis
of potential hazard. In the 1982 eruption of El Chichón, pyroclastic flow
deposits accounted for only 30 per cent of the total pyroclastic movement
(Sigurdsson *et al.*, 1987). Most material spread out as a gravity current
unaffected by topography and only the low-level flow was channelled
into gorges and valleys. But discrimination between pyroclastic flows and
surge deposits can be difficult. Thus the blast deposits produced by the
18 May 1980 eruption of Mount St Helens has been identified as a
pyroclastic surge (Moore and Sisson, 1981), a combination of flow and
surge (Hoblitt *et al.*, 1981; Hoblitt and Miller, 1984; Waitt, 1981; 1984),
and a pyroclastic flow (Walker and McBroome, 1983).

The difficulty in differentiating the various flow types can be
illustrated with reference to events on the north-east side of Mount St

Helens following the eruption in 1980. Pyroclastic surges and lahars produced several catastrophic flows into tributaries on the north-east flank of the volcano. These tributaries channelled flows to the Smith Creek Valley. Evidence suggests that a dry pyroclastic surge preceded a very wet lahar. The lahar must have originated as snowmelt and not as ejected water-saturated debris. Six metres of snow was removed from Shoestring Glacier (Brugman and Meier, 1981), and ten to fifteen metres of snow and ice was lost from Toutle and Talus glaciers. Three sequences of deposits have been recognised in Smith Creek Valley: a complex valley-bottom facies of pyroclastic surge and related pyroclastic flow, followed by an unusual hummocky diamict caused by complex mixing of lahars with dry pyroclastic debris, followed in turn by the deposition of secondary pyroclastic flows.

Lahars

'Lahar' is a Javanese term for what is essentially a volcanic mudflow or wet avalanche. Lahars may form during eruptions or by the later reworking of volcanic debris. Anderson (1933) has provided a classification of lahars according to their mode of origin. Some of the most destructive lahars have been created by the sudden draining of crater lakes. They may originate as a nearly clear water flow but become a debris flow as they incorporate sediment (Waitt *et al.*, 1983; Pierson, 1985; Pierson and Scott, 1985). In 1919, 38×10^6 m^3 of water were blown out of Kelut volcano in Java. Lahars created by this volume of water buried 131 km^2 of cultivated land, destroyed 104 villages and killed more than 5000 persons. Considerable erosion was effected on the upper slopes and the most mobile flows travelled 38 km. Catastrophic lahars can also be initiated by rapid melting of snow and ice during eruptions. The initial eruptions of the Kamchatka volcano, Bezymianny, in 1956 caused rapid melting of snow and generated bouldery lahars that travelled more than 85 km (Gorschkov, 1959). Similar lahars have been caused by the passage of glowing avalanches across glaciers around the summit of Cotopaxi (Fenner, 1948). Extensive lahars are also associated with Mount Rainier, Washington (Crandell and Waldron, 1959; Crandell and Mullineaux, 1967). In East Africa, lahars are extensive south of Mount Meru and about 12 km south-west of Mount Rungwe in Tanzania. Lahar deposits are 8 m thick on the north and east slopes of Sabinio in south-west Uganda.

Lahars may also be initiated by eruptions during heavy tropical rains. Ash falls will blanket the slopes and bury the vegetation, which normally acts as a filter for the rainfall. Without this protective cover the rain will run across the surface, taking with it large amounts of highly erodible ash which soon develops into highly mobile mudflows. These mudflows are concentrated in drainage channels and torrents of bouldery mud sweep down the slopes and spread across the adjacent lowlands. Such lahars, initiated by the prolonged eruption of fine ash from Irazu volcano, destroyed part of the city of Cartago in Costa Rica. Taylor (1958, p. 56) has vividly described the mudflows, which continued for six months after the eruption of Mount Lamimgton in 1951:

They rose with dangerous rapidity, reaching full flood within a few minutes and, at times, advancing down the valley as a wall . . . The flows rarely lasted longer than an hour or two and subsided rapidly, leaving on the margins of the stream and the adjacent valley floor stranded boulders and a deep layer of mud.

Lahars deposit levees and lobes as their water content drops and mobility decreases (Crandell, 1971). It has often been suggested that eruptions themselves produce heavy rains that then cause lahars but much of the evidence is contradictory (Finch, 1930).

The eruption of the Nevada del Ruiz volcano, Colombia in November 1985, in which up to 20 000 lives were lost, is a devastating reminder of the danger of living on the flanks of large volcanoes. The eruption melted snow and ice on the mountain, creating catastrophic mudslides and mudflows. Torrential rain also added to the havoc. Nevada del Ruiz, which stands 4900 m high in the northern Andes, had been dormant since 1845, with its last major eruption being four centuries ago. A minor eruption was recorded six hours before the big blast at 0200 GMT on 14 November 1985, when tons of ash and rocks were hurled 6000 m into the air. The melted snow and ice soon turned into a mass of mud and boulders, which swept down the mountain burying four towns and seriously effecting five villages. The La Lagunilla river burst its banks and became a rushing wall of mud. The river town of Armero, with a population of 28 000, was 85 per cent destroyed and buried in up to 12 m of mud. In the village of Chinchina three neighbourhoods were swept away, with an estimated loss of 1000 lives.

A variant of the normal lahar is the Icelandic jökulhlaup, a catastrophic flood caused by volcanic eruptions or the rise of geothermal heat under ice caps. Thirty to forty floods have been traced to the geothermal area of Grímsvötn on Vatnajökull (Björnsson 1974; 1977), and seventeen have been ascribed to the volcano Katla under the ice cap of Myrdalsjökull. Similar floods have also been created by eruptions at Öræfajökull and Eyjafjallajökull. The association of jökulhlaups with volcanic activity is well documented. The Öræfajökull eruptions of 1362 and 1727 have left vast expanses of sediment. Blocks associated with the 1362 hlaup are still visible and the largest are over 50 m^3. The eruption of Katla in 1357 released a jökulhlaup that flooded the Sólheimasandur in southern Iceland. The flood lines are about 10 m higher than the present riverbed and the flow deposited tens of millions of cubic metres of ash, thickening the sandur plain by several metres.

Floods emanating from the Grímsvötn area of Vatnajökull have been the most spectacular. The water originates in a subglacial lake situated

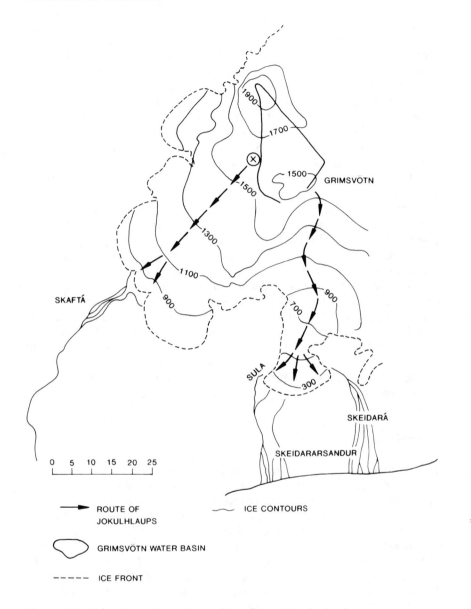

Figure 7.3. Volcanic outburst floods from Grímsvötn, Iceland.

in a volcanic caldera of about 30–40 km³ in the interior of the accumulation area of Vatnajökull. The presence of a high-temperature geothermal area is the key factor in the development of the floods. The melting of the ice affects the surface slope of the ice cap and directs ice and water flow into the caldera (Figure 7.3). Over a period of five or six years the water level in the lake rises about 100 m. Then the glacier is either floated off a subglacial ridge east of the lake or the temperature rises sufficiently to melt the ice in contact with the ridge. Water is then forced subglacially for 50 km to emerge from about ten tunnels, although the main water volume comes from three to four main rivers. The lake level then gradually rises again until conditions are suitable for the next flood event.

Irrespective of how they originate, all jökulhlaups produce distinctive hydrographs. There is an initial period of slow leakage which may take place for many weeks prior to the main flood. This is followed by a steep limb rising to a peak flow which is usually impossible to measure accurately. The final element is a very steep falling limb. Maximum discharges can be enormous. It has been estimated that jökulhlaups from the Katla region may have reached a maximum of 100 000 m³ s⁻¹ (equivalent to the flow of the River Amazon). The Skeidharárhlaup in March 1938 was estimated at 45 000 m³ s⁻¹, and the total quantity of water discharged at about 7 km³. It is very difficult to estimate the amount of sediment carried by these floods. It is thought that 86 million tonnes of suspended sediment was carried by the Skeidharárlhaup of 1938 (Tomasson *et al.*, 1980).

Volcanic debris avalanches

Large horseshoe-shaped craters, open at one end, with hummocky deposits of volcanic debris at their bases, have been widely observed on volcanic mountains. But it is only recently that they have been recognised as being the result of large-scale mass movement. The eruption of Mount St Helens on 18 May 1980 focused attention on the importance of such large-scale collapse events and resulting debris avalanches. Previously unknown avalanches have since been identified on Mount Shasta, USA (Crandell *et al.*, 1984); Popocatepetl (Robin and Boudal, 1984) and Colima (Luhr and Carmichael, 1982), Mexico; Lastarria (Naranjo and Francis, 1987) and Socompa (Francis *et al.*, 1985), Chile; and in the Central Andes (Francis and Wells, 1988). Such avalanches have occurred at roughly four per century in historic time.

A number of names have been used for these avalanches such as 'dry mudflow' (Murai, 1961), 'debris flow' (Cattermole, 1982; Sigurdsson, 1982), 'volcanic dry avalanche' (Ui, 1983), 'rockslide avalanche' (Voight *et al.*, 1981) and 'rockslide-debris avalanche' (Siebert, 1984). 'Volcanic debris avalanche' seems to be the term most commonly used at the

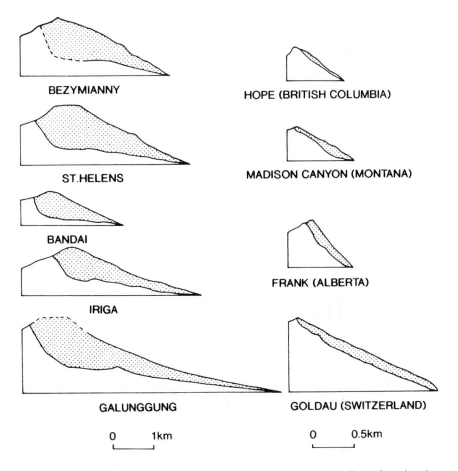

Figure 7.4. Comparison of profiles of large landslides with profiles of avalanche calderas.

Source: Siebert (1984).

present time. Avalanche calderas, rather than being subcircular or elliptical, are horseshoe-shaped with one end of the caldera widely breached. They often range from 1 to 3 km in width, whereas collapse calderas have a median diameter of 6 km and are often much larger (Siebert, 1984).

Siebert (1984) has provided an interesting comparison between some historic landslide scars and avalanche calderas (Figure 7.4). The Bezymianny profile is after Borisova and Borisov (1962); St Helens from Moore and Albee (1982); and Bandai-san from Sekiya and Kikuchi (1889). The Hope, Madison Canyon and Frank profiles are from Voight (1978), and Goldau from Zaruba and Mencl (1969). Avalanche calderas

Table 7.1. Characteristics of Andean debris avalanches.

Volcano	Summit height (m)	Max. summit slope (degrees)	Vertical descent (H) (m)	Horizontal runout (L) (km)	H/L	Area (km²)	Azimuth (degrees)
Parinacota	6348	35	1900	23	0.08	156	253
Tata Sabaya	5430	35	1730	25	0.07	331	166
Tittivilla	5050	30	1350	20	0.07	189	68
Caiti	4900	32	1200	12	0.10	67	313
Aucanquilcha	6176	25	2100	17	0.12	59	311
Ollague	5863	35	2171	16	0.14	113	286
San Pedro	6145	45	2845	16	0.18	118	275
Laguna Verde	5412	35	912	3*	0.30	4	11
Socompa	6051	35	3020	37	0.08	606	312
Llullaillaco	6723	36	2500	25	0.10	197	114
Rosado	5100	–	960	6	0.16	10	96
Lastarria	5700	–	1200	9	0.13	18	105
Cajaro	5790	32	1300	17	0.08	45	160

*The extent of the avalanche debris is controlled by a neighbouring volcano.

Source: Francis and Wells (1988).

are deeper relative to the length of the depression, with relatively high sidewalls, gently sloping floors and sharp breaks of slope where the headwall rises from the floor. The greater depth of volcanic debris avalanches may be due to the presence of large amounts of pyroclastic and altered material at the volcano's summit. Also, explosive activity, often associated with the major failures, brecciates the cone, over-steepens the headwalls and promotes additional collapse.

Debris avalanches have often been described as lahars, but they possess characteristics of landslide deposits, such as large blocks of homogeneous composition surrounded by crushed fragments forming a matrix of similar composition. The deposits also show evidence that water was not a major component in their movement. The major characteristic of such movements is their great mobility. Four-fifths of the 119 debris avalanches listed by Siebert (1984) travelled over 10 km from their source. Velocities of 160 km h^{-1} have been estimated at Chaos Crags (Crandell *et al.*, 1974), 124 km h^{-1} at Meru (Cattermole, 1982) and 180–288 km h^{-1} at Mount St Helens (Voight *et al.*, 1981).

Remote sensing studies of the Central Andean volcanic province at latitudes 18–27° with the Landsat Thematic Mapper have identified twenty-eight breached volcanic cones and fourteen major debris avalanches (Francis and Wells, 1988). These studies have provided valuable information on the nature of the processes involved. Characteristics of the debris avalanches are provided in Table 7.1. The avalanche on Socompa volcano, on the border of Chile and Argentina, is one of the most impressive and, with an area greater than 600 km^2, is the biggest exposed in the Andes (Francis *et al.*, 1985). Collapse of the original volcano involved a 70° sector of the cone and created an amphitheatre 10 km wide at its mouth. Large parts of the mass slid as coherent blocks and came to rest at the mouth of the amphitheatre. The avalanche also involved rocks from the basement below the composite cone and initially travelled at high velocity towards the north-west across ground that sloped towards the north-east. The avalanche came to rest temporarily before moving in a secondary flow at right angles to the primary flow direction. Even 35 km from the source the deposit has a steep front over 40 m thick. Radio-carbon dates from charcoal beneath the deposits indicate that the event took place about 7200 years ago.

The other avalanches exhibit similar features. At Parinacota the largest hummocks of the deposits are 400–500 m in largest dimension and up to 80 m high. Lake Chungara (elevation 4550 m) was formed by an avalanche damming the westward drainage from the Cerro de Quisiquisini and the Nevados de Quimsachata. Water seeping through the avalanche deposits supplies numerous small lakes trapped in the hummock topography. The hummock topography associated with the avalanche of Tata Sabaya is the best in the Andes.

It is difficult to establish triggering mechanisms for the avalanches. At Socompa fresh magmatic material was involved in the avalanche, therefore the growth of a lava dome may have been involved. In addition, Socompa is constructed over a major fault zone and seismic triggering might have been involved (Francis and Self, 1987). The bimodal pattern of debris avalanche and breach azimuths for the Andean examples indicate that sector collapses occur perpendicular to the regional fault trend.

The remote sensing studies have established that such collapses and avalanches are extremely common in the Central Andes and this begs the question why. Francis and Wells (1988) offer some explanations. The erosion rates in the Central Andes are lower than in almost any other major volcanic area in the world. Both the northern and southern volcanic zones of the Andes have much higher precipitation and rates of erosion. Also, these areas are heavily glaciated. Where erosion is rapid, it is difficult for a volcano to construct a steep cone. As a result of the arid environment of the Central Andes, lava flows erupted from the summits are not rapidly eroded away and may build steep cones. The second important factor is that the Central Andean region is dominated by the eruption of andesitic and dacitic lavas rather than basaltic material. Thus the volcanoes are composed of thicker, shorter lava flows. Dacitic lavas especially tend to pile up in thick accumulations around the vents. Summit slopes greater than 45° have been measured on some volcanoes. From an analysis of 578 composite volcanoes in the Central Andean volcanic province, Francis and Wells (1988) have established a relationship between sector collapse and cone height. Collapse events have occurred on 42 of 578 cones over 500 m high, 24 of 239 cones over 1000 m high, 11 of 74 cones over 1500 m high, 4 of 14 cones over 2000 m high, and 3 of 4 cones over 2500 m high. These data suggest that once composite cones reach a threshold height, somewhere between 2000 and 3000 m, sector collapse becomes almost inevitable. The extreme mobility of volcanic debris avalanches can be assessed by examining a number of morphometric indices and comparing them with non-volcanic debris avalanches, pyroclastic flows and lahars. One of the simplest relationships is that between travel distance (L) and vertical drop (H). The ratio H/L has been equated with an apparent coefficient of friction (see Table 7.1 for values for the central Andean avalanches discussed above). In Figure 7.5 volcanic debris avalanches are compared with other types of movement. The H/L ratios for debris avalanches (median H/L is 0.11) are similar to some pyroclastic flows but are intermediate between those of non-volcanic debris avalanches and large volume pyroclastic flows. Hsu (1975) has noted that the apparent coefficient of friction for non-volcanic debris avalanches decreases with increasing volume. A similar trend has been obtained by Ui (1983).

An index called the excessive travel distance (L_e) has been devised by Hsu (1975). It is defined as the travel distance beyond the point obtainable by sliding of a body with an 0.62 friction coefficient. The value of 0.62 for the friction coefficient is the value applicable to the

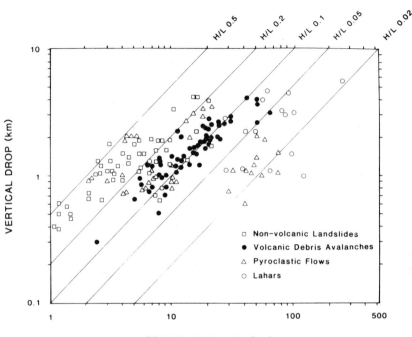

Figure 7.5. Comparison between various types of mass movement in terms of travel distance.

Source: Hsu (1975).

sliding of rigid blocks. The relation is expressed by $L_e = L - H/0.62$. Volcanic debris avalanches have a larger excess travel distance compared with non-volcanic avalanches of the same volume (Figure 7.5).

It is clear from this analysis that mass failure of volcanic mountains is somewhat different from large-scale mass failure on other types of mountain and is additional justification for the view that volcanic mountains should be treated as separate landforms. Volcanoes are always undergoing adjustments. Gravitational adjustments will lead to the slow settling of the volcanic cone, possibly with deformation at the base. Also, segments of volcanoes may break off along arcuate faults and slide slowly downslope. The accumulation of large volumes of material around central vents may result in failure if triggered by explosive activity or earthquakes. The collapse of craters and the production of large debris avalanches seems to be related to the height and slope angle of the volcanic cone (Figure 7.6) Catastrophic failure can occur on cone heights as low as 500–1000 m but is more likely when slope angle exceeds 20°. The analysis by Siebert (1984) suggests that there are many large

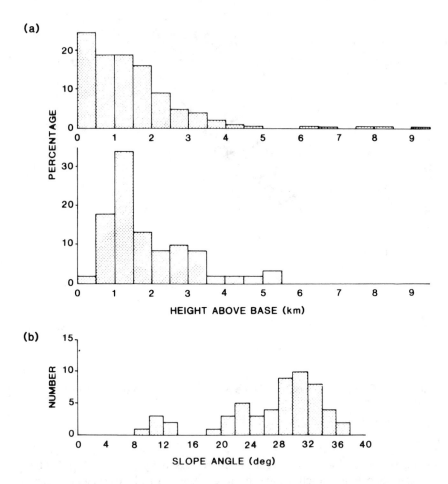

Figure 7.6. a) Elevation above base for 845 Holocene volcanoes compared with data for 62 volcanoes having undergone avalanche caldera formation. b) Slope angle of Quaternary volcanoes which have undergone major slope failure.

Source: Siebert (1984).

volcanoes potentially at risk from such avalanches, given the right triggering mechanisms such as a major earthquake.

The main factors that separate volcanic mountains from other mountain types in terms of potential instability have been listed by Siebert (1984) as follows:

(a) the dilational effect of the intrusion of parallel dike swarms;
(b) massive lavas overlying weak pyroclastic rocks;
(c) hydrothermal alteration weakening the volcanic cone;

(d) the migration of vents in a direction parallel to the axis of the avalanche caldera;

(e) reduction in strength of the volcanic mass due to saturation by hydrothermal waters.

HAZARD ASSESSMENT AND MITIGATION

The problems of predicting the timing, nature and intensity of volcanic action can be seen by examining the events associated with the eruption of Mt St Helens on 18 May 1980. The Cascade Ranges of the Pacific Northwest of North America have, for a long time, been seen as a likely site for volcanic eruptions (Coates, 1985). Subduction zones, as noted in Chapter 1, are also often associated with volcanic activity. Mt St Helens is situated almost directly over the zone where the Juan de Fuca oceanic plate is being subducted by the North American continental plate. In 1975 the geologists Crandell *et al.* stated that on past behaviour there was a strong probability that Mt St Helens would erupt again even though there had been no appreciable activity for over a hundred years. They also suggested that potential volcanic hazards should be considered in planning for future uses of the land on the mountain.

A large number of micro-earthquakes began on 20 March 1980, followed by a small eruption of smoke and ash on 27 March. A crater formed, followed by another on the following day which gradually merged with the first one. Seismic activity developed into harmonic tremors on 3 April and by 9 April the crater was 300 m wide, 500 m long and 250 m deep. Intermittent activity then ensued until the eruption of 18 May, the suddenness of which surprised everyone. An earthquake measuring 5 on the Richter scale triggered landslides in the area that had previously exhibited marked surface bulging followed by the eruption of an ash column and a sub-horizontal blast of light-grey clouds of incandescent gas and ash. This blast devastated an area of 500 km^2 and the mountain's peak was reduced from 2950 m to 2438 m. Mudslides cascaded into Spirit Lake then down the North Fork of the Toutle River filling it to a depth of 60 m. The main channels of all the nearby rivers were silted up with the 180 m wide channel of the Columbia being reduced from its normal depth of 12 m to 4.3 m.

The variable nature of the processes involved can be seen in a simplified map of the mountain (Figure 7.7). Ash flows, mudflows and debris flows were all involved, superimposed on the devastation caused by the blast. Hazard assessment would need to be able to take all these processes into consideration. In general the flows and avalanches were channelled along pre-existing channels and depressions. But, as seen earlier, many large pyroclastic flows and debris avalanches are powerful enough to override the effects of topography. This demonstrates the nature of the problems associated with the development of hazard and

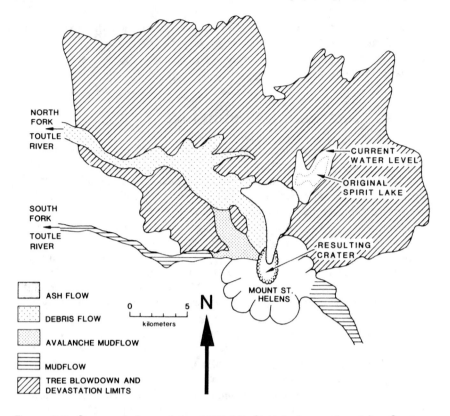

Figure 7.7. Geomorphology of the 1980 Mt. St Helen's eruption (after Coates, 1985).

risk maps. Coates (1985) has listed four main tasks involved in the creation of such maps: identifying the disaster process that will create the hazard, i.e. lava flow, avalanching, lahars etc.; determining the timing of the event; estimating the magnitude of the event; and locating the areas that will be affected. Fortunately there are a number of ways in which this task can be made easier.

Prediction of eruptions

There are often signs that a volcano is building up to an eruption. It is rare for a major eruption to occur without some prior warning. The main problem is predicting not that an eruption is likely but when it will occur. The following phenomena can be used to predict eruptions.

Seismicity
As mentioned in the case of Mt St Helens, most major volcanic eruptions

are preceded by shallow-depth seismic shocks. In some cases these shocks increase up to the eruption whereas in other instances the shocks increase, then suddenly stop and eruption occurs sometime later. This is why it is important to 'fingerprint' individual volcanoes. It is not possible to apply general rules to all volcanoes.

Ground tilting
Many volcanoes swell before an eruption because of the upwelling of lava. Detailed and repeated ground survey will detect such changes as will tiltmeters which can be linked up to an automatic warning system.

Temperature changes
The temperature of hot springs, crater lakes and fumaroles often rises before an eruption. Regular measurement may provide evidence of a future eruption.

Gas composition
The composition of gases erupted from craters and fumaroles often changes before eruption. There is likely to be an increase in the amount of HCl. Amounts of fluorine, chlorine and sulphur have also been shown to increase.

Gravity and magnetism
Movements of lava at depth cause small changes in local gravity and magnetism which may be detected. Prediction of volcanic activity is difficult because changes in gravity and magnetism are not always followed by eruption.

Behaviour of volcanoes
Many volcanoes exhibit patterns in the timing of eruptions which can be used to assess the likelihood of an eruption occurring within specified time periods (Wickman, 1966a). Stochastic models have been developed involving statistical analysis of eruptive and repose periods. Wickman (1966b) has produced what is called a 'survivor function', $F(t)$, defined as the number of reposes which lasted longer than time t divided by the total number of reposes. The survivor function gives the probability that a repose period has not ended after the time (t) has elapsed since the last eruption. Two types of volcano can be identified. Volcanoes characterised by random repose time intervals exhibit a straight survivor function line. Their probability of eruption is constant. Popocatepetl is of this type. The second type are characterised by non-random repose time intervals and exhibit a more complex survivor function line. Mts Pelee and Hekla, Iceland, belong to this type.

Hazard zonation

The development of accurate hazard zone maps requires a thorough knowledge of the nature of the individual volcano, its past activity patterns and its current geomorphological development. It is often necessary to identify specific hazards such as lakes. Large bodies of water on volcanic mountains are a major hazard and the likelihood of sudden draining needs assessing, as does the path taken by the water if drainage does occur. Lake Nyos, a deep maar lake in north west Cameroon, is a good example. It is 208 m deep, contains 132 million cubic metres of water, and is in imminent danger of collapse (Lockwood *et al.*, 1988). In August 1986 there was a catastrophic release of carbon dioxide gas killing 1700 persons (Kling *et al.*, 1987).

Typical hazard maps for surge or surge-like eruptions often display risk zones as circles centred at the principal vent (Westercamp, 1980). This may be unrealistic, and more sophisticated models including specific topographic characteristics are required. Sheridan and Malin (1983) developed model-based hazard maps for various vent locations, eruption types and mass production rates which matched reasonably well pyroclastic surge deposits from several recent eruptions on Vulcano, Lipari and Vesuvius. Ruapehu (2797 m), a very active andesitic composite volcano and the highest mountain in North Island, New Zealand, is an interesting example of the benefits to be gained from hazard zonation (Houghton *et al.*, 1987). There have been 54 small eruptions in the last 37 years, with explosive events in 1969, 1971, 1975, 1977, 1978, 1979, 1980, 1981-2, 1985 and 1987 (Latter, 1986). Destructive lahars were associated with the 1969, 1971, 1975 and 1977 eruptions. Skifield facilities, road and rail bridges and portions of a major hydroelectric power scheme have been destroyed or damaged. Destruction of a rail bridge by a lahar in 1953 caused the loss of 151 lives. Other hazards include Strombolian and Sub-Plinian explosive eruptions, lava extrusion from summit or flank vents and collapse of portions of the volcano. Three hazard types have been recognised: inner zones of extreme risk from ballistic blocks and surges;, an outer zone of fall deposits; and, risk from lahars down major river valleys. Early warning schemes have been installed to deal with risks from lahars.

CONCLUSIONS

This chapter has revealed that volcanic mountains share many of the characteristics and processes of other mountain types. However, it has also shown that volcanoes possess many attributes not shared by other mountains. The interaction between large-scale constructional and destructional processes is unique to volcanoes. They also present specific and potentially catastrophic problems for human activity. The 1980 eruption of Mt St Helens has provided a major impetus for applied studies. The scale and nature of the volcano's collapse has led to a reassessment of the landforms on many other large volcanoes. Immense volcanic avalanches are now known to be far more common than originally supposed and it is clear that the process of reassessment is going to be a major one.

Chapter 8

MOUNTAINS UNDER PRESSURE: APPLIED PHYSICAL GEOGRAPHY

INTRODUCTION

Mountains world-wide are coming under increasing pressure from a variety of human activities. In developed countries the main pressure is being created by tourism and recreation. In developing countries pressure on land in the lowlands is leading to encroachment on increasingly marginal land in the uplands and mountains. But tourism pressure is also becoming a problem in such areas. Thus between the early 1970s and 1980s the annual number of tourists visiting Nepal's Khumbu region rose from a few hundred to several thousand (Pawson *et al.*, 1984). Damage to the region's already denuded forests has only been avoided by requiring visitors to import their own fuel for cooking purposes. In general, land degradation, soil erosion and landsliding have increased (see, for example, Hurni, 1982; Messerli and Ives, 1984a), and hazards, of one form or another, have become a major focus of study.

Mountain hazards are construed as being restricted to natural processes such as the mass movement of water, snow, ice, earth and rock. Such processes are avalanches, debris floods and landslides, soil erosion, earthquakes, and so on. They are considered hazardous when they directly endanger life and property, causing death, injury, destruction of property and ecological damage. These are natural hazards, although their extent and seriousness can be increased by human activity. The relief energy embodied in mountains is the ultimate cause of all these geomorphic hazards (Kienholz, 1984). A potential failure or movement is built up in a slow but steady manner by weathering, debris accumulation, snow accumulation and metamorphoses, many of the processes examined in this book. Eventually a sudden, discontinuous event occurs and potential energy is converted into kinetic energy. This conversion can be divided into two basic types:

(a) The discharge is triggered immediately and no potential is accumulated over time. Such directly triggered discharges usually occur after torrential rainstorms, earthquakes, etc. The potential energy of water and material on slopes is immediately converted into kinetic energy of moving material.

(b) In an indirectly triggered discharge the potential for the seemingly spontaneous event exists over a period of time. A change in conditions causes the event. A snow avalanche illustrates this concept. The potential energy of the accumulated snow is converted into kinetic energy only when a change in snow conditions causes the formation of a failure surface in the snow.

These ideas are very similar to those put forward by Brunsden and Thornes (1979). They regard external shocks to landscape systems as being either pulsed or ramped inputs. With pulsed inputs, the imposed disturbance is short in relation to the time-scale being considered. This type of change is typical of extreme, episodic events. The impact of such extreme events will vary in different climates and geomorphological zones, depending on the relative efficiency of more frequent events. Impact is also a function of reinforcing or restorative processes. In the ramped type of disturbance the changes in inputs are sustained at the new level as a result of permanent shifts in the controlling variables such as climatic or land-use changes.

The response of the landscape will also depend on its sensitivity. Some parts of the landscape will be more sensitive than others. Mobile, fast-responding systems have a high sensitivity to externally generated pulses. Slow-responding, insensitive areas tend to be far removed from the zones where changes are propagated. Thus interfluves and plateaux, in general, are insensitive areas.

The significance of this analysis is that the sensitive areas are the areas where natural hazards are most likely to occur. Sensitivity is the result of many processes, enhanced in many instances by human action. This entire book has really been a demonstration of the sensitivity of mountain landscapes — steep slopes, unstable materials and high energy processes. Debris deposition by glaciers is often in an unstable, highly sensitive position. There are many examples where hazards have been induced or made more likely by human action, some of which will be examined in greater detail later. In the European Alps, widespread deforestation in the seventeenth, eighteenth and nineteenth centuries has had a negative effect on the landscape. Many floods and debris flows would have been less extensive if deforestation had not occurred. Many of the flash floods that occurred in 1875, 1907, 1938 and 1968 on the Liembach, near Frutigen in the Bernese Oberland, may have been enhanced by deforestation. The felling of trees near the timber line has also increased the avalanche hazard. A useful classification of mountain hazards has been produced by Messerli and Ives (1984a) and is

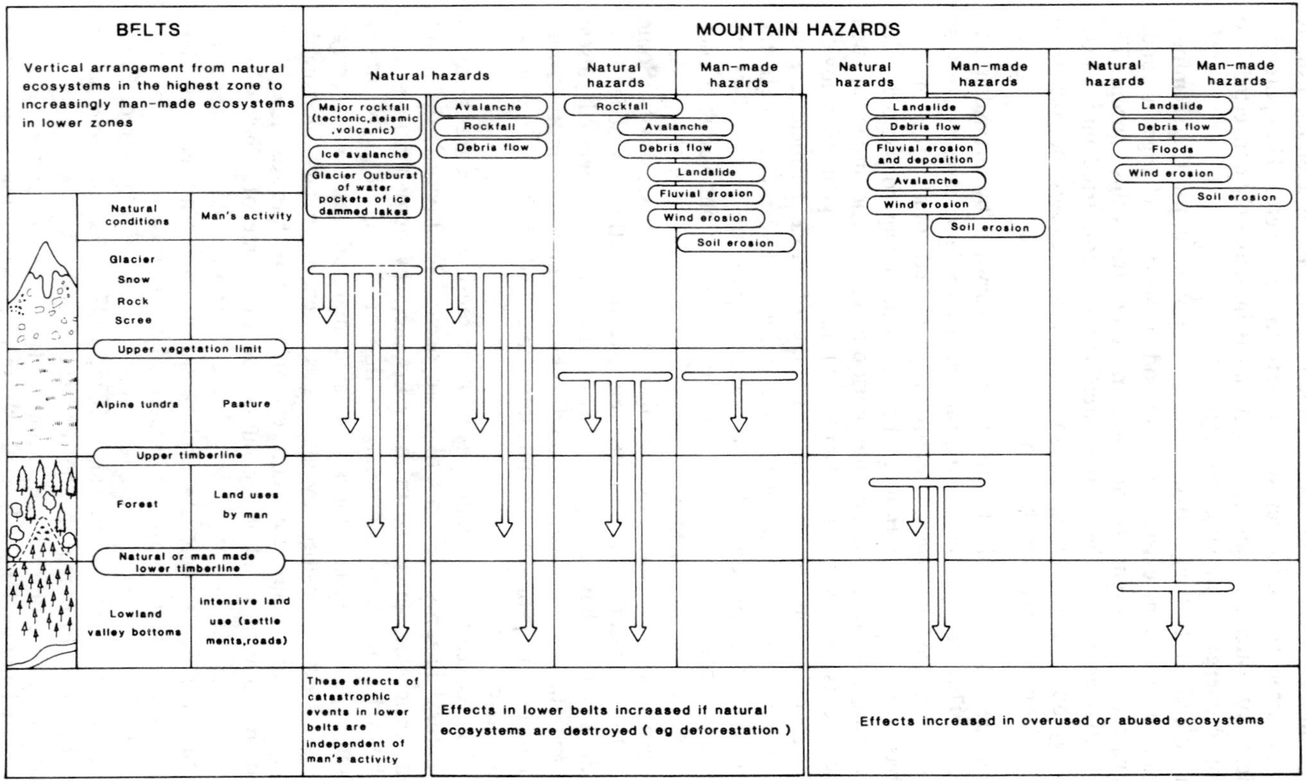

Figure 8.1. Classification of mountain hazards.

Source: Messerli and Ives (1984a).

reproduced in Figure 8.1. This emphasises the zonation characteristic of most mountain landscapes and the interaction between natural processes and human activity.

An important distinction has been made between mountains in developed and those in developing countries. Mountains in industrialised countries are experiencing a decrease in their agricultural populations and an increase in populations dependent on tourism. The mountains of developing countries are experiencing rapid growth in a rural and largely subsistence population which forces the increasingly heavy use of marginal land (Kienholz, Hafner and Schneider, 1984). Both types of mountain system are under stress and need to achieve a new stability. Mauch (1983) has also emphasised the contrast between developed and developing countries in his comparison of Nepal and Switzerland. In Nepal the problems are characterised by strong ecological and relatively weak economic interactions. In contrast, in the Swiss Alps, there are strong primary economic interactions leading to secondary ecological repercussions in the highlands.

A synthesis of the processes operating in the mountains of the developed and developing countries is shown in Figure 8.2. As Messerli (1983) has stressed, it demonstrates the need for scientific cooperation and interdisciplinary approaches on a world scale. It also needs the establishment of a number of principles that can be used in a variety of environments such as those proposed by Gigon (1983) and Winiger (1983). Landscapes can be classified as:

(a) *stable* where they are subject to little change, are capable of sustaining the present use and of returning to equilibrium after having been disturbed;
(b) *conditionally unstable* where they are liable to irreversible change; and
(c) *unstable* where they are easily disturbed and not capable of returning to equilibrium following disturbance.

It is not possible to examine all the interactions between human activity and natural processes present in mountains nor all the types of landscape. All that is attempted is to highlight some of the problems and to demonstrate the various approaches that have been adopted to tackle those problems.

SNOW AVALANCHES

Snow avalanches are common features in high alpine mountains and present a considerable hazard in areas that are being developed for tourism. Thus, most research has been concentrated in the mountains of developed countries. Conventional avalanche forecasting was based on general weather and snow conditions. Heavy reliance was also placed on

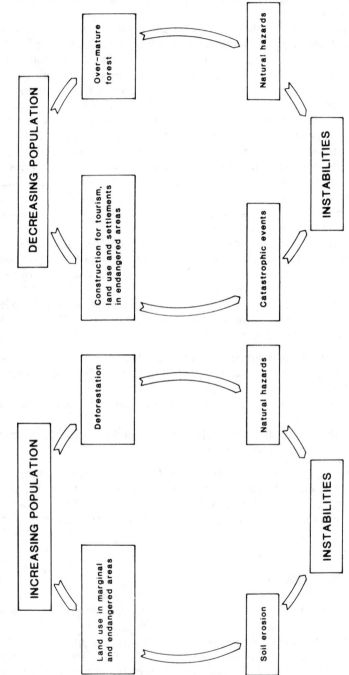

Mountains in Developing Countries

Mountains in Developed countries

Figure 8.2. Pressures on mountain environments in developed and developing countries.
Source: Messerli (1983).

actual avalanche activity, as this was one of the best available indicators of instability. Gradually more information became available with details of precipitation, wind and snowpack parameters. Forecasts based on this type of information can be very flexible in that they can be rapidly adjusted to local conditions. The disadvantages are that the reasoning is subjective and the memory capacity of a forecaster is less than that of a computer (Fohn *et al.*, 1977). Initial predictions attempted to indicate the general circumstances when avalanches could occur, rather than the location and extent of the area to be affected. Meteorological data were comparatively easy to obtain and fortunately are a major controlling factor (Perla, 1970; Schaerer, 1972; Bovis, 1977; Hartline, 1979). However, such forecasts were often inaccurate and additional information was required concerning the actual snow and slope conditions.

Snow is a complex substance with many unique properties. Its density, viscosity, strength, compressibility and thermodynamic instability can vary substantially, which leads to several problems in avalanche prediction. New snow changes quickly by destructive metamorphism where density increases and water vapour moves from the crystal points to the centre. Constructive metamorphism causes the vertical transfer of water vapour which can result in depth hoar. Varying wind strengths and direction during and after snowfall also influence the surface characteristics, providing a variety of bonding surfaces for future snow.

The importance of these processes is that a complex assortment of layers are produced which have a variety of strengths that can change relatively quickly. The shovel test allows a potential sliding surface to be identified. A pit is dug into the lower layers of the snow and a test block is isolated. The block is then loosened using an ice-axe, shovel or heel of a ski. The ease of separation is a measure of the adhesion of the surface layer to the layer below and provides useful first-hand evidence of the likelihood of a slab avalanche. The snow profile can be examined to identify any weak layers, such as depth hoar, a buried layer of dry unconsolidated powder snow or a layer of wet snow. It is especially important to look for sharp discontinuities between adjacent layers, as these represent boundaries where the adhesion between layers is low and consequently are the likely focal planes for avalanches. Snow profiles, if taken at regular intervals throughout the winter, can provide a record of the development of the total snow cover and the progress of individual layers within it. If this record is coupled with the careful recording of all observed avalanches, it may be possible to establish patterns of association which will assist in the recognition of dangerous conditions in future years.

In order to forecast accurately when an avalanche will occur on a particular slope, pre-failure signals are required. Acoustic and seismic emissions have been studied, as it was considered that when a failure occurred it might be detected by the noise or movement (St Lawrence *et al.*, 1973; St Lawrence and Bradley, 1977; Gubler, 1979; 1980). The emissions which occurred were due to small displacements, deformation of ice crystals and the breaking of bonds between individual grains of snow. But problems were encountered because snow is an excellent muffler of acoustic energy (Hartline, 1979) and there was a great amount of background noise.

Numerous researchers have attempted to design statistical models to predict avalanche activity. The basis for many of these are high correlations between certain meteorological and snowpack conditions and avalanche occurrence. Discriminant analysis is one such technique that has been used (Khomenyk, 1978; Obled and Good, 1980) where the information is processed to provide the best variables which distinguish between avalanche and non-avalanche days. Such techniques were improved by the use of antecedent data, especially using a two-day record (Hartline, 1979). Cluster analysis (Fohn *et al.*, 1977) and stochastic transfer-function time-series methods (Salway, 1979) have also been used. The stochastic method suggested that the product of snowpack depth, precipitation and relative humidity was the best index. A simulation process-orientated model was developed by Judson *et al.* (1980) but it indicated avalanche potential on 50 per cent of non-avalanching days. In a later study a new approach was used (Judson, 1983). The idea was to use key or index paths to provide warnings of avalanches in particular areas. Although the system seemed to work for one winter it was totally inappropriate for the next. Attempts have also been made to establish periodicity of large avalanches (Fitzharris and Schaerer, 1980). At Rogers Pass, British Columbia, Kahn (1966) and Loup and Lovie (1967) suggest that major avalanching occurs every eighteen to thirty years and Tushinskiy (1966) and Kahn (1966) have shown a correlation between the eleven-year sunspot cycle and avalanche activity.

Other workers have concentrated on predicting runout distances once an avalanche had occurred (Martinelli *et al.*, 1980). This often involves calculating avalanche friction coefficients. Fitzharris (1985) attempted to estimate the runout distances for five avalanche paths in the Mount Cook region of New Zealand by comparing the Norwegian model of Lied and Bakkehoi (1980), the regression model of Bovis and Mears (1976), the numerical model of Perla *et al.*, (1980) and Voellmy's (1955) model. The Norwegian model appeared to correlate best with the observed distances, whereas the regression model gave a variety of results. All except the regression model offer promise for predicting maximum runout distances.

Snow avalanche management is of two main types: controlling the snow, and controlling those who might be affected by it. The former can be sub-divided into stopping the avalanche forming, artificially releasing it, and trying to arrest or divert its motion once it has started. Many of these techniques have been reviewed by La Chapelle (1977a; 1977b). Avalanche management in respect of people can be carried out in several ways. In Switzerland, Avalanche Hazard Maps (AHMs) are used for advice and avalanche defence planning. They are then used to produce the Avalanche Zone Plan (AZP) which is used for land-use planning (Frutiger, 1980). Often colour maps are produced in relation to the perceived avalanche danger. White areas are presumed to be without danger. Red indicates the most hazardous areas and that no building permits will be granted. In blue areas, licences may be refused but if they are issued certain restrictions are stipulated. Buildings have to be reinforced and the roofs must meet specific design standards. In France, town plans are constructed so that building orientations are aligned to allow avalanches to pass easily between buildings (De Crecy, 1980). Similar techniques have been used by Ives and Bovis (1978) in the San Juan Mountains of Colorado, USA.

In New Zealand information for producing avalanche maps has been obtained for several areas including parts of Mt Cook National Park, the Craigieburn Range, Arthur's Pass National Park, the Milford Track and the Milford Road. No detailed mapping of zones of different frequencies of avalanche occurrence has been undertaken. This is probably not necessary because of low pressure for development in avalanche prone areas. All that seems to be required are maps showing potential runout areas. Such areas should not be built on and should be avoided by transport routes. More detailed planning and a wider range of adjustments to the avalanche hazard must be employed on ski-fields.

Where transport routes cross avalanche terrain methods have been developed in New Zealand for assessing the degree of hazard and the most appropriate methods of managing the hazard. Assessment of degree of hazard involves average size, frequency, length of road affected and traffic volume and speed so as to calculate an encounter probability for both moving traffic and traffic halted by one avalanche and endangered by adjacent ones. Summation of these probabilities for all paths provides a realistic hazard index. Milford Road has been assessed as possessing a moderate degree of hazard. Management measures include occasional road closures and signposting of no-stopping and safe-stopping areas. If the level of road use increases the index can be recalculated and appropriate additional measures undertaken. Milford Track possesses over fifty dangerous avalanche paths but relocation of the track beyond avalanche runout zones is more feasible than for a road.

It is clear that avalanche prediction is not a precise science. The actual forecast depends on whether general warnings are required or whether particular predictions of the extent and timing are needed for certain slopes. Total management is dependent on accurate prediction and sometimes this is vital for evacuation purposes in extreme cases. Avalanche prediction requires a stress increase or a strength decrease in the snow cover to be forecast, whereas avalanche management requires substantial forward planning. Avalanche hazard analysis therefore involves a substantial element of uncertainty. The general uncertainty problem in mountain areas is examined in greater detail in the next chapter.

LANDSLIDES AND DEBRIS TORRENTS

Landslides pose considerable problems in mountain areas. Two specific areas, British Columbia and the Austrian Alps, are chosen to illustrate these problems.

British Columbia

Instability is a common feature of the slopes in the mountains of western North America, and British Columbia is no exception. Heavily jointed and faulted igneous, metamorphic and sedimentary rocks are subject to frequent earthquakes. Also, bedrock slopes have been oversteepened by glacial erosion and are mantled by pockets of till, fluvio-glacial and solifluction sediments. Heavy rainfall, high snowfall amounts and undercutting by rivers and marine action combine with the other factors to create optimum conditions for mass movements of many types. These conditions have made it extremely difficult and dangerous to build settlements and lines of communication. Since 1865 approximately 365 people have been killed by landslides and direct damage is counted in hundreds of millions of Canadian dollars. The greatest loss of life occurred in the winter of 1981–2 in the areas between Squamish and Vancouver when ten people were killed by debris torrents and rockfall. Mass movement is also a threat to hydroelectric projects such as the Downie slide in the Revelstoke Dam pondage and the failures taking place upstream from the Seven Mile Dam in the Pend'Oreille Valley.

A number of large rockslides have occurred in British Columbia, although for some reason not with the same frequency as further east in the Rocky Mountain area of Alberta. The most famous rockslide in British Columbia occurred at Hope in the lower part of the Fraser basin in 1965 (Figure 8.3). The slide descended into the valley of the Nicolum Creek, destroying a 3 km stretch of the Hope–Princeton Highway and filling the bottom of the valley to a depth of up to 80 m (Figure 8.4). The slide consisted of 91 megatonnes of earth and rock which crashed down in seconds from the 2000 m high ridge forming the north side of the valley. The rocks consist of fine-grained lavas, or greenstones, interbedded with cherts, and they have been folded and faulted. A major fault area exists in the valley 5 km to the northeast. A strong set of joints divides the rock into polygonal blocks which break away to form extensive talus sheets. On the morning of the slide an earthquake was recorded on local seismographs at 6.58 a.m. The slide followed two minutes later. Since earthquakes are relatively common in British

Figure 8.3. The Hope landslide, British Columbia. (Photo: author.)

Figure 8.4. Valley floor infilled with debris from the Hope landslide. (Photo: author.)

Columbia, the possibility of another slide like that at Hope must be ever present.

A similar event was the Pandemonium Creek rock avalanche which occurred in the Coast Mountains in 1959. A slab of gneiss, estimated at 5×10^6 m^3, descended into the main valley of Pandemonium Creek at speeds up to 100 m s^{-1}. Debris ran 300 m up the opposite side of the valley, then turned and travelled through a series of bends to run out on a fan surface at the head of Knot Lakes (Evans, 1989).

Mudflows are comparatively rare in British Columbia but, given the right conditions, can be spectacular and catastrophic. The largest recorded mudflow occurred near Pemberton in 1975 where a debris avalanche developed into a mudflow involving 29 million cubic metres of material. The lower Thompson and Fraser valleys are also prone to mudslides and mudflows, a result of the combination of valley infill composed of lake sediments, outwash sands, gravels and silts, and incision by the river. Big Slide, on the Frazer River north of the town of Quesnel, is one such example. This is basically a retrogressive mudflow in Pleistocene lacustrine sediments. Rates of retrogression have varied in the range 6–12 m yr^{-1}. Yearly rates of movement have been up to 271 m yr^{-1}. Volume calculations estimate that between 1973 and 1982, 5.6×10^6 m^3 of material was moved and 1.9×10^6 m^3 was deposited resulting in a net yield into the Fraser River of 3.7×10^6 m^3.

Of a slightly different nature are large, slow-moving earthflows that occur in the Interior Plateau. Elongated masses of hummocky debris, some totalling more than 30×10^6 m^3, occupy minor tributary valleys incised into the steep margins of the plateau. In some cases movement has been occurring for at least 6500 years. Such earthflows are notably concentrated within the Camelsfoot and Clear Ranges, which form the western boundary of the Interior Plateau. The large earthflows at Drynoch and Pavilion are developed in Upper Cretaceous sediments containing bentonitic shales, sandstone and conglomerate. Earthflows are common within the Eocene dacite-to-rhyolite volcanics at Gibbs Creek–Fountain Valley, in the Hat Creek Graben and along the Fraser Valley upstream of Big Bar Creek, particularly where thick interbedded sediments occur.

Debris slides and debris avalanches are common features on the steep mountain slopes especially in the Coast and Cascade Ranges (Church and Miles, 1987). On the highest mountains failure tends to occur in talus and leads to the extension of the talus cones with very little transfer of material to lower levels. Potential damage is also slight. However, this is not the case in the Coast Mountains, where slides commonly occur when the soil mantle becomes saturated. This results from high precipitation or snow melt acting on weathered rock or till which acts as an impermeable barrier to water flow. Slopes underlain by argillite, slates, sandstones and schists are especially prone to instability. Several studies have shown that the magnitude and frequency of debris avalanches and torrents within a small area vary considerably. Major torrents have been triggered in some basins by rainstorms with return periods of less than five years, whereas in adjacent basins no known torrents have occurred over a twenty- to thirty-year period. This differing response appears to be related to the varying availability and stability of debris within the basins. Slope failures in rock slope dominated basins are usually decoupled from any streamflow influences whereas in areas dominated by colluvium or glacial drift, slope failures tend to be linked more directly to undercutting by streamflow. In general, the three attributes of debris texture, debris storage, and the potential for loading of channels by rockfall and landslide events account for the high incidence of debris torrents in the plutonic rocks of the Coast Mountains.

The Austrian Alps

Debris torrents are also a major hazard in the Austrian Alps. A few examples are chosen to illustrate the scale of the problem. Every major tributary valley that drains and channels water and material from the mountain valley sides into the main valley systems has been subjected to catastrophic events. The Enterbach of Inzing, a southern tributary of the Inn River west of Innsbruck, is one such example. It has a drainage area of 12.3 km^2 and water emanating from two cirques and valley slopes is concentrated in a small basin. The Village of Inzing, on the alluvial fan at the point where the Enterbach enters the main valley, has often been affected by debris flow disasters. The last disaster was on 26 July 1969 which took three lives, destroyed twelve houses, a chapel and a new recreation centre. Four thousand metres of road and 500 metres of railway were also blocked. During this episode the Enterbach deposited about 400 000 m^3 of debris on the fan. The highest stated discharge was 4,000 cumecs (cubic metres per second) and the highest stated speed of the debris was more than 100 km h^{-1}.

The 1969 debris flow destroyed forty-two of the forty-four concrete locks that had been built since 1897 to protect the lower part of the valley. The destruction was caused by the undercutting of the valley slopes. Since 1969 new check dams have been built higher up the valley and a major dam has been constructed lower down to protect the village. This dam will allow 150 000 m^3 of debris to be deposited. The total cost of rebuilding the roads, bridges and check dams was about £1.8 million.

The Schesa Gully near Bludenz, Vorarlberg, illustrates the same problems. Between 1796 and 1975, 45.2 million cubic metres of material has been removed from the gully systems, 20 million cubic metres of this material being deposited in an alluvial fan on the lower slopes. The catchment area of the Schesa Gully systems is only about 14.7 km^2, with the majority of the gullies being cut into weak Quaternary deposits.

The gully systems appear to have developed around 1776 when there were major changes in the catchment area. In 1776 the former commune of Bürs was divided into two new communes; Bürs and Bürserberg. The new commune of Bürserberg faced financial difficulties and the forest area of the Schesa area was entirely felled by the commune. This was when the Schesa Gully began to develop extremely rapidly. Disasters were reported in 1802, 1806, 1810, 1811, 1820, 1823, 1868, 1879, 1880 and 1885. The development of the gully is shown in Table 8.1. The 1892 Austrian–Swiss agreement concerning the regulation of the River Rhine led the torrent control service to build a series of check dams starting in 1899. However, further disasters occurred in 1904, 1907, 1924, 1966, 1967 and 1969. There is little that can be done to halt erosion in the upper part of the gully; all that can be done is to control the debris as it works its way down through the system.

Somewhat different problems are faced by the Upper Ötz Valley. During periods of glacier advance the Vernagt Glacier in the Rofen Valley and the Gurgler Glacier in the Gurgl Valley have often blocked their valleys, thus damming streams to form lakes. Such dams have often burst creating flooding in the Ötz Valley. A comprehensive summary of the Vernagt Glacier has been provided by Grove (1988). Periods of ice advance, with dam and lake formation have occurred in 1599–1601, 1677–82, 1770–4 and 1845–8. It appears that the lake emptied slowly six times and rapidly seven times in 1600, 1678, 1680, 1773, 1845, 1847 and 1848. The worst disaster occurred on 16 July 1678 because the dam burst at a time when all the tributaries were flooding as a result of heavy precipitation and melting snow in the mountains. All bridges and many houses and villages were destroyed. Längenfeld in the central Ötz Valley was badly hit, with the valley floor being flooded and the village buried by a debris flow from the Fischbach. The Fischbach itself has produced eighteen catastrophic flows since 1700.

The Gurgl Glacier near Obergurgl advanced rapidly in the period 1717–18, continuing until 1724. Further advances occurred around 1770 and from 1845 to 1855. At the glacier's maximum extent in 1855 the lake behind the ice contained 11.7 million cubic metres of water (Patzelt, 1986). The Gurgler ice lake has tended to empty gradually, however, as the glacier began to retreat in 1867 the lake did burst, flooding the valley and destroying the bridges in the Gurgl Valley.

A somewhat different flood event occurred in the Ötz Valley following summer precipitation in 1987. The late spring and early summer period was comparatively cold, thus maintaining a greater snow cover on the mountains. July and August reversed the trend and were particularly warm, producing rapid snow melt and high river discharges. Heavy precipitation in late August created a devastating flood, with about 25 per cent of the houses in Solden being affected. Seventeen people lost their lives and damage amounted to $21 millions. Most of the flooded houses had been built since 1918 on the lowest part of the valley floor

Table 8.1. Development of the Schesa Gully, 1899-1975.

Period	Number of years	Eroded area (m²)	(m² yr⁻¹)	Eroded volume (m³)	(m³ yr⁻¹)	Volume of debris stopped in the gully by check dams (m³)	(m³ yr⁻¹)	Volume of material deposited outside of the gully (m³)	(m³ yr⁻¹)	Ratio between checked and unchecked debris movement
1899-1927	28	37660	1345	28670000	102393	955000	34107	19120000	68286	0.49
1928-1934	7	15900	2271	1358000	194000	655000	93571	703000	100429	0.93
1935-1946	12	22400	1867	1435000	119583	425000	35417	1010000	84167	0.42
1947-1965	19	12900	679	959000	50474	520000	27368	439000	23105	1.18
1966-1975	10	21180	2118	700000	70000	420000	42000	280000	28000	1.50
Total/average		110040	1448	7319000	96303	2975000	39145	4344000	57158	

Source: Prucker (1976).

(Aulitzky, 1988). The magnitude of the disaster was thus increased by ill-advised rapid expansion of the village of Solden. Hazard zones have now been defined to try to prevent similar disasters happening in the future.

HIGHWAY ENGINEERING IN MOUNTAINS

Problems of accessibility place considerable constraints on techniques used in highway engineering in mountains. Emphasis is placed on the use of rapid methods which require little in the way of materials or field support (Fookes *et al.*, 1985). A mountain road project usually involves five stages:

1. The *feasibility* stage involves the initial project planning, the collection of terrain data, regional investigations using mainly remote sensing techniques and the selection of the route corridor.
2. The *reconnaissance* stage consists of terrain studies within the route corridor and the selection of a provisional road alignment.
3. The *site-investigation* stage involves a detailed study, using mainly ground techniques, of a narrow zone on either size of the proposed alignment.
4. The *construction* stage is often accompanied by additional site investigation at problem sites. Road design is finalised and remedial measures are designed for slope failures created during construction.
5. The *post-construction* stage involves regular maintenance and continued observation of unstable slopes and road structures.

The aim of the feasibility stage is to make an overall appraisal of the terrain conditions within the project area. This will involve use of background information, such as geological and climatic data, as well as remote sensing techniques. Field mapping is the main technique used in the reconnaissance stage, combining both engineering geology and geomorphology. This field mapping should place specific slope sites in an overall terrain context. The engineer is then able to assess potentially dangerous features adjacent to the proposed route alignment. The post-construction stage is often neglected, but a critical assessment of construction techniques, slope-erosion control and gully protection measures is invaluable in building up a series of case studies of use to future projects. The methodology used in choosing the final alignment of the Dharan–Dhankuta Road Project illustrates many of these considerations (Brunsden *et al.*, 1975a; 1975b; 1975c).

Some landforms present more problems than others. If a road has to cross scree, then alignment across the lower part is to be preferred. The toe slopes generally possess the lowest slope angles and the largest rock fragments. These factors provide the greatest strength available for excavations. In addition, rockfall hazard will usually be at a minimum at the toe of the scree. A road alignment across the head of a scree will encounter smaller fragments at a greater angle with a greater rockfall hazard. However, if the scree is especially active with a high angle an alignment across the toe may lead to undercutting and greater scree activity. In this case a higher line has the advantage that the scree will be at its thinnest. O'Dell (1956) has described such problems where the Peruvian Central Railway crosses a 1000 m high scree. Rockfalls and scree movement were so frequent that during the wet season a bulldozer was used to clear material from the track each morning.

Mudflows and mudslides present the most difficult type of terrain for road alignment. If a road has to cross a mudflow zone it should be as high as possible where the mudflow is thinnest. There is also a greater chance of draining the flow in such a situation. Mass movement catchments, where all types of slope instability may occur in close association, should be avoided. Alluvial fans, as noted above in the Austrian Alps, cause both obstacles and hazards. As long ago as 1908, Anderson was describing the problems created by the constant changing of river courses across alluvial fans experienced during the construction of the Dooars railway at the foot of the Bengal Himalaya.

Many of these problems can be illustrated by examining the Karakoram Highway along the valley of the Hunza River (Jones *et al.*, 1983). The Hunza Valley is a deep, steep-sided, sediment-floored, glaciated valley set between high ice-covered mountain masses reaching up to 6500 m. Highly dynamic glacier tongues often reach near the main valley floor. The character of the valley, as is normal in mountain regions, is highly variable due to structural, lithological and geomorphological factors. In the lower sections the valley is broad and floored by low-angled debris flows and terraces. The upper section is similar except that morainic ridges and outwash fans are more important. The middle section is more complex. The eastern third consists of a series of long, narrow, winding gorges where the river has cut through igneous rocks. The remainder consists of short gorges through resistant lithologies separating broad embayments filled with till, outwash and debris flow deposits. The river is usually flanked by unstable, near-vertical terrace cliffs.

The catchment area of the Hunza is approximately 13 000 km^2, with a calculated mean runoff of 900 mm yr^{-1} or 11 965 $\times$ 10^6 m^{3}s^{-1}. Minimum flows occur in winter and maximum flows in July and August. Peak flows exceed 2000 m^3 s^{-1}. Sudden increases in discharge are quite common, caused by sudden releases of meltwater or the collapse of natural dams created by glacier extension, landslide or catastrophic debris flows. In 1858 the Phungurh landslide dammed the river for six months, eventually leading to a flood wave that caused the river level at Attock on the Indus lowlands to rise 9 m in ten hours, suggesting drainage in excess of 30 000 m^{3}s^{-1}. Ice-dam bursts in the Shimstal Valley led to flood surges in 1884, 1893, 1905, 1906, 1927, 1928 and 1959. The 1906 flood raised the river level at Chalt by 15 m and the 1959 wave destroyed Pasu Village. Catastrophic debris flows occasionally dam

Table 8.2. *Relative significance of different terrain and material types along the Karakoram Highway.*

Road section (km)	Fan	Rockfall and scree	Rock	Terrace	River	Till
0- 33.60	17.25	7.10	5.45	3.80	0.00	0.00
%	(51.34)	(21.13)	(16.22)	(11.31)	(0.00)	(0.00)
33.6-65.75	11.40	5.30	5.50	5.05	0.00	4.85
%	(35.5)	(16.51)	(17.13)	(15.73)	(0.00)	(15.10)
66.75-93.0	11.60	4.15	4.10	1.20	0.20	6.06
%	(42.49)	(15.20)	(15.02)	(4.40)	(0.73)	(22.16)
93.0-128.8	12.10	10.10	8.50	2.05	0.30	2.75
%	(33.79)	(28.21)	(23.74)	(5.73)	(0.84)	(7.68)
Total km	52.35	26.65	23.55	12.10	0.50	13.65
%	(40.64)	(20.69)	(18.28)	(9.39)	(0.39)	(10.60)

Source: Jones *et al.* (1983).

the river. In August 1974, an alluvial fan near Batura Glacier created a fan 300 m by 200 m by 4 m in ten minutes, following rainfall of 10.3 mm in two days.

This general background demonstrates that road construction problems must be considerable. Jones *et al.* (1983) have assessed the Karakoram Highway with respect to the following factors: the occurrence of debris on the road and damage to the road; the nature of the materials traversed by the road; the slope angles (characteristic, maximum, minimum) both upslope and downslope of the road; the dip and orientation of joints in road cuts; the occurrence and nature of drainage across the road; potential hazards; and possible remedial measures. Five generalised terrain units were recognised: fluvial and outwash terraces; till-dominated sediment accumulations; alluvial fans; screes (including rockfall cones); and rock outcrops. The relative importance of these units in different sections of the valley is shown in Table 8.2.

Most of the roadline has been constructed across unlithified Pleistocene and Holocene deposits. Only 39 per cent of the road traverses rock. One of the major problems is created by outwash from glaciers such as the Batura, Pasu and Ghulkin. Batura Glacier is one of the eight longest extra-polar glaciers in the world, with a length of 59 km. In 1980 its terminus was within 700 m of the Hunza River, thus confining the Highway between the glacier and the river.

The major problem is meltwater with sudden changes in discharge and in the direction of meltwater release. The original 1970 road possessed two bridges, 30 m and 8 m long, where it crosses the meltwater streams of the Batura Glacier. The 30 m bridge soon developed subsidence cracking due to settlement of unconsolidated young moraine and was destroyed by meltwaters in 1972. During the following year a major shift in the pattern of meltwater release damaged the 8 m bridge, necessitating a realignment of the road. Glaciological investigations have suggested that the Pasu Glacier poses a relatively minor short-term threat. But the same cannot be said of the Ghulkin Glacier. The terrain on the west (Ghulkin) side is generally preferable to that on the east, where active debris flow fans and unstable scree slopes reach down to the river. But the western route has to pass across a narrow corridor of dissected outwash material concentrated between the snout of the Ghulkin and the Hunza River. This snout has fluctuated by 625 m during the period 1885–1980, and so the road has been constructed as near to the river as possible. Three main meltwater paths occur. When the road was constructed the magnitude of meltwater discharge decreased from south to north. This was clearly incorporated in bridge design because the three channels were spanned by bridges 15 m and 5 m long, and a 1 m culvert respectively. But

meltwater paths are notoriously transient features under the prevailing glaciological conditions, for changes in sub-glacial water routeways due to pressure variations, sediment deposition, the creation and disruption of sub-glacial lakes, the collapse of ice blocks and the compaction of sub-glacial debris can all lead to altered patterns of meltwater release (Jones *et al.*, 1983, p. 344).

In the summer of 1980 the southern meltwater flow ceased and a newly developed exit in the snout fed water to the central channel. This led to scour in the upper sections and a new boulder fan which destroyed the bridge and buried a 300 m stretch of the highway.

The general assessment, by Jones *et al.* (1983), of the route taken by the Karakoram Highway was that it had been chosen carefully with respect to geomorphological constraints. It is inevitable that there will be problem sections and that the road will be periodically disrupted and occasionally destroyed by catastrophic events such as earthquakes, debris flows and flood surges. Several conclusions emerged from this investigation which reinforce the points made earlier in this section. There is a clear need for extensive areal surveys in mountainous terrain so that planned roadlines can be examined in the context of geomorphological units. Road construction near glacier snouts requires careful glaciological investigations, and the crossing of debris flow fans requires detailed analysis of upslope catchment characteristics. Finally, scree slopes require careful management to minimise disturbance.

TIMBER HARVESTING AND EROSION IN MOUNTAINS

Many forested mountain slopes are being exploited commercially for their timber and this is having a tremendous influence on the stability of those slopes and overall sediment yields from mountain basins. Nowhere is this more apparent than in the western parts of the North American Cordillerra. Forest vegetation is generally believed to play an important role in stabilising slopes and reducing the movement and rate of mass erosion processes. When such forests are removed a temporary increase in erosion is likely. Accelerated erosion due to timber harvesting may result in reduced productivity of forest soils, damage to roads and bridges and adverse impacts on the stream environment downstream (Swanston and Swanson, 1980). There have been many examples of slumps and earth flows being reactivated by timber operations. Where massive, deep-seated failures are involved, anchoring by tree-root systems is likely to be negligible but destruction of the forest cover will alter the hydrological balance. Increased moisture availability will allow water to pass through the soil to deeper levels possibly reactivating earth flows. In the H.J. Andrews Experimental Forest, western Cascade Range, Oregon, even nine years after clearcutting, runoff in a small watershed still exceeded by 50 cm the estimated yield for the watershed in a forested condition (Rothacher, 1971).

Clearcutting often results in an increase of debris avalanche erosion by a factor of 2–4 (Swanson and Dryness, 1975). Roads have a more profound impact, with debris avalanche activity increased by factors of up to 300. Clearcut slopes appear to remain susceptible to debris avalanching for up to twelve years after cutting. The duration of impact of road construction may be twice as long as clearcutting impacts. The increase in the frequency of debris avalanches also leads to an increase in the occurrence of debris torrents. In the H.J. Andrews Experimental Forest, clearcutting appeared to increase the occurrence of debris torrents by between 4.5 and 8.8 times (Swanston and Swanson, 1980). Increased debris torrents are the result of increased debris avalanches which trigger most debris torrents. Soil creep, slumping, earthflows, debris avalanches and debris torrents form interlocking systems. Thus changes to any one of these systems will have widespread repercussions on the others. This can be illustrated by the results of experiments conducted at Coyote Creek in the South Umpqua Experimental Forest and in the H.J. Andrews Experimental Forest in Oregon.

The Coyote Creek research area is situated in the western Cascade Range. Four research watersheds have been monitored since 1966 to determine water discharge and sediment yield under forested and clearcut conditions. Two access roads were built in the upper half of one of the basins in 1971 and the watershed was logged by clearcutting in 1972. Between 1966 and 1970 sediment movement was much less than 0.01 m^3 per metre active stream channel per year. In 1971, as a result of heavy winter precipitation and increased runoff from the construction of logging roads, sediment yield increased to 0.05 m^3 per metre per year. Material was mainly produced by debris avalanching and rotational slumping along stream banks. In 1972, the first year after the watershed was clear felled and a year of exceptional rainfall, bedload movement tripled to 0.16 m^3 per metre per year.

The H.J. Andrews Experimental Forest is situated in the western

Cascade Range on lava flow and volcaniclastic bedrock with an average annual precipitation of 230 cm and Douglas Fir–Western Hemlock forest vegetation (Swanson and James, 1975). Detailed investigation has shown that although slump-earthflow movement only occurs on 3.3 per cent of the forest area, it appears to have contributed to conditions leading to 40 per cent of the total volume of debris avalanche erosion. Slightly more than half of this earthflow-generated debris avalanche erosion occurred in streamside areas where earthflow movement constricted channels initiating stream undercutting and rapid shallow mass movement. In clearcut areas and where roads have been constructed, debris avalanche erosion has been increased by about five times (Swanson and Dyrness, 1975).

Soil erosion on hillslopes and sedimentation in streams have also been accelerated by timber harvesting in Queen Charlotte Islands, British Columbia (Roberts and Church, 1986). The sediment budget was estimated for four severely disturbed small drainage basins in the Queen Charlotte Ranges. These results indicated that sediment production and delivery to stream channels may increase following logging, but not invariably. Significant stream bank erosion following logging resulted in the formation of large wedges of coarse sediment in the stream channel. Sediment transport through the channels increased by up to ten times but the residence time for the increased volume of sediment increased by up to 100 times. These detailed studies clearly substantiate the increase in mass movement and soil erosion following timber operations and should serve as a warning to future such operations in mountain areas.

MOUNTAIN HAZARDS MAPPING

The previous examples have demonstrated the need for integrated research into mountain problems. Integrated research is well to the fore in schemes for mapping mountain hazards. There have been numerous schemes devised to map and assess the risk and hazard from particular processes such as landsliding (see, for example, Ives *et al.*, 1976; Kienholz, 1977; 1978; Moser, 1978; Malgot and Mahr, 1979; Jones *et al.*, 1983; Hansen, 1984). Attempts are also being made to devise all-purpose mapping schemes to evaluate the range of hazards encountered in mountain environments. However,

Mountain hazards mapping is far from being a widely recognised component of either regional or site specific land-use planning . . . This may be because many of the large-scale mountain developments are very recent and because of the relatively new awareness that manipulation of one or a few components of a landscape affects all other components (Dow *et al.*, 1981, p. 56).

One of the most significant is the Mountain Hazards Mapping (MHM) Project in Nepal. This project is used here as an illustration of the problems involved in such a project. The MHM Project is sponsored by the Nepal National Committee for Man and the Biosphere and the United Nations University (UNU). It is part of the UNU Development Studies Division. The main objectives of the project are (Kienholz *et al.*, 1984):

(a) the development of prototype hazard maps for type areas in three Himalayan altitudinal belts — Middle Hills, High Mountain–Upper Timberline and Terai–Siwalik Interface (Ives and Messerli, 1981);
(b) the evaluation of methods for assessing and mapping mountain hazards and slope stability (Kienholz *et al.*, 1983);
(c) examination of the means of transferring the experience derived from (a) and (b) into the regional planning process.

The High Mountain area was chosen because it provided a site representative of human impacts from a complex of processes such as tourism, overgrazing and deforestation. The area selected was a transect from the vicinity of Lukla up the Dudh Kosi Valley to the snowline in the Sagarmatha (Mount Everest) area, centred on Namche Bazar. This is also the area known as Khumbu-Himal. The Middle Mountain area (Kakani-Kathmandu) contains Nepal's greatest population densities with deforestation, agricultural terracing on steep slopes, and soil erosion. The Siwalik-Gangetic Plain (Terai) interface is where the major rivers leave the mountains and deposit their coarse sediment load. Such an area possesses problems of heavy population in-migration, deforestation, wet season flooding and dry season drought. The mapping scales selected were based on available topographic maps and air photographs: Khumbu–Himal, 1:50 000; Kakani–Kathmandu, 1:10 000; Siwalik–Terai, 1:25 000).

A number of reports have been published concerned with the Project (see Ives and Messerli, 1981; Caine and Mool, 1981; 1982; Johnson *et al.*, 1982; Kienholz *et al.*, 1983; 1984; Peters and Mool, 1983) The research design for the Kathmandu-Kakani area is shown in Figure 8.3. The hazards map should fulfil several functions. It must provide a general impression of which areas are subject to various degrees of hazard and which are relatively stable. It must also provide detailed information of the distribution of the different types of hazard and the individual degrees of hazard. In order to achieve this differentiation four groups of colours were used: reds and browns indicate a high degree of hazard or slope instability; yellows indicate an intermediate state; greens indicate stability or freedom from hazard, and blue is used for streams.

A second level of detail is provided by the differentiation within the reds and browns as to whether human life and hard structures such as roads and buildings are endangered (reds) or whether the hazard is confined to loss of arable land and soils (brown). At the third level, differentiation of suspected hazards or instability of a higher degree and for special conditions within important forested areas is indicated by the use of screens on colours. A fourth level of detail is achieved by letter symbols and a fifth level introduces information about all types of hazard

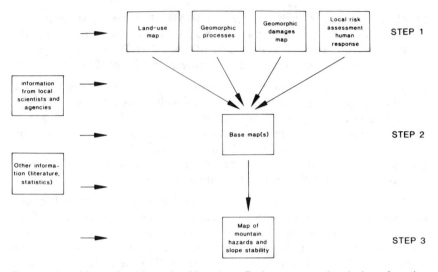

Figure 8.5. Mountain Hazards Mapping Project research design for the Kathmandu-Kakani area.

Source: Kienholz *et al.* (1981).

that occur in a particular area.

The type of evidence provided is shown in Figure 8.6. Each type of hazard may possess a specific disposition, trigger mechanism process and effect and may require particular preventive measures. A hazard is classed as confirmed if there is actual damage or silent witness of former or actual movement in the area under consideration. A hazard is inferred if there is no actual damage within the area but actual damage observed in adjacent areas of a comparable nature. A hazard is classed as supposed if it is neither confirmed nor inferred but if it is assumed based on experience in other areas. Inaccessible areas are mapped indirectly, based on a model involving all the factors thought to engender instability.

The resultant map indicates a concentration of hazards in the intensely cultivated Kolpu Khola watershed and in the Kathmandu Valley (Kienholz *et al.*, 1984). The Kolpu Khola watershed possesses long, steep slopes with high relative relief developed on metamorphic rocks which are highly susceptible to erosion and landslides. In certain areas the risk of death or injury to the population is high, especially along streams and on slopes with gradients greater than 25°. The gentler slopes and the ridge spurs are prone only to less severe shallow erosion and landsliding. This pattern appears to be representative of the Nepal Middle Mountains as a whole.

The Kathmandu Valley area is underlain by alluvium and old lake deposits. The major hazards are associated with the tar slopes where slumps, river erosion and rill wash are common. Flooding and siltation are also a problem close to stream channels. The major forested areas are much less subjected to hazards, although there is a potential for hazard

Type of hazard		Evidence and corresponding degree			
		confirmed more than 50% of unit area	inferred in less than 50% of unit area	suspected in more than 50% of unit area	less than 50% of unit area
E	deep erosion by water (> 2 m) deep gullies	E 2a	E 1 & 2a	*(E)* 1 (2a)	(E) 1 & 1 (2a)
e	shallow erosion by water (< 2 m) shallow gullies	e 2c	e 1 & 2b	*(e)* 0 (2b)	(e) 0 & 0 (2b)
e'	surficial erosion rills	e' 0 (2b)	e' 0 & 0 (2b)	*(e')* 0	(e) 0
L	deep landslide (> 2 m)	L 2a	L 1 & 2a	*(L)* 1 (2a)	(L) 1 & 1 (2a)
l	shallow landslide (< 2 m)	l 2b (2a)	l 1 & 2b (2a)	*(l)* 1 (2a)	(l) 1 & 1 (2a)
l'	collapsed terraces	l' 1 (2b)	l' 0 & 0 (1)		
D	debris flow	D 2a	D 1 & 2a	*(D)* 1 (2a)	(D) 1 & 1 (2a)
Ad	accumulation of debris flow material	Ad 2a	Ad 1 & 2a	*(Ad)* 1 (2a)	(Ad) 1 & 1 (2a)
F	flooding	F 2a	F 1 & 2a	*(F)* 1 (2a)	(F) 1 & 1 (2a)
Aw	accumulation of water-transported material	Aw 2a	Aw 1 & 2a	*(Aw)* 1 (2a)	(Aw) 1 & 1 (2a)
T	major torrent activity		T 1 & 2a		
t	minor torrent activity		t 1 & 2b		
R	rockfall source area	R 2a	R 1 & 2a	*(R)* 1 (2a)	(R) 1 & 1 (2a)
Ar	accumulation of rockfall material	Ar 2a	Ar 1 & 2a	*(Ar)* 1 (2a)	(Ar) 1 & 1 (2a)

NOTE	Example	Meaning	
		Hazard or instability is:	occurrence in:
	E	Confirmed/inferred	more than 50% of unit area
	E	Confirmed/inferred	less than 50% of unit area
	(E)	Suspected	more than 50% of unit area
	(E)	Suspected	less than 50% of unit area
	E, E, *(E)*, (E)	Letters in italics indicate forested areas which have been delimited by interpretation of aerial photographs and assessed by comparison of geological, geomorphological and hydrological parameters with the pattern of actual damages in unforested areas.	
	2a	The whole unit area ia endangered by a definite hazard of degree 2a.	
	1 & 2a	The unit area is endangered by a definite hazard partly of degree 1, partly of degree 2a.	
	1 (2a)	The whole unit area is endangered by a definite hazard of degree 1 and a suspected hazard of degree 2a.	

Figure 8.6. Type, evidence and corresponding degree of hazard or instability.

Source: Kienholz *et al.* (1984).

occurrence with deforestation. Protection and long-term forest maintenance are necessary to maintain stability.

Kienholz *et al.* (1984), in their conclusions, note two important reservations which are not only true for the MHM Project but also for mountain environment studies in general. First, the landscapes under investigation are dynamic, with natural and anthropogenic changes occurring rapidly Each change in landscape conditions creates a new set of conditions and affects the operation of the natural systems. Stable landscapes will be capable of returning to equilibrium after having been disturbed. Conditionally unstable landscapes are liable to irreversible change and are vulnerable to disturbance. Unstable landscapes are easily disturbed and are not capable of returning to equilibrium following disturbance without technological input or reclamation. The second reservation concerns situations where it is very difficult to predict future events, and errors, especially in hazard assessment, are likely to occur. This emphasises the need for continuing research into such high-energy rapidly changing environments and has been the rationale behind this book. It also stresses the great deal of uncertainty that exists concerning what is happening in mountains. This problem is analysed in the next chapter.

Chapter 9

INTEGRATION OF SPATIAL AND TEMPORAL MOUNTAIN SYSTEMS — THE UNCERTAINTY DILEMMA

A great many topics have been examined in this book in varying degrees of detail. It is hoped that some impression has been provided concerning the current state of knowledge of the workings and linkages of natural systems in mountain environments. But it is probably just as important to emphasise how little is known of such systems and of mountains in general. This lack of knowledge is especially important when interactions between human and physical systems are considered. Topics have been examined at a variety of temporal and spatial scales which makes integration difficult. Thus detailed freeze-thaw mechanisms at the timber line in the Colorado Front Range of the Rocky Mountains are difficult to reconcile with long-term advance and retreat of valley glaciers. Geoecological relationships also vary from the generalised approach of Troll to that of detailed soil, topography and vegetation characteristics of small drainage basins. Some attempt to combine various spatial and temporal scales is present in sediment budget studies, but there are few such studies in mountain regions and those that exist have been conducted in somewhat restricted geographical zones. However, such studies as do exist provide a valuable insight into the nature of integrated mountain systems.

SPATIAL AND TEMPORAL VARIATIONS

More is known of mid- to high-latitude mountains than is known of tropical mountains. In addition, the environments above the tree line, the so-called Alpine region, have been more intensively studied than areas below the tree line. An interesting review of the geomorphology of the Indian Peaks Area, Front Range, Colorado Rocky Mountains, by Thorn and Loewenherz (1987) indicates the sort of uncertainty that exists. The region in question has probably been more intensively studied than any other equivalent area. Most peaks rise to 3900–4100 m in elevation and the north–south trending Continental Divide and valley bottoms possess classic glacial features with relative relief up to 700 m. The climate and general arrangement of ecosystems have been studied in detail by many workers (such as Marr, 1967; Marr, Clark, Osburn and Paddock, 1968; Marr, Johnson, Osburn and Knorr; 1968; Barry, 1973).

The available information on the linking of geomorphic systems can be assessed using Caine's (1974) model combining sediment fluxes with landform units. The landform units are interfluve, free-face, talus, talus foot and valley floor. Within and across these landform units three sediment fluxes — coarse debris (greater than 8 mm), fine sediment (less than 8 mm) and geochemical — move material. Research on interfluves in the Indian Peaks region has been centred almost exclusively on Niwot Ridge. Little information is available on the coarse debris flux. Blockfields dominate much of the landscape but appear to be relict and have received little attention. Coarse debris only appears to be moved at the present time by frost creep and sorting to produce patterned ground is restricted to wet sites on high cirque floors.

Study of fine sediment fluxes has revealed some interesting results, one of the most important being the significance of aeolian input (Thorn and Darmody, 1980; 1985a; 1985b). This influx is of uniform character, with mean grain size in the 7.38–3.21 phi (0.006–0.108 mm) range, is not highly weathered and has a heterogeneous mineralogy which makes determination of its provenance difficult. Caine (1974) estimated annual aeolian input to be 0.005 mm. Interesting work on soil loss by Bovis (1978; 1982) and Bovis and Thorn (1981) has shown that snow patches occupy only 3 per cent of the area but yield about 59 per cent of the sediment total, and that tundra meadow occupies about 50 per cent of the area but yields only about 5 per cent of the sediment total. Geochemical fluxes have, in general, been thought to be relatively insignificant, though work undertaken in the Indian Peaks area has led to this opinion being modified. Dixon (1986) has shown that alpine geochemical processes are similar to those in temperate regions and occur quite rapidly. Also Thorn's (1976) and Caine's (1982) identification of the relative importance of chemical solution suggests that the 'geochemical situation on the interfluves merits serious reappraisal unfettered by traditional preconceptions' (Thorn and Loewenherz 1987, p. 243).

Free faces have received little direct attention but it is believed that structural controls, hydrothermal activity and glaciation explain free face formation. The widespread assumption that freeze-thaw weathering is the dominant bedrock failure mechanism is being challenged by a number of workers (White, 1976b; Thorn, 1982).

The movement of fines and water through the talus slope is little understood but the behaviour of coarse debris is quite well known. Although rockfall, alluvial and avalanche-type talus occur, most of the research has been conducted on rockfall talus. This may be unfortunate, bearing in mind Caine's (1974) suggestion that alluvial talus is important because it seems able to deliver material all the way to the valley bottom.

Compared to the talus slope, little is known about sediment flux on the talus foot zone. Lobate rock glaciers are abundant in this zone in the Front Range and have generally received considerable attention. But a great deal more is known about the superficial layer of coarse debris on their surface than about the mixture of ice, fines and coarse debris that forms the bulk of their mass. Large rock glaciers move at rates in the range 50–60 cm yr^{-1} while smaller ones move at lower rates of 5–10 cm yr^{-1}. Mass wasting forms in valley bottoms have received little attention, apart from rock glaciers.

THE UNCERTAINTY DILEMMA

This account of the sediment fluxes through the five landform elements — interfluve, free-face, talus, talus foot and valley floor — in the Indian Peaks Region, Colorado, has highlighted a number of methodological problems and a great deal of uncertainty. This level of uncertainty is increased when processes and fluxes are examined at the basin level (Caine 1982; 1984; 1986). The crest of Niwot Ridge has been estimated to be lowering at a rate of less than 0.01 mm yr^{-1}, whereas cliff retreat is of the order of 0.76 mm yr^{-1}. The coarse debris flux on valley sides is the dominant process with respect to mass moved over distance, but is internal to the slope system. Such debris is released only by exceptional catastrophic events or by valley glaciation. As mentioned earlier, movement of aeolian material is probably very important, which means calculations of erosion rates from lake sediments must be adjusted for this input. Lake sedimentation rates suggest surface-lowering rates of 4–10 μm yr^{-1} without adjustment (Caine, 1984). Such values are very similar to a surface lowering rate of about 10 μm yr^{-1} estimated for rock weathering. These results indicate that the Indian Peaks region is a relatively inactive alpine region. Zones of high activity are surrounded by an inactive matrix with poorly developed linkages within basins leaving interfluves, valley sides and the stream networks as independent entities with respect to coarse debris movement.

Uncertainty concerning these systems operates at two levels. There is uncertainty concerning the accuracy and appropriateness of the data. Have the right observations been made within the most efficient methodological framework? This uncertainty, which concerns the small spatial scales and short temporal scales studied, is shared with all other alpine regions where field experiments have been conducted. But this uncertainty is much less than for those areas where detailed observations are lacking. The other uncertainty level concerns how this information can be used to examine larger spatial systems over longer time-spans.

Clearly there cannot be a simple extrapolation or upgrading of the data to these, more general, systems. As Thorn and Loewenherz (1987) stress, this problem will not be rectified by merely accumulating more data from diverse regions. This will only create a more comprehensive but still contemporary picture. Also, another twenty or thirty years of the same type of records will not necessarily produce a significant improvement. The sample will still be too small and just as unrepresentative. Thus, 'extension to macroscale or regional perspectives will require not only additional data, but more importantly, a concerted effort to develop new methodological schemes' (Thorn and Loewenherz, 1987, p. 247).

The previous chapter has, indirectly, stressed the uncertainty just examined. Applied problems can be tackled satisfactorily only if the interactions between human and natural systems are assessed accurately and realistically. Unfortunately this is rarely the case and assumptions are too often made which later are shown to be false or for which there is little or no evidence. This uncertainty is partly technical, in that there is an absence of certainty, but it is also structural in the sense that some workers have succeeded in imposing their desired certainties on the issues in question (Ives and Messerli, 1989). The latter can be called unwarranted certainties. Nowhere is this more obvious than in attempts to examine what has become known as the Himalayan dilemma. The widespread temptation to extrapolate or generalise, as noted above, must be resisted. The most obvious generalisation is that the Himalayan region is so varied and so complex that generalisation is counterproductive (Ives and Messerli, 1989).

The complex series of interactions between human and natural systems in the Himalayas has led to much speculation concerning the probable course of events. One of the main problems concerns the apparent increase in landsliding, soil erosion and land degradation as a response to increased population pressure, deforestation and land-use changes. This chain of events poses a number of fundamental questions. However, the number of uncertainties that exist may require a totally different approach than is normally adopted. The Himalayan landscape is one of the most dynamic landscapes in the world in terms of natural processes. Even casual observations indicate numerous examples of active landslides and mudflows, rockfall events and erosion in general. The rivers also appear to be carrying great quantities of material. The vigour of erosion and the frequency of large erosional events produces extremely high rates of erosion. Superimposed on this high rate of natural activity is the effect of human activity in a variety of forms and with varying intensity. It is this interaction that causes most of the uncertainty in Himalayan landscapes. The fundamental questions concern the relationships between the naturally occurring processes and human impact. Thus, 'A complex series of interactions between man and nature in the Himalayan region has many experts from various disciplines speculating about the probable course of events in the area' (Thompson and Warburton, 1988, p. 1).

HIMALAYAN ENVIRONMENTAL DEGRADATION THEORY

The possible interlinking of the various components has been called by Ives and Messerli (1989) 'Himalayan Environmental Degradation Theory'. The Theory has eight components to it:

1. The introduction of modern health care, new medicines and malaria suppression since 1950 has led to an unprecedented population growth. In Nepal the growth rate for the 1971–81 decade was 2.6 per cent per annum.
2. This population explosion has been augmented by migration from the plains to the mountains in India and by illegal immigration from India into Nepal. This has led to rapidly increasing demands for fuelwood, construction timber, fodder and agricultural land.
3. This increasing population exerts increased pressure on the forest cover, leading to massive deforestation.
4. The deforestation, especially on the steeper and more marginal mountain slopes has led to a catastrophic increase in soil erosion and loss of productive land through landsliding. The normal hydrological cycle has also been disrupted.
5. This change in the nature of the slopes has led to increased runoff during the summer monsoon with increases in flooding and siltation in the plains of the Ganges and Brahmaputra rivers. Springs and wells dry up during the dry season, rivers change course, reservoirs become silted up and bridges buried. Agricultural land is lost on the lowland plains.
6. Increased sediment load of the major rivers from the Himalayas are extending the Ganges and Brahmaputra delta.
7. The continued loss of agricultural land in the mountains leads to greater pressure and more deforestation. Eventually the walking distances involved to collect fuelwood become so great that it is easier to use animal dung as fuel.
8. This leads to another vicious circle with soils, deprived of natural fertilizer, becoming less fertile. This lowers crop yield and the soil structure may be weakened, leading to further soil erosion and landslides.

This general sequence appears deceptively simple, with a number of self-reinforcing loops which lead to a worsening of the situation. But there are many levels of interdependent cause and effect relationships to consider in the context of extremely fluid and complicated ecological and social systems. Uncertainty is the key issue, and it has been stated that the Himalayas constitute a region where there is nothing but mess and uncertainty. Thompson and Warburton (1988) have summarised the problem in the following way. Analysis in terms of the physical facts has first to identify all the components of the Himalayan system and all the connections between those components. The result is a qualitative model made up of boxes representing the various components of the system. But before it is possible to tell whether the system is being sustained or transformed and the direction of the transformation, it is necessary to know the relative rates of all the processes. At the very least it is necessary to know the rates of operation of the key processes. This is where the problems start and uncertainties begin to creep in. All the components and the dynamic processes that link those components may not have been identified. Also, the key variables may not have been identified and sufficiently accurate measurements of the rate of operation of the processes may not be available.

The main problem concerns not only the linkages between the eight components of the Himalayan Environmental Degradation Theory discussed earlier but the accuracy and rationality of the components themselves. Thus a number of fundamental questions need to be asked. What is the nature and rate of deforestation? What is the per capita fuelwood consumption rate and the sustainable yield from forest production? How does deforestation affect the incidence of landslides? Is there a link between what is happening in the mountains and what is happening in the plains?

Ives and Messerli (1989) provide an exhaustive review and critical discussion of the deforestation problem, when it occurred, what the perceived pressures on the forests are, and the nature of their role as an environmental shield. They also examine mountain slope instability, both natural and caused by human intervention, and the Himalaya–lowland interactive system. Their conclusions are that in many places there are major problems concerning all these elements but that there is little evidence of widespread occurrence and linkage of all the elements in the Himalayan Environmental Degradation Theory.

One of the key problems concerns statistics of the rate of deforestation. Thus, it is not uncommon to find vast areas of open ground which support only shrubs, although they were formerly forested and are still counted as forests in official statistics because of unrecorded deforestation. However, it is distinctly possible that the shrub layer on the cleared land is a more effective barrier to erosion than the original forest cover. It is a question not just of forest statistics but of statistics couched in terms that are applicable to the land degradation problem. Also, 'deforestation' is too ambiguous a term because it can be used to describe fuelwood cutting, commercial logging, shifting cultivation, and forest clearance for continuous annual cropping or for grazing (Hamilton and Pearce, 1988). The subsequent land use is vitally important. It is necessary to know the precise land-use change before it is possible to estimate its effect on soil and hydrology. In a similar way, simple statistics of population density are not necessarily a good index of potential land degradation because the best-managed lands are often found where population densities are high.

The consequences of deforestation are location-specific. It is difficult to infer the losses, whether of species diversity, wood production potential, soil fertility or watershed protection, from an analysis of the overall rate of deforestation (Allen and Barnes, 1985). Thus conversion of forests to agriculture might not lead to deleterious effects if the lands are properly managed. It is possible for farming systems to combine food cropping with wood energy production and environmental protection (Dunkerley *et al.*, 1981; Eckholm, 1979; Ranganathan, 1979).

Perhaps the greatest uncertainty concerns estimates of per capita rate of fuelwood consumption. This is one of the most crucial variables for understanding the whole system and estimates have been used to demonstrate the environmental impact of an expanding population. Estimates of this nationwide rate in Nepal vary by a factor of 67 (Thompson and Warburton, 1988). Estimates of the sustainable yield of firewood production from Himalayan forests also vary considerably, thus the balance between the two estimates is going to be even more uncertain. But this balance is crucial to the long-term stability of the slopes.

Uncertainty also exists within the physical or natural systems. Thus Laban (1978), working in Nepal, has shown that soil erosion rates of 10–20 t ha^{-1} yr^{-1} are generally within the tolerance limits of the landscape. In other areas such a rate would lead to catastrophic changes. Laban (1979) has also shown that 50 per cent of landsliding in Nepal is due to natural causes and that the least affected areas are the most densely populated because terraces are well maintained. However he does stress the smaller landslides may have been missed. Starkel (1972), working in the Darjeeling region of India, argues that major catastrophies may occur on average four or five times a century with precipitation exceeding 400 mm in twenty-four hours. He estimated that 100–200 mm in twenty-four hours was usually sufficient to cause local failures. But Caine and Mool (1982) found no direct correlation between twenty-four-hour rainfall and landsliding in Nepal. Heavy rainfalls early in the season did not produce landslides. Landsliding was delayed until ground water recharge had brought the water table to within less than 1 metre of the surface. Piping in the soil was a major factor, but unfortunately the effect of piping is extremely difficult to predict. The general conclusions from a number of studies are that rates of land loss and landslide initiation are generally less than anticipated. Local farmers have developed an understanding of the hazard and have been able to mitigate some of the worst degradation effects. It has even been suggested that moving population from the hazard areas may increase the problem, because without some remedial measures, however primitive these may be, erosion will accelerate.

The problems of conceptualising and quantifying such complex interacting systems can be illustrated by examining a model developed by Rieger (1981). A number of assumptions are made in order to generate the model and predict future events. It is assumed that population growth is of the order of 2 per cent per annum. If the average size of households is five persons and the average size of agricultural holding is 0.5 ha, the amount of purely agricultural land required per capita is 0.1 ha. But, additional land is required for housing, grazing land, tracks, and so on, so that a land requirement of 0.2 ha per capita is deemed to be more realistic. The extraction of fire-wood, construction timber and leaves and twigs for cattle fodder, etc., is estimated at about 1400 kg per capita per annum. Natural forest density is about 360 tonnes per hectare of timber, twigs and leaves, with natural forest growth at about 5 per cent per decade.

On the basis of these assumptions a simple ecological model of the deforestation process was set up. However, the uncertainty problem is stressed when each of these estimates can be challenged. There is little certainty in any of the figures. If the model is run, because of the exponential growth of population, extraction from the forest increases so rapidly that before year 120 is reached, it exceeds natural regeneration. Within a few decades the remaining forests are depleted to the point of complete destruction.

As Rieger (1981) stresses, the model simulation should not be taken as a statement or prediction of what will happen in the future, but as an indication of what is happening at the present time. Also, the model is clearly simplistic in that it ignores many of the relationships discussed earlier. For instance, there is no indication of the loss of land by soil erosion and landsliding as deforestation progresses. The techniques of agriculture and forestry are regarded as constant in the model, although they will change with the changing environment. The purpose of the model is to 'indicate the dangers inherent in a static view of the ecological system and to show the way in which future research could be geared to the construction of a more realistic model' (Rieger, 1981, p. 368). Such a model could provide guidelines for policy-makers indicating the ecological carrying capacity of a given area.

It is important to return to the simple spiral of cause and effect in the Himalayas mentioned earlier and to examine a number of fundamental issues this raises. The spiral moves from population pressure through deforestation, soil erosion and landsliding, to increased runoff leading to high sediment loads in the rivers, siltation and flooding often in areas far removed from the source of the sediment. Once the sequence has been established it is self-reinforcing and the rates of all the processes increase. The questions that need to be addressed are a mixture of universal and site-specific problems. The most important would seem to be:

1. What is the nature of the population pressure in terms of numbers, social make-up, location, type of terrain, etc?
2. What is the rate of deforestation?
3. Where is the deforestation occurring and what types of forest are involved?
4. What is the forest replaced with and how does this replacement

compare with the original forest in terms of affecting soil and slope hydrology and land degradation?

5. Underlying the specific question of deforestation is the more general one as to how forests protect, if they do, the slopes from erosion and how the balance is affected following deforestation. It is apparent from the previous chapter and from studies from around the world that this question is far from straightforward.

6. What forms of soil erosion and landsliding are present?

7. Are soil erosion and landsliding restricted to certain slope types or slope positions?

8. What is the rate of landsliding, and is this rate increasing, decreasing or static?

9. Are the rivers carrying increased sediment loads?

10. If rivers are carrying greater loads, is this the result of human action or is it part of a naturally occurring cycle?

11. Are the rivers more liable to flood under current conditions?

These questions are only a few of those that need to be asked. Finding answers to each of them is extremely difficult and establishing links between them is even more difficult. Since only that part of a process that is caused by human agency can be altered by altering human behaviour, the whole question of how much of the damage inflicted by these processes is due to human activity and how much is due to events which would occur whether man were present or not has become increasingly important to policy-makers. Answers to some of these questions will be attempted with reference to the Darjeeling region of northeast India.

LAND DEGRADATION IN THE DARJEELING AREA

Darjeeling is the northernmost and smallest district of West Bengal. Its total area, which includes a substantial portion of lowland plains beyond the foothills of the Himalayas, is 3106 km^2 with a total population in 1981 of 1 006 434. The three hill sub-divisions of Darjeeling, Kurseong and Kalimpong comprise an area of 2157 km^2 with a 1981 population of approximately 400 000.

The geological framework of the region varies from unaltered sedimentary rocks in the south to various grades of metamorphic rock and some intrusive rocks in the middle and northern zones. The mountains are composed of rocks folded and thrust over one another by a series of north–south compressive movements as well as tangential thrusts. The complex multiple folding and faulting have sheared, folded, crumpled, fractured and jointed the rocks, thus weakening them and making them more susceptible to water percolation and the build-up of potentially high pore pressures. Four main geological zones can be identified: the

Terai and the plains at the foot of the Himalayas; the frontal range of Siwalik rocks; an intermediate belt of the Damuda Series; and the main zone of rocks belonging to the Daling and Darjeeling Series.

The Terai is composed of horizontal beds of unconsolidated sand, silt, gravel and cobbles, the result of intense erosion of the mountains during the Quaternary Period. The Tertiary Siwalik beds are similar but more compacted and cemented and have been affected by the mountain building episodes. They are mainly arkosic and micaceous sandstones and siltstones interbedded with unconsolidated conglomerates. These rocks dip generally northwards at angles of 0–50° under the overthrusts of the Damuda Series. Rocks of the Damuda Series are generally coarse sandstones and quartzites with a number of impersistent coal seams. North of the Damuda rocks the hills are composed of low-grade metamorphic sediments represented by quartzites, slates, schists, phyllites and gneiss. The Daling Group (phyllites and schists) covers the Kalimpong hills east of the River Tista while the Darjeeling gneiss dominates the western part of the region.

The rivers are characterised by high runoff fed either from glaciers or by monsoon rainfall. The Tista is the most important river rising from a glacier in north Sikkim at 6300 m and joined by the Rangpu and Great Rangit before leaving the hills at Sevok and eventually joining with the Brahmaputra in Bangladesh. The discharge of the Tista varies from 10 000 to 100 000 cusecs and there have been major floods in 1950 and 1968. The total runoff of the rivers draining the Darjeeling District is over 37×10^9 m^3, of which the Tista contributes 24.66×10^9 m^3. Rainfall amounts and timing are governed by the monsoon systems, with most of the rain falling in the period June to September. Darjeeling receives about 3000 mm of rainfall, Kalimpong 2000 m and Kurseong 4000 mm. As noted in Chapter 2, there is a distinct altitudinal zonation of vegetation. Deciduous forests of sal, champ and simul dominate the areas between 300 m and 1000 m, forests of alder, walnut and poplar occur between 1000 m and 2000 m, with coniferous forest at higher levels.

Earthquake-induced landslides are an additional complicating factor when trying to assess land degradation. The Darjeeling region suffers from persistent tectonic activity associated with large active faults such as the Main Boundary Fault and the Himalayan Front Fault, as well as a number of minor faults extending parallel to the main fault zones. Continuing tectonic activity is visible in the landscape in features such as convex-up terrace profiles, many generations of alluvial fans at different altitudes and eroded fault scarps. The drainage systems are essentially antecedent, extending north–south across the main orogenic trends, and the Tista, especially, has created several narrow gorge sections at sites of major uplift.

A number of severe earthquakes have occurred over the last 100 years, such as the Bengal earthquakes (14 July 1885), Assam earthquakes (12

Table 9.1. *Recorded major landslides of Darjeeling area.*

Year	Details	Rainfall (mm)	Effect on landscape	Casualties
1899	24-25 September	610	Innumerable landslips	Heavy
1934	15 January	–	Associated with Bihar earthquake	–
1950	10-14 June	842	Widespread landslides, communication badly affected	Total disruption of civil life
1968	3-5 October	1266	Communication links badly affected	Heavy toll of human life
1980	27 August 3-4 September	– –	2966 acres of cultivated land and 30 acres of plantations damaged	–
1984	1-5 September	450-700	150 metre stretch of main road washed off	

Source: Chattopadhyay (1987).

June 1897, 8 July 1918, 10 July 1969), north Bihar earthquake (15 January 1934) and the Tibet/Assam earthquake (15 August 1950). Earthquakes such as these can lead to severe flooding and landslide damage. The 1950 Assam earthquake almost completely denuded the Dibang valley and tributaries of forest cover (Kingdom-Ward, 1955). Many of the tributaries were blocked and many of the hillsides reduced to bare rock. The severity of the destruction depended less on the type of forest than on angle of slope and nature of the rock The luxuriant subtropical forest was no better than the pine forest in resisting the earthquake and many scars were still pouring material six weeks after the earthquake. Every uprooted tree left behind a hole full of loose earth and stones which often produced a gully head. This example illustrates the devastation and instability created by a single natural event; an instability which was to remain for many years totally unrelated to human activity.

One of the most important issues is to examine the nature and incidence of landsliding and to try to assess the factors involved. It might then be possible to decide whether landsliding is an increasing hazard as a result of human activity. The incidence of major landslide events in the Darjeeling District since 1899 has been admirably summarised by Anis Chattopadhyay (1983) and Guru Prasad Chattopadhyay (1987). The timing and nature of the events are shown in Table 9.1. Although landslides prior to 1899 are not recorded in detail it is clear from travellers' accounts that landslide scars have always been prominent features. Indeed Joseph Hooker mentioned such scars in his memoirs for 1854. It must be stressed that the landslide events listed in Table 9.1 have been the most devastating; landslides occur each year but most go unrecorded. This is reflected in the Geological Survey of India's investigation between 1900 and 1914 of frequent landslips at the Limbu Jhora and Happy Valley areas of Darjeeling.

All except the landslips associated with the 1934 Bihar earthquake were caused by rainfall of above average intensities. The landslips of 24 and 25 September 1899 were caused by 60.96 cm (24 in.) of rain, with the township of Darjeeling being most affected. It is clear that excessive rainfall was the major cause but in combination with defective drainage, badly constructed revetments and the undercutting of steep slopes for the formation of paths, roads and houses. Accounts of the landslide damage inflicted by the 1934 Bihar earthquake are conflicting but there seems little doubt that many areas of Darjeeling town were badly affected.

Apart from the effect of the earthquake, there is little indication of major landsliding between 1899 and 1950. This reintroduces the uncertainty issue raised earlier. Either there were no major events during this period or the landsliding occurred in remote rural areas and escaped recording. Rainfall induced landsliding was again excessive in June 1950, when 84.20 cm (33.15 in.) of rain fell in Darjeeling between 10 and 14 June, with heavy intensities of 45.42 cm (17.88 in.) and 25.81 cm (10.16 in.) on 12 and 13 June, respectively. Areas worst affected in 1899 were also badly affected in 1950 as well as a number of new areas. These landslips were investigated by Ghosh (1950) and Dutta (1951).

Similar events occurred in 1968 as a result of heavy rainfall between 3 and 5 October 1968. The Hill Cart Road between Kurseong and Darjeeling was blocked at eighteen different points. The landslides of August and September 1980 were even more devastating, following heavy and continuous rain on 27 August and 3–4 September. In all, an area of 160 km², comprising Darjeeling town, Darjeeling-Pulbazar Block II, Gorubathan Block, Jorebunglow-Sukhiapokhri Block, Kalimpong Block I, and 84 825 people were affected by landslides (Chattopadhyay, 1983). Landslides following heavy rainfall between 1 and 5 September 1984 removed a 150 m stretch of the main Darjeeling road.

Considerable insight into the nature of the landsliding in 1968 is provided by Starkel (1972). This study is especially important because it provides an analysis of landslides and slopes away from settlements and lines of communication. Landslide types are of various characters but mudflows or debris flows were the most common. Starkel differentiates two types. The first, most frequent in tea gardens, started in small depressions on slopes of 25–40° at a considerable distance from the watersheds. Subsurface water had clearly collected in these depressions and then burst through the surface, initiating a small 1–2 m scar and a long, narrow debris track. These are identical to those described by Temple and Rapp (1972) in the Uluguru Mountains of Tanzania created by torrential rain. Deforestation and land-use changes were also implicated in those events. The flows described by Starkel (1972) often

Figure 9.1. Debris torrent cut through dense forest in the Darjeeling Himalaya.
(Photo: author.)

Figure 9.2. Damage to Siliguri–Darjeeling road in 1988 caused by severe land-sliding. (Photo: author.)

extended for up to 1000 m, some reaching the valley bottoms before coming to rest on river terraces. Some of this debris was eventually washed into the rivers. A second type of mudflow, involving coarser material, was initiated by slumps on steep slopes usually on structural breaks, edges of destructional levels or terraces or on undercuts of roads. Many of these movements commenced at the crest of ridges.

Flows also occurred in pre-existing gullies and were directed very quickly downslope. Considerable velocities were achieved which were capable of transporting quite large boulders and considerable damage was done where gullies crossed roads. In forest areas only water carrying fine gravels flowed in gullies formed earlier, while the few flows reaching wooded areas were impeded by the vegetation (Figure 9.1). Roads were most affected, with over 200 slips occurring on the Siliguri to Darjeeling road (Figure 9.2).

Large deep-seated rock and debris slides occurred in some locations. These were due either to deep infiltration of water or undercutting by flooded rivers. Many of the rock slides were 0.5–1 km long with an area exceeding 20 ha. Undercutting by the Tista and Little Rangit rivers created large earth and rock slides on slopes overlain with thick covers of decomposed rocks. In the Tista valley the shales of the Darjeeling Series were especially vulnerable to movement. Jain (1966) has also noted that landslides east of the Tista were twice as common on Daling Series as on other rock types.

The role of vegetation and land use was also examined by Starkel (1972). Tea gardens were badly affected, with about 20 per cent on average of the area of each garden destroyed. Terraced slopes were even more affected with damage of some 30–50 per cent. Destruction in the forests was one-twentieth to one-tenth of that on the tea slopes and tended to be concentrated along the roads. It appeared that forests inhibited the shallower processes but had little effect on deep landslides (Figure 9.3). Similar results have been reported for the Mussoorie–Tehri road section in the front range of the Lesser Himalaya of Uttar Pradesh by Haigh (1984). In 1978, 148 slides had their origins on slopes with a tree cover of 40 per cent or less and 118 had their origins on slopes which retained 60 per cent or more of their tree cover. The average size of slides on deforested slopes was greater (26 m^3 as against 12 m^3). Slides generated from slopes which retain 60 per cent or more of their tree cover on average have longer, wider, deeper and steeper outfalls.

The general conclusion from this brief survey of the Darjeeling area is that it is difficult to establish detailed links between the various processes and factors involved. There is little doubt that landsliding and soil erosion are major problems. Landsliding occurs on two temporal scales: there is an annual problem represented by slope failure induced by normal monsoonal rainfall and a more catastrophic, spasmodic problem created by exceptional rainfall that seems to occur five or six times a century. The annual problem is concentrated in areas most affected by

Figure 9.3. Debris slide on a forested slope in the Ballason Valley, Darjeeling. (Photo: author.)

Figure 9.4. Protection of road from landslides induced by undercutting by the River Tista, Darjeeling Himalayas. (Photo: author.)

human activity such as along highways and in villages and towns where houses are casually constructed and located (Figure 9.4). Thus landslides have been called 'the nightmare of the hill roads' (Bansal and Mathur, 1976). Landslides also occur on slopes altered by cultivation practices, but not to the extent first supposed. The catastrophic events are also concentrated in these areas but major failures occur on other zones, even in forested areas.

Detailed links between land-use pressure, population increase, deforestation and landsliding are difficult to establish and it is even more difficult to establish links between slope failure and increased sediment load in the major rivers. The majority of the slope failures come to rest on the hillslopes and material very rarely reaches the rivers. Thus, if there is increased sediment load in the major rivers it is related either to events that occurred some time ago, and material is only now reaching the rivers or to active undercutting of the rivers themselves — a natural process. The natural time-lag in Himalayan slope systems is quite substantial, which poses considerable management problems and stresses the uncertainty problem which has been a major theme of this chapter.

CONCLUSIONS

This book has attempted to summarise the current state of knowledge concerning the physical geography of mountains and to emphasise the progress that has been made since the early 1970s. But, as Ives (1987) has stressed and this last chapter has shown, the extreme variability in space and time of mountain processes remain as severe problems. This book has also demonstrated that research has tended to concentrate substantially on a small number of high-altitude mountain areas in mid- to high latitudes. More study is required of tropical, vegetation-covered mountain systems. The last two chapters have also attempted to show that knowledge of the interaction between human activity and physical systems is fundamental to an understanding of landscape change in mountains over the next 50–100 years. Traditional divisions between physical and human geography need to be breached to provide a fuller understanding of mountain environments. Ives (1987, p. 248) has stressed the dilemma in the following way:

With the contemporary pressures and resource-use conflicts facing the mountain lands, much high-mountain research . . . begins to appear effete, the perpetuation of an elitism that is a hold-over from the gentlemen adventurers of a previous era who were privately funded.

The challenge is to overcome this reputation and to place mountain physical geography on a firmer basis. It is hoped that this book has gone some way to achieving this.

BIBLIOGRAPHY

Addison, K., 1981, The contribution of discontinuous rock-mass failure to glacier erosion, *Annals of Glaciology*, **2**: 3–10.

Ahlmann, H.W., 1935, Contribution to the physics of glaciers, *Geographical Journal*, **86**: 97–113.

Ahlmann, H.W., 1948, *Glaciological research on the North Atlantic coasts*, Royal Geographical Society Research Series, 1, 83 pp.

Ahnert, F., 1970, Functional relationships between denudation, relief and uplift in large mid-latitude drainage basins, *American Journal of Science*, *268*: 243–63.

Ahuja, P.R. and Rao, P.S., 1958, Hyrological aspects of floods, *Proceedings of the Symposium on Meteorological and Hydrological Aspects of Floods and Droughts in India*, New Dehli.

Albjar, G., Rehn, J. and Stromquist, L., 1979, Notes on talus formation in different climates, *Geografiska Annaler*, **61A**: 179–185.

Alger, C.S. and Ellen, S.D., 1987, Zero-order basins shaped by debris flows, Sunol, California, USA. In *Erosion and Sedimentation in the Pacific Rim*, International Association of Hydrological Sciences Publication, **165**: 111–19.

Allen, J.C. and Barnes, D.D., 1985, The causes of deforestation in developing countries, *Annals of the Association of American Geographers*, **75** (2): 163–84.

Anderson, C.A., 1933, The Tuscan formation of Northern California, *University of California Bulletin of the Department of Geological Science*, **25**: 347–422.

Anderson, F.P., 1980, River-control by wire net-work, *Selected paper no. 3748, Mining Proceedings of the Institution of Civil Engineers*, CLXXIII, 244–58.

Anderson, R.S., Hallett, B., Walker, J. and Aubry, B.F., 1982, Observations in a cavity beneath Grinnel Glacier, *Earth Surface Processes and Landforms*, **7**: 63–70.

Andrews, E.D., 1983, Entrainment of gravel from naturally sorted riverbed material, *Bulletin of the Geological Society of America*, **94**: 1224–31.

Andrews, J.T., 1961, The development of screes in the English Lake District and central Quebec-Labrador, *Cahiers Géographie de Québec*, **10**: 210–30.

Andrews, J.T., 1965, The corries of the northern Nain-Okak section of Labrador, *Geographical Bulletin*, **7**: 129–36.

Andrews, J.T., 1972, Glacier power, mass balances, velocities and erosion potential, *Zeitschrift für Geomorphologie*, **13**: 1–17.

Andrews, J.T. and Dugdale, R.E., 1971, Quaternary history of northern Cumberland Peninsula, Baffin Island, NWT. Part V. Factors affecting corrie glacierization in Okoa Bay, *Quaternary Research*, **1**: 532–51.

Andrle, R. and Abrahams, A.D., 1989, Fractal techniques and the surface roughness of talus slopes, *Earth Surface Processes and Landforms*, **14**: 197–209.

Anon, 1966, Mass balance terms, *Journal of Glaciology*, **8** (52): 3–7.

Anon, 1978, *Atlas on Seismicity and Volcanism*, Swiss Reinsurance Co., Zurich.

Aniya, M. and Naruse, R., 1985, Structure and morphology of Solar Glacier. In Nakajima, C. (ed.), *Glaciological Studies in Patagonia Northern Icefield 1983–84*, Data Center for Glacier Research, Japanese Society of Snow and Ice, 70–9.

Aniya, M. and Welch, R., 1981, Morphological analyses of glacial valleys and estimates of sediment thickness on the valley floor: Victoria Valley system, Antarctica, *The Antarctic Record*, **71**: 76–95.

Antevs, E., 1932, *Alpine zone of Mt. Washington Range, Auburn, Maine*, Merrill and Webber, Boston.

Aramaki, S., 1963, Geology of Asama volcano, *Journal of the Faculty of Science, University of Tokyo*, Section II, **14**: 229–443.

Aramaki, S. and Yamasaki, M., 1963, Pyroclastic flows in Japan, *Bulletin Volcanologique*, **26**: 89–99.

Argand, E., 1922, La Tectonique de l'Asie, *Comptes Rendus, XIIIe Congrès International Géologique*, **1**: 171–372.

Ashley, G.H., 1935, Studies in Appalachian mountain sculpture, *Bulletin of the Geological Society of America*, **46**: 1395–1436.

Aulitzky, H., 1988, Sommerhochwasser 1987 in Tirol – Naturkatastrophen oder fehlende Vorbeugung, *Österreichische Wasserwirtschaft*, **40**: 5/6, 122–8.

Avsiuk, G.A., 1955, Temperaturnoe Sostoianie Lednikov, *Iszestiya Akademii Nauk SSR, Seriya Geograficheskaya*, **1**: 14–31.

Bagchi, A.K., 1982, Orographic variation of precipitation in a high-rise Himalayan basin. In Glen, J.W. (ed.), *Hydrological Aspects of Alpine and High Mountain Areas*, International Association of Hydrological Sciences Publication, **138**: 3–9.

Bakker, J.P. and Le Heux, J.W.N., 1946, Projective-geometric treatment of O. Lehmann's theory of the transformation of steep mountain slopes, *Koninklijke Nederlandsche Akademie van Wetenschappen Series B*, **49**: 533–47.

Bakker, J.P. and Le Heux, J.W.N., 1947, Theory of central rectilinear recession of slopes, *Koninklijke Nederlandsche Akademie van Wetenschappen Series B*, **50**: 959–66, 1154–62.

Bakker, J.P. and Le Heux, J.W.N., 1950, Theory on central rectilinear recession of slopes, *Koninklijke Nederlandsche Akademie van Wetenschappen Series B*, **53**: 1073–84, 1364–74.

Bakker, J.P. and Le Heux, J.W.N., 1952, A remarkable new geomorphological law, *Koninklijke Nederlandsche Akademie van Wetenschappen, Series B*, **55**: 399–410, 554–71.

Bansal, R.C. and Mathur, H.N., 1976, Landslides, the nightmares of the hill roads, *Soil Conservation Digest*, **4** (1): 36–7.

Barry, R.G., 1973, A climatological transect on the east slope of the Front Range, Colorado, *Arctic and Alpine Research*, **5**: 89–110.

Barry, R.G., 1981, *Mountain Weather and Climate*, Methuen, London.

Barsch, D., 1969, Studien und Messungen an Blockgletschern in Macun, Unterengadin, *Zeitschrift für Geomorphologie*, Supplementband, **8**: 11–30.

Barsch, D., 1988, Rockglaciers. In Clark, M.J. (ed.), *Advances in Periglacial Geomorphology*, J. Wiley, Chichester, 69–90.

Barsch, D. and Caine, N., 1984, The nature of mountain geomorphology, *Mountain Research and Development*, 4: 287–98.

Barsch, D. and Treter, U., 1976, Zur Verbreitung von Periglazial-Phänomenen in Rondane/Norwegen, *Geografiska Annaler*, 58A: 83–93.

Bathurst, J.C., 1978, Flow resistance of large-scale roughness, *American Society of Civil Engineers, Journal of the Hydraulics Division*, 104 (12): 1587–1603.

Bathurst, J.C., 1987a, Measuring and modelling bedload transport in channels with coarse bed materials. In Richards, K., (ed.), *River Channels: Environment and Process*, Institute of British Geographers Special Publication, 18: 272–94.

Bathurst, J.C., 1987b, Critical conditions for bed material movement in steep, boulder-bed streams. In *Erosion and Sedimentation in the Pacific Rim*, International Association of Hydrological Sciences Publication, 165: 309–18.

Bathurst, J.C., Leekes, G.J.L. and Newson, M.D., 1986, Field measurements for hydraulic and geomorphological studies of sediment transport — the special problems of mountain streams. In *Measuring Techniques in Hydraulic Research*, Proceedings of the International Association of Hydrology Research, Delft Symposium, Balkema, Rotterdam, 137–51.

Battey, M.H., 1960, Geological factors in the development of Veslgjuv-botn and Veslskautbreen. In Lewis, W.V. (ed.), Norwegian cirque glaciers, *Royal Geographical Society Research Series*, 4: 5–10.

Battle, W.R.B., 1960, Temperature observations in bergschrunds and their relationship to frost shattering. In Lewis, W.V. (ed.), Norwegian cirque glaciers, *Royal Geographical Society Research Series*, 4: 83–95.

Bauer, A., 1955, The balance of the Greenland ice-sheet, *Journal of Glaciology*, 2: 456–62.

Beecroft, I., 1983, Sediment transport during an outburst from Glacier de Tsidjiore Nouve, Switzerland, 16–19 June 1971, *Journal of Glaciology*, 29: 185–90.

Beltaos, S., 1982, Dispersion in tumbling flow, *American Society of Civil Engineers, Journal of Hydraulics Division*, 108 (4): 591–612.

Benda, L. and Dunne, T., 1987, Sediment routing by debris flow In *Erosion and Sedimentation in the Pacific Rim*, International Association of Hydrological Sciences Publication, 165, 213–23.

Benedict, J.B., 1970, Downslope soil movement in a Colorado alpine region: rates of processes and climatic significance, *Arctic and Alpine Research*, 2: 165–226.

Billings, W.D., 1969, Vegetational pattern near alpine timberline as affected by fire-snow drift interactions, *Vegatio Acta Geobotanica*, 19 (1–6): 192–207.

Bird, J.B., 1980, *The Natural Landscapes of Canada*, 2nd edn, Wiley, Toronto.

Birkeland, P.W., 1967, Correlation of soils of stratigraphic importance in western Nevada and California and their relative rates of profile development. In R.R. Morrison and H.E. Wright (eds) *Quaternary Soils*, INQUA Congress, Desert Research Institute, University of Nevada, Reno, vol. 9, VII, 71–91.

Bjerrum, L. and Jorstad, F., 1963a, Correspondence on: stability of steep slopes on hard unweathered rock by K. Terzaghi, *Géotechnique*, 13: 171–3.

Bjerrum, L. and Jorstad, F., 1963b, Correspondence on: an approach to rock mechanics by K.W. John, *American Society of Civil Engineers, Journal of Soil Mechanics and Foundation Engineering*, 89, SM1: 300–2.

Bjerrum, L. and Jorstad, F., 1968, Stability of rock slopes in Norway, *Norwegian Geotechnical Institute Publication* 79: 1–11.

Björnsson, H., 1974, Explanation of jökulhlaups from Grímsvötn, Vatnajökull,

Iceland, *Jokull*, **24**: 1–24.

Björnsson, H., 1977, The course of jökulhlaups in the Skafta River, Vatnajökull, *Jokull*, **27**: 71–7.

Blake, D.H. and Loffler, E., 1971, Volcanic and glacial landforms on Mount Giluwe, Territory of Papua New Guinea, *Bulletin of the Geological Society of America*, **82**: 1605–14.

Blong, R.J., 1966, Discontinuous gullies on the volcanic plateau, *Journal of Hydrology (NZ)*, **5**: 87–99.

Boesch, H., 1961, Beiträge zur Kenntnis der Blockströme, *Die Alpen*, **27**: 1–5.

Bonell, M. and Gilmour, D.A., 1978, The development of overland flow in a tropical rainforest catchment, *Journal of Hydrology*, **39**: 365–82.

Bones, J.G., 1973, Process and sediment size arrangement on high arctic talus, south west Devon Island, NWT, Canada, *Arctic and Alpine Research*, **5**: 29–40.

Bonnefond, J., 1977, Le Néotectonique et sa traduction dans les paysages géomorphologiques de l'île de Crète, *Revue de Géographie Physique et de Géologie Dynamique*, **19**: 93–108.

Borisova, V.N. and Borisov, O.G., 1962, Observations on the crater of the Bezymianny volcano in the summer of 1960, *USSR Academy of Sciences, Bulletin of Volcanological Station*, **32**: 13–19.

Borland, W.M., 1961, Sediment transport on glacier-fed streams in Alaska, *Journal of Geophysical Research*, **66**: 3347–50.

Boulton, G.S., 1970, On the origin and transport of englacial debris in Svalbard glacier, *Journal of Glaciology*, **9** (56): 213–29.

Boulton, G.S., 1974, Processes and patterns of glacial erosion. In Coates, D.R. (eds) *Glacial Geology*, State University of New York, Binghampton, 41–87.

Boulton, G.S., 1978, Boulder shapes and grain-size distribution of debris as indicators of transport paths through a glacier and till genesis, *Sedimentology*, **25**: 773–99.

Boulton, G.S., 1979, Processes of glacier erosion on different substrata, *Journal of Glaciology*, **23**: 15–38.

Boulton, G.S., Morris, E.M., Armstrong, A.A. and Thomas, A., 1979, Direct measurements of stress at the base of a glacier, *Journal of Glaciology*, **22**: 3–24.

Bousquet, B. and Pechoux, P.Y., 1977, La Sismicité du Bassin égéen pendant l'Antiquité, *Bulletin Société Géologique de France*, **19**: 679–84.

Bousquet, B. and Pechoux, P.Y., 1980, Seismotectonique et conjecture géographiques dans l'évolution recente des paysages de la rive nord de la Mediterranée orientale, *Bulletin de l'Association de Géographie France*, **466–7**: 13–19.

Bovis, M.J., 1977, Statistical forecasting of snow avalanches, San Juan Mountains, southern Colorado, *Journal of Glaciology*, **18**: 87–100.

Bovis, M.J., 1978, Soil loss in the Colorado Front Range, *Zeitschrift für Geomorphologie* Supplementband 29: 10–21.

Bovis, M.J., 1982, Spatial variation of soil loss controls. In Thorn, C.E. (ed.) *Space and Time in Geomorphology*, George Allen & Unwin, London, 1–24.

Bovis, M.J. and Mears, A.I., 1976, Statistical prediction of snow avalanche runout from terrain variables in Colorado, *Arctic and Alpine Research*, **8**: 115–20.

Bovis, M.J. and Thorn, C.E., 1981, Soil loss variation within a Colorado alpine area, *Earth Surface Processes and Landforms*, **6**: 151–63.

Brayshaw, A.C., Frostick, L.F. and Reid, I., 1983, The hydrodynamics of particle clusters and sediment entrainment on coarse alluvial channels, *Sedimentology*, **30**: 137–43.

Bretz, J.H., 1935, Physiographic studies in east Greenland. In Boyd, L.A. (ed.), *The Fiord Region of East Greenland*, American Geographic Society, Special Publication **18**: 161–266.

Broscoe, A.J. and Thompson, S., 1969, Observations of an alpine mudflow, Steele Creek, Yukon, *Canadian Journal of Earth Sciences*, **6**: 219–29.

Brown, M. and Powell, J.M., 1974, Frost and drought in the Highlands of Papua New Guinea, *Journal of Tropical Geography*, **38**: 1–6.

Brown, R.J.E., 1969, Factors influencing discontinuous permafrost in Canada. In Pewe, T.L. (ed.) *The Periglacial Environment: Past and Present*, McGill-Queen's University Press, Montreal, 11–53.

Brown, R.J.E., Johnston, G.H., Mackay, J.R., Morgenstern, N.R. and Shilts, W.W., 1981, Permafrost distribution and terrain characteristics. In Johnston, G.H. (ed.), *Permafrost: Engineering Design and Construction*, Wiley, Toronto, 31–72.

Brown, W.H., 1925, A possible fossil glacier, *Journal of Geology*, **33**: 464–6.

Brugger, E.A., Furrer, G., Messerli, B. and Messerli, I.P. (eds), 1984, *The Transformation of Swiss Mountain Regions*, Verlag Paul Haupt, Berne and Stuttgart.

Brugman, M.M. and Meier, M.F., 1981, Response of glaciers to the eruptions of Mount St. Helens. In Lipman, P.W. and Mullineaux, D.R. (eds), *The 1980 Eruption of Mount St. Helens*, United States Geological Survey Professional Paper **1250**: 743–56.

Brundall, J.A., 1966, Recent debris flows and related gullies in the Cass basin. Unpublished MA thesis, University of Canterbury, New Zealand.

Brunsden, D. and Allison, R.J., 1986, Mountains and highlands. In Fookes, P.G. and Vaughan, P.R. (eds), *A Handbook of Engineering Geomorphology*, Surrey University Press, Guildford, 150–65.

Brunsden, D., Doornkamp, J.C., Fookes, P.G., Jones, D.K.C. and Kelly, J.M.H., 1975a, Geomorphological mapping techniques in highway engineering, *Journal of the Institute of Highway Engineering*, **22** (12); 35–41.

Brunsden, D., Doornkamp, J.C., Fookes, P.G., Jones, D.K.C. and Kelly, J.M.H., 1975b, Large scale geomorphological mapping and highway engineering design, *Quarterly Journal of Engineering Geology*, **8**: 227–53.

Brunsden, D., Doornkamp, J.C., Fookes, P.G., Jones, D.K.C. and Kelly, J.M.H., 1975c, Geomorphological mapping and highway design, *Proceedings 6th Regional Conference for Africa Soil Mechanics and Foundation Engineering*, Durban, 3–9.

Brunsden, D. and Jones, D.K.C., 1984. The geomorphology of high magnitude – low frequency events in the Karakoram Mountains. In Miller, K.J. (ed.), *The International Karakoram Project*, Vol. I, Cambridge University Press, Cambridge, 383–8.

Brunsden, D., Jones, D.K.C. and Goudie, A.S., 1984, Particle size distribution on the debris slopes of the Hunza valley. In Miller, K.J. (ed.), *The International Karakoram Project*, vol. II, Cambridge University Press, Cambridge, 536–79.

Brunsden, D., Jones, D.K.C., Martin, R.P. and Doornkamp, J.C., 1981, The geomorphological character of part of the Low Himalaya of eastern Nepal, *Zeitschrift für Geomorphologie*, Supplementband, **37**: 25–72.

Brunsden, D. and Thornes, J.B., 1979, Landscape sensitivity and change, *Transactions of the Institute of British Geographers*, new series, **4** (4): 463-84.

Bull, C. and Marangunic, C., 1968, Glaciological effects of debris slide on Sherman Glacier. In *The Great Alaska Earthquake of 1964*, Hydrology volume, National Academy of Sciences Publication 1603, Washington, DC, 309-17.

Bullard, F., 1962, Volcanoes of Southern Peru, *Bulletin Volcanologique*, **24**: 443-53.

Bullard, F., 1984, *Volcanoes*, 2nd edn, University of Texas Press, Austin.

Burns, S.F., 1980, Alpine soil distribution and development, Indian Peaks, Colorado Front Range. Unpublished PhD dissertation, University of Colorado, Boulder.

Burns, S.F. and Tonkin, P.J., 1982, Soil-geomorphic models and the spatial distribution and development of alpine soils. In Thorn, C.E. (ed.), *Space and time in Geomorphology*, Binghampton Symposium in Geomorphology 12, George Allen & Unwin, London, 25-43.

Butler, B.E., 1959, *Periodic Phenomena in Landscapes as a basis for Soil Studies*, CSIRO, Australian Soil Publication 14.

Caine, N., 1969, A model for alpine talus slope development by slush avalanching, *Journal of Geology*, **77**: 92-101.

Caine, N., 1974, The geomorphic processes of the alpine environment. In Ives, J.D. and Barry, R.G. (eds), *Arctic and Alpine Environments*, Methuen, London, 721-48.

Caine, N., 1976. The influence of snow and increased snowfall on contemporary geomorphic processes in alpine areas. In Steinhoff, H.W. and Ives, J.D. (eds), *Ecological Impacts of Snowpack Augmentation in the San Juan Mountains, Colorado*, Colorado State University, Fort Collins, 145-200.

Caine, N., 1982, Water and sediment fluxes in the Green lakes Valley, Colorado Front Range. In Halfpenny, J.C. (ed.) *Ecological Studies in the Colorado Alpine: A Festschrift for John W. Marr*, Institute of Arctic and Alpine Research, University of Colorado, Boulder, Occasional Paper 37, 1-12.

Caine, N., 1983, *The Mountains of Northeastern Tasmania: A Study of Alpine Geomorphology*, Balkema: Rotterdam.

Caine, N., 1984, Elevational contrasts in contemporary geomorphic activity in the Colorado Front Range, *Studia Geomorphologica Carpatho-Balcanica*, **18**: 5-31.

Caine, N., 1986, Sediment movement and storage on alpine slopes in the Colorado Rocky Mountains. In Abrahams, A.D. (ed.), *Hillslope Processes*, George Allen & Unwin, London, 115-37.

Caine, N. and Mool, P.K., 1981, Channel geometry and flow estimates for two small mountain streams in the Middle Hills, Nepal, *Mountain Research and Development*, **1**: 231-43.

Caine, N. and Mool, P.K., 1982, Landslides in the Kolpu Khola Drainage, Middle Mountains, Nepal, *Mountain Research and Development*, **2** (2): 157-73.

Carniel, P. and Scheidegger, A.E., 1974, Morphometry of an alpine scree cone, *Rivista Italiana di Geofisica*, **23**: 95-100.

Carson, M.A. and Griffiths, G.A., 1985, Tractive stress and the onset of bed particle movement in gravel stream channels: different equations for different purposes, *Journal of Hydrology*, **79**: 375-88.

Carson, M.A. and Kirkby, M.J., 1972, *Hillslope Form and Process*, Cambridge University Press, London.

Catalano, L.R., 1972, Datos hidrológicos del desierto de Atacama, *Decion Minas y Geológica Publication*, **35**: 10-16.

Cattermole, P., 1982, Meru — a rift valley giant, *Volcano News*, **11**: 1-3.

Chamberlin, T.C. and Chamberlin, R.T., 1911, Certain phases of glacial erosion, *Journal of Geology*, **19**: 193-216.

Chandler, R.J., 1973, The inclination of talus, arctic talus, terraces and other slopes composed of granular material. *Journal of Geology*, **81**: 1–14.

Chattopadhyay, A., 1983, A survey on the occurrence of landslides and their impact on the economy of Darjeeling District. In Sarkar, R.L. and Lama, M.P. (eds), *The Eastern Himalayas: Environment and Economy*, Atma Ram and Sons, New Delhi, 74–87.

Chattopadhyay, G.P., 1981, Landslide phenomena in the Darjeeling Himalaya: some observations and analysis. In Datye, V.S., Diddee, J., Jog, S.R. and Patil, C. (eds), *Explorations in the Tropics*, University of Poona, Pune, 198–205.

Cheng, G. and Wang, S., 1982, On the zonation of high-altitude permafrost in China, *Journal of Glaciology an Cryopedology*, **4**: 1–17.

Chinn, T.J.H., 1981, Use of rock weathering-rind thickness for Holocene absolute age dating in New Zealand, *Arctic and Alpine Research*, **13**: 33–45.

Church, M. and Miles, R.J., 1987, Meteorological antecedents to debris flow in southwestern British Columbia: Some case studies, *Geological Society of America, Reviews in Engineering Geology*, VII, Boulder, Co., 63–79.

Church, M., Stock, R.F. and Ryder, J.M., 1979, Contemporary sedimentary Environments on Baffin island NWT, Canada: debris slope accumulations, *Arctic and Alpine Research*, **11**: 371–402.

Clapperton, C.M., 1987, Glacial geomorphology, Quaternary glacial sequence and palaeoclimatic inferences in the Ecuadorian Andes. In Gardiner, V. (ed.), *International Geomorphology 1986*, Pt II, Wiley, London, 843–70.

Clark, M.J., Gurnell, A.M., Milton, E.J., Seppala, M. and Kyostila, M., 1985, Remotely sensed vegetation classification as a snow depth indicator for hydrological analysis in sub-arctic Finland, *Fennia*, **163**: 195–225.

Clark, S.P., Jr. and Jager, E., 1969, Denudation rate in the Alps from geochronologic and heat flow data, *American Journal of Science*, **267**: 1143–60.

Clément, P. and Vaudour, J., 1968, Observations on the pH of melting snow in the southern French Alps. In Osburn, W.H. and Wright, H.E. (eds), *Arctic and Alpine Environments*, Indiana University Press, Bloomington, 205–16.

Coates, D.R., 1985, *Geology and Society*, Chapman and Hall, London.

Collier, E.P., 1957, Glacier variation and trends in runoff in the Canadian Cordillera, *International Association of Scientific Hydrology Special Publication*, **46**: 344–57.

Collins, B.D., Dunne, T. and Lehre, A.K., 1983, Erosion of tephra-covered hillslopes north of Mount St. Helens, Washington: May 1980–May 1981, *Zeitschrift für Geomorphologie*, Supplementband, **46**: 103–21.

Collins, D.N., 1979, Sediment concentration in melt waters as an indication of erosion processes beneath an Alpine glacier, *Journal of Glaciology*, **23**: 247–57.

Coltorti, M., Dramis, F. and Pambianchi, G., 1983, Stratified slope waste deposits in the Esino River basin, Umbria, Marche Apennines, Central Italy, *Polarforschung*, **53**: 59–66.

Conacher, A.J. and Dalrymple, J.B., 1977, The nine-unit landsurface model: An approach to pedogeomorphic research, *Geoderma*, **18**: 1–154.

Corbato, C.E., 1965, Theoretic gravity anomalies of glaciers having parabolic cross sections, *Journal of Glaciology*, **5**: 255–8.

Corbel, J., 1959, Vitesse de l'erosion, *Zeitschrift für Geomorphologie*, **3**: 1–28.

Corbel, J., 1964, L'erosion terrestre, étude quantitative, *Annales de Géographie*, **398**: 385–412.

Corner, G.D., 1980, Avalanche impact landforms in Troms, North Norway,

Geografiska Annaler, **62A**: 1–10.

Corrado, G. and Luongo, G., 1981, Ground deformation measurements in active volcanic areas using tide gauges, *Bulletin Volcanologique*, **44** (3): 505–11.

Corte, A.E., 1976, Rock glaciers, *Biuletyn Peryglacjalny*, **26**: 175–97.

Corte, A.E., 1978, Rock glaciers as permafrost bodies with a debris cover as an active layer. A hydrological approach, Andes of Mendoza, Argentina. In *Proceedings of the Third International Permafrost Conference, Edmonton, Canada*, National Research Council, Ottawa, 163–9.

Costin, A.B., Jennings, J.N., Black, H.P. and Thom, B.G., 1964, Snow action on Mount Twynam, Snowy Mountains, Australia, *Journal of Glaciology*, **5**: 219–28.

Costin, A.B., Jennings, J.N., Bautovich, B.C. and Wimbush, D.J., 1973, Forces developed by snowpatch action, Mt. Twynam, Snowy Mountains, Australia, *Arctic and Alpine Research*, **5** (2): 121–6.

Cotton, C.A., 1922, *Geomorphology of New Zealand*, New Zealand Board of Science and Art, Wellington, New Zealand.

Cotton, C., 1942, *Climatic Accidents*, Whitcombe and Tombs, Christchurch, New Zealand.

Cotton, C.A., 1944, *Volcanoes as Landscape Forms*, Whitcombe and Tombs, Christchurch, New Zealand.

Cotton, C.A., 1960, The origin and history of central Andean relief: divergent views, *Geographical Journal*, **125**: 476–8.

Court, A., 1957, The classification of glaciers, *Journal of Glaciology*, **3** (21): 3–7.

Crandell, D.R., 1971, Postglacial lahars from Mount Rainier volcano, Washington, *United States Geological Survey Professional Paper*, **667**: 1–75.

Crandell, D.R., Miller, C.D., Glicken, H., Christiansen, R.L. and Newhall, C.G., 1984, Catastrophic debris avalanche from ancestral Mount Shasta, California, *Geology*, **12**: 143–6.

Crandell, D.R. and Mullineaux, D.R., 1967, *Volcanic Hazards at Mount Rainier, Washington*, United States Geological Survey Bulletin, **1238**, 26 pp.

Crandell, D.R., Mullineaux, D.R., Sigafoos, R.S. and Rubin, M., 1974, Chaos crags eruptions and rockfall-avalanches, Lassen Volcanic National Park, California, *United States Geological Survey Journal of Research*, **2**: 45–59.

Crandell, D.R., Mullineaux, D.R. and Rubin, M., 1975. Mount St Helens volcano: recent and future behavior, *Science,* **187**, 438–441.

Crandell, D.R. and Waldron, H.H., 1956, A recent volcanic mudflow of exceptional dimensions from Mt. Rainier, Washington, *American Journal of Science*, **254**: 349–62.

Cruden, D.M., 1976, Major rock slides in the Rockies, *Canadian Geotechnical Journal*, **13**: 8–20.

Curray, J.R. and Moore, D.G., 1971, Growth of the Bengal deep-sea fan and denudation in the Himalayas, *Bulletin of the Geological Society of America*, **82** (3): 563–72.

Curry, R.R., 1966, Observation of alpine mudflows in the Tenmile Range, Central Colorado, *Bulletin of the Geological Society of America*, **77**: 771–6.

Dahl, R., 1965, Plastically sculptured detail forms on rock surfaces in northern Nordland, Norway, *Geografiska Annaler*, **47**: 83–140.

Dahl, R., 1966a, Blockfields and other weathering pits and tor-like forms in the Narvik Mountains, Nordland, Norway, *Geografiska Annaler*, **48**: 55–85.

Dahl, R., 1966b, Blockfields and other weathering forms in the Narvik Mountains, *Geografiska Annaler,*. **48**: 224–7.

Dalrymple, J.B., Blong, R.J. and Conacher, A.J., 1968, A hypothetical nine unit landsurface model, *Zeitschrift für Geomorphologie*, **12**: 60–76.

Daly, C., 1984, Snow distribution patterns in the alpine krummholz zone, *Progress in Physical Geography*, **8**: 157–75.

Daly, R.A., 1905, Summit levels among alpine mountains, *Journal of Geology*, **13**: 105.

Das, K.N., 1968, Soil erosion and the problem of silting in the Kosi catchment, *Journal of Soil and Water Conservation in India*, **16**: 2–4, 60–7.

Davies, J.L., 1969, *Landforms of Cold Climates*, MIT Press, Cambridge, MA.

Davis, G.H., 1962, Erosional features of snow avalanches, Middle Fork, Kings River, California, *United States Geological Survey Professional Paper*, **450D**: 122–5.

Davis, W.D., 1916, The Mission Range, Montana, *Geographical Review*, **2**: 267–88.

Davis, W.M., 1889, The rivers and valleys of Pennsylvania, *National Geographic Magazine*, **1**: 183–253.

Davis, W.M., 1900, Glacial erosion in France, Switzerland and Norway, *Proceedings of the Boston Society of Natural History*, **29**: 273–322.

Davis, W.M., 1906, The sculpture of mountains by glaciers, *Scottish Geographical Magazine*, **22**: 76–89.

Davis, W.M., 1909, *Geographical Essays*, ed. D. Johnson, Ginn and Co., Boston.

Davis, W.M., 1923, The cycle erosion and the summit level of the Alps, *Journal of Geology*, **31**: 1–41.

Davoran, A. and Mosley, M.P., 1986, Observations of bedload movement, bar development and sediment supply in the braided Ohau River, *Earth Surface Processes and Landforms*, **11** (6): 643–52.

Day, T.J., 1972, The channel geometry of mountain streams. In Slaymaker, O. and McPherson, H.J. (eds), *Mountain Geomorphology, Geomorphological Processes in the Canadian Cordillerra*, Tantalus Press, Vancouver, 141–9.

Decker, R.W., 1986, Forecasting volcanic eruptions, *Review of Earth and Planetary Sciences*, 274–91.

De Crecy, L., 1980, Avalanche zoning in France — regulation and technical bases, *Journal of Glaciology*, **26**: 325–30.

Derbyshire, E., 1964, Cirques, Australian landform example no. 2, *Australian Geographer*, **9**: 178–9.

Derbyshire, E., 1968, Cirques. In Fairbridge, R.W. (ed.), *The Encyclopedia of Geomorphology*, Reinhold, New York, 119–23.

Derbyshire, E. and Love, M.A., 1986, Glacial Environments. In Fookes, P.G. and Vaughan, P.R. (eds), *A Handbook of Engineering Geomorphology*, Surrey University Press, London, 66–81.

Derrau, M., 1968, Mountains. In Fairbridge, R.W. (ed.), *The Encyclopedia of Geomorphology*, Reinhold, New York, 737–9.

Dewey, J.F. and Bird, J.M., 1970, Mountain belts and the new global tectonics, *Journal of Geophysical Research*, **75**: 2625–47.

DeWolf, Y., 1988, Stratified slope deposits. In Clark, M.J. (ed.), *Advances in Periglacial Geomorphology*, Wiley, Chichester, 91–110.

Dickinson, W.T. and Whitelet, H., 1970, Watershed areas contributing to runoff, *International Association of Scientific Hydrology Publication*, **96**: 12–26.

Dietrich, W.E. and Dunne, T., 1978, Sediment budget for a small catchment in mountainous terrain, *Zeitschrift für Geomorphologie*, Supplementband, **29**: 191–206.

Dingwall, P.R., 1972, Erosion by overland flow on an alpine debris slope. In Slaymaker, O. and McPherson, H.J. (eds), *Mountain Geomorphology: Geomorphological Processes in the Canadian Cordillerra*, Tantalus Press, Vancouver, 113–20.

Dixon, J.C., 1986, Solute movement on hillslopes in the alpine Environment of the Colorado Front Range. In Abrahams, A.D. (ed.), *Hillslope Processes*, George Allen & Unwin, London, 139–59.

Dixon, J.C., Thorn, C.E. and Darmody, R.G., 1984, Chemical weathering processes on the Vantage Peak Nunatak, Juneau Icefield, southern Alaska, *Physical Geography*, 5 (2): 111–31.

Dollfus, O., 1960, Etude d'un bassin torrentiel dans la vallée de Rimac, Andes Centrales Peruviennes, *Revue de Géomorphologie Dynamique*, 11: 159–63.

Dollfus, O., 1964, L'Influence de l'exposition dans le modèle des versants des Andes central Peruviennes, *Zeitschrift für Geomorphologie*, Supplementband, 5: 131–5.

Doornkamp, J.C. and King, C.A.M., 1971, *Numerical Analysis in Geomorphology*, Arnold, London.

Dow, V., Kienholz, H., Plam, M. and Ives, J.D., 1981, Mountain Hazards Mapping: the development of a prototype combined hazards map, Monarch Lake Quadrangle, Colorado, USA, *Mountain Research and Development*, 1 (1): 55–64.

Downie, C., 1964, Glaciations of Mount Kilimanjaro, northeast Tanganyika, *Bulletin Geological Society of America*, 75: 1–16.

Dresch, J., 1941, *Recherches sur l'evolution du relief dans le Massif Central du Grand Atlas, Le Haouz et le Sous*, Arrault, Tours.

Dresch, J., 1952, Le haut atlas occidentale, *XIX International Geological Congress, Regional Monograph*, 3: 107–21.

Drewry, D.J., 1972, The contribution of radio echo sounding to the investigation of Cenozoic tectonics and glaciation in Antarctica. In Price, R.J. and Sugden, D.E. (eds), *Polar Geomorphology*, Institute of British Geographers, Special Publication 4: 43–57.

Dreyer, N.N., Nikolayeva, G.M. and Tsigelnaya, I.D., 1982, Maps of streamflow resources of some high-mountain areas in Asia and North America. In Glen, J.W. (ed.), *Hydrological Aspects of Alpine and High Mountain Areas*, International Association of Hydrological Sciences, Publication 138: 11–20.

Dunkerley, J., Ramsay, W., Gordon, L. and Cecelski, E., 1981, *Energy Strategies for Developing Nations*, John Hopkins University Press for Resources for the Future Inc., Baltimore, MD, USA.

Dunn, J.R. and Hudec, P.P., 1966, Water, clay and rock soundness, *Ohio Journal of Science*, 66: 153–68.

Dutta, K.K., 1951, Report on the landslides in Darjeeling and neighbouring hillslopes in June 1950, *Records of the Geological Survey of India*, 29.

Dyson, J.L., 1937, Snowslide striations, *Journal of Geology*, 45: 549–57.

Dyson, J.L., 1938, Snowslide erosion, *Science*, 87: 365–6.

Eckholm, E., 1979, *Planting for the Future: Forestry for Human Needs*, Worldwatch Institute, Washington, DC.

Ekblaw, W.E., 1918, The importance of nivation as an erosive factor and of soil flow as a transporting agency in northern Greenland, *Proceedings of the United States National Academy of Sciences*, 4: 288–93.

Elliston, G.R., 1973, Water movement through the Gornergletscher. In *Proceedings of the Symposium on the Hydrology of Glaciers*, International Association of Scientific Hydrology Publication 95: 79–84.

Embleton, C. and King, C.A.M., 1975, *Glacial Geomorphology*, Arnold, London.

Evans, I.S., 1969, The geomorphology and morphometry of glacial and nival areas. In Chorley, R.J. (ed.) *Water, Earth and Man*, Methuen, London, 369–80.

Evans, I.S., 1972a, Inferring process from form: the asymmetry of glaciated mountains, *International Geography*, 1: 17–19.

Evans, I.S., 1972b, General geomorphometry: derivations of altitude and descriptive statistics. In Chorley, R.J. (ed.), *Spatial Analysis in Geomorphology*, Methuen, London, 17–90.

Evans, I.S. and Cox, N., 1974, Geomorphometry and the operational definition of cirques, *Area*, 6 (2): 150–3.

Evans, S., 1989, Landslides and related processes in the Canadian Cordillera, *Landslide News*, 3: 3–6.

Eyles, N., 1983, The glaciated valley landsystem. In Eyles, N. (ed.), *Glacial Geology: An Introduction for Engineers and Earth Scientists*, Pergamon, Oxford, 91–110.

Eyles, N. and Menzies, J., 1983, The subglacial landsystem. In Eyles, N. (ed.), *Glacial Geology: An Introduction for Engineers and Earth Scientists*, Pergamon, Oxford, 19–70.

Eyles, N., Sasseville, D.R., Slatt, R.M. and Rogerson, R.J., 1982, Geochemical denudation rates and solute transport mechanisms in a maritime temperate glacier basin, *Canadian Journal of Earth Sciences*, 18: 1570–81.

Fahey, B.D., 1973, An analysis of diurnal freeze-thaw and frost heave cycles in the Indian Peaks Region of the Colorado Front Range, *Arctic and Alpine Research*, 5 (3): 269–81.

Fahey, B.D., 1983, Frost action and hydration as rock weathering mechanisms on schist: a laboratory study, *Earth Surface Processes and Landforms*, 8: 535–45.

Fahey, B.D. and Dagesse, D.F., 1984, An experimental study of the effect of humidity and temperature variations on the granular disintegration of argillaceous carbonate rocks in cold climates, *Arctic and Alpine Research*, 16 (3): 291–8.

Fahnestock, R.K., 1963, *Morphology and Hydrology of a Glacial Stream, White river, Mount Rainier, Washington*, United States Geological Survey Professional Paper 422-A, 70 pp.

Fairbridge, R.W., 1968, Mountain and Lilly terrain, mountain systems; mountain types. In Fairbridge, R.W. (ed.), *Encylopedia of Geomorphology*, Reinhold, New York 745–61.

Fakuda, M., 1971, Freezing-thawing process of water in pore space of rocks, *Low Temperature Science, Series A*, 29: 225–9.

Fakuda, M., 1987, Freezing-thawing process of water in pore space of rocks II, *Low Temperature Science, Series A*, 30: 183–9.

Fenn, C.R., 1987, Sediment transfer processes in alpine glacier basins. In Gurnell, A.M. and Clark, M.J. (eds), *Glacio-fluvial Sediment Transfer*, Wiley, Chichester, 59–85.

Fenneman, N.M., 1928, Physiographic divisions of the United States, *Annals of the Association of American Geographers*, XVIII: 261–353.

Ferguson, R.I., 1984, Sediment load of the Hunza River. In Miller, K.J. (ed.), *International Karakoram Project*, Vol. II, Cambridge University Press, Cambridge, 581–98.

Fenner, C.N., 1948, Incandescent tuff flows in southern Peru, *Bulletin of the Geological Society of America*, 59: 879–93.

Finch, R.H., 1930, Rainfalls accompanying explosive eruptions of volcanoes, *American Journal of Science*, **19**: 147–50.

Finch, V.C. and Trewartha, G.T., 1936, *Elements of Geography*, McGraw-Hill, New York.

Fisher, O., 1866, On the disintegration of a chalk cliff, *Geological Magazine*, **3**: 354–6.

Fisher, R.V., 1979, Models for pyroclastic surges and pyroclastic flows, *Journal of Volcanology and Geothermal Research*, **6**: 305–18.

Fisher, R.V., 1983, Flow transformations in sediment gravity flows, *Geology*, **11**: 272–4.

Fisher, R.V. and Heiken, G., 1982, Mt. Pelee, Martinique: May 8 and 20 1902 pyroclastic flows and surges, *Journal of Volcanology and Geothermal Research*, **13**: 339–71.

Fisher, R.V. and Waters, A.C., 1970, Base surge bedforms in maar volcanoes, *American Journal of Science*, **268**: 157–80.

Fitzgibbon, J.E. and Dunne, T., 1981, Land surface and lake storage during snow-melt runoff in a subarctic drainage system, *Arctic and Alpine Research*, **16**: 291–8.

Fitzharris, B.B., 1985, Estimation of avalanche runout distances in New Zealand. In Church, M. and Slaymaker, O. (eds), *Field and theory, Lectures in Geocryology*, University of British Columbia Press, Vancouver, 57–73.

Fitzharris, B.B. and Schaerer, P.A., 1980, Frequency of major avalanche winters, *Journal of Glaciology*, **26**: 45–52.

Flint, R.F., 1971, *Glacial and Quaternary Geology*, Wiley, Chichester.

Fohn, P., Good, W., Bois, P. and Obled, C., 1977, Evaluation and comparison of statistical and conventional methods of forecasting avalanche hazard, *Journal of Glaciology*, **19**: 375–87.

Fookes, P.G., Sweeney, H., Manby, C.N.D. and Martin, R.P., 1985, Geological and geotechnical engineering aspects of low cost roads in mountainous terrain, *Engineering Geology*, **21**: 1–152.

Ford, D.C., Schwarcz, H.P., Drake, J.J., Gascoyne, M., Harmon, R.S. and Latham, A.G., 1981, Estimations of the age of the existing relief within the southern Rocky Mountains of Canada, *Arctic and Alpine Research*, **13** (1): 1–10.

Francis, P., 1976, *Volcanoes*, Penguin, London.

Francis, P. and Self, S., 1987, Collapsing volcanoes, *Scientific American*, **256**: 90–7.

Francis, P.W., Gardeweg, M., Ramirez, C.T. and Rothery, D.A., 1985, Catastrophic debris avalanche deposit of Socompa volcano, northern Chile, *Geology*, **13**: 600–3.

Francis, P.W. and Wells, G.L., 1988, Landsat thematic mapper observations of debris avalanche deposits in the Central Andes, *Bulletin Volcanologique*, **50**: 258–78.

Francis, P.W., Roobol, M.J., Walker, G.P.L., Cobbold, P.R. and Coward, M., 1974, San Pedro and San Pablo volcanoes of north Chile and their avalanche deposits, *Geologische Rundschau*, **63**: 357–88.

Francis, S.C., 1987, Slope development through the threshold slope concept. In Anderson, M.G. and Richards, K.S. (eds), *Slope Stability*, Wiley, Chichester, 601–24.

Francou, B., 1982, Chutes de pierres et éboulisation dans les parois de l'étage periglaciaire. Observations faites d'octobre 1979 à juin 1981 dans la Combe de Laumchard (Hautes-Alpes), *Revue de Géographie Alpine*, **70** (3): 279–300.

Francou, B., 1984, Géodynamique des dépôts de pieds de paroi dans l'étage periglaciaire des Alpes internes, *Revue de Géographie Physique et de Géologie Dynamique*, **24** (5): 411–24.

Frank, R.C. and Lee, R., 1966, *Potential Solar Beam Irradiation on Slopes*, United States Department of Agriculture Forest Service Research Paper, RM **18**, 116 pp.

Frank, T.D. and Thorn, C.E., 1985, Stratifying Alpine Tundra for geomorphic studies using digitized aerial imagery, *Arctic and Alpine Research*, **17** (2): 179–88.

French, H.M., 1976, *The Periglacial Environment*, Longman, New York.

Frutiger, H., 1980, History and actual state of legislation of avalanche zoning in Switzerland, *Journal of Glaciology*, **26**, 313–24.

Fryxell, F.M. and Horberg, L., 1943, Alpine mudflows in Grand Teton National Park, Wyoming, *Bulletin Geological Society of America*, **54**: 457–72.

Fujii, Y. and Higuchi, K., 1972, On the permafrost at the summit of Mt. Fuji, *Seppyo*, **13**, 175–86.

Gage, M., 1966, Franz Josef Glacier, *Ice*, **20**: 26–7.

Galibert, G., 1965, *La Haute Montagne Alpine*, Thèse Lettres, Toulouse.

Galloway, R.W., Hope, G.S., Loffler, E. and Peterson, J.A., 1973, Late Quaternary glaciation and periglacial phenomena in Australia and New Guinea, *Palaeoecology of Africa and the Antarctic*, **8**: 127–38.

Gardner, J., 1967, Notes on avalanches, icefalls and rockfalls in the Lake Louise district, July and August, 1966, *Canadian Alpine Journal*, **50**: 90–5.

Gardner, J., 1969a, Snow patches: their influence on mountain wall temperatures and the geomorphic implications, *Geografiska Annaler* **51**: 114–20.

Gardner, J., 1969b, Notes on avalanches, icefalls and rockfalls in the Lake Louise district, July and August 1966. In Nelson, J.G. and Chambers, M.H., (eds), *Geomorphology*, Methuen, Toronto, 195–201.

Gardner, J., 1969c, Observations on surficial talus movement, *Zeitschrift für Geomorphologie*, **13**: 318–23.

Gardner, J., 1970a, Rockfall — a geomorphic process in high mountain terrain, *Albertan Geographer*, **6**: 15–20.

Gardner, J., 1970b, Geomorphic significance of avalanches in the Lake Louise Area, Alberta, Canada, *Arctic and Alpine Research*, **2**: 135–44.

Gardner, J., 1972, Recent glacial activity and some associated landforms in the Canadian Rocky Mountains. In Slaymaker, O. and McPherson, H.J. (eds), *Mountain Geomorphology: Geomorphological Processes in the Canadian Cordillerra*, Tantalus Press, Vancouver, 55–62.

Gardner, J., 1977, High-magnitude rockfall-rockslide: frequency and geomorphic significance in the Highwood Pass area, Alberta, Great Plains, *Rocky Mountain Geography Journal*, **6**: 228–38.

Gardner, J., 1983a, Observations on erosion by wet snow avalanches, Mount Rae area, Alberta, Canada, *Arctic and Alpine Research*, **15** (2): 271–4.

Gardner, J., 1983b, Accretion rates on some debris slopes in the Mt. Rae area, Canadian Rocky Mountains, *Earth Surface Processes and Landforms*, **8**: 347–55.

Garner, H.F., 1959, Stratigraphic-sedimentary significance of contemporary climate and relief in four regions of the Andes Mountains, *Bulletin Geological Society of America*, **70**: 1327–68.

Garner, H.F., 1974, *The Origin of Landscapes*, Oxford University Press, New York.

Garnier, B.J. and Ohmura, A., 1968, A method of calculating the direct short-wave radiation income of slopes, *Journal of Applied Meteorology*, **7**: 796–800.

Garstka, W.V., Love, L., Goodell, B.C. and Bertle, F.A., 1959, *Factors Affecting Snowmelt and Streamflow*, US Department of Interior Bureau of Reclamation and US Department of Agriculture Forest Service, 187 pp.

Geikie, A., 1903, *Text-book of Geology*, Macmillan, London, 2 vols.

Geikie, J., 1898, *Earth Sculpture, or the Origin of Land-forms*, Putnam, London.

Geikie, J., 1914, *Mountains, Their Origin, Growth and Decay*, Van Nostrand and Co., Princeton, NJ.

Gerber, E. and Scheidegger, A.E., 1969, Stress induced weathering of rock masses, *Ecologia Geologia Helvetica*, **62**: 401–14.

Gerber, E. and Scheidegger, A.E., 1973, Erosional and stress-induced features on steep slopes, *Zeitschrift für Geomorphologie*, Supplementband, **18**: 38–49.

Gerber, E. and Scheidegger, A.E., 1975, Geomorphological evidence for the geophysical stress field in mountain massifs, *Revista Italia Geofisica Scientia Affini*, **2**: 47–52.

Gerrard, A.J., 1981, *Soils and Landforms*, George Allen & Unwin, London.

Gerrard, A.J., 1985, Soil erosion and landscape stability in southern Iceland: A Tephrochronological approach. In Richards, K.S., Arnett, R.R. and Ellis, S. (eds), *Geomorphology and Soils*, George Allen & Unwin, London, 78–95.

Gerrard, A.J., 1988a, *Rocks and Landforms*, Unwin Hyman, London.

Gerrard, A.J., 1988b, Periglacial modification of the Cox Tor – Staple Tors area of western Dartmoor, England, *Physical Geography*, **9** (3): 280–300.

Ghosh, A.M.N., 1950, Observation on landslides of 11th and 12th June 1950 in the Darjeeling, Himalayas, *Report Geological Survey of India*.

Giardino, J.R., Shroder, J.F. and Lawson, M.P., 1984, Tree-ring analysis of movement of a rock-glacier complex on Mount Mestas, Colorado, USA, *Arctic and Alpine Research*, **16** (3): 299–309.

Gibbs, H.S., 1980, *New Zealand Soils: An Introduction*, Oxford University Press, Wellington, New Zealand.

Gibbs, H.S. and Wells, N., 1966, Volcanic ash soils in New Zealand, *Bulletin Volcanologique*, **29**: 669–70.

Gibbs, R.J., 1967, The geochemistry of the Amazon River system: Part 1, The factors that control the salinity and composition and concentration of suspended solids, *Bulletin Geological Society of America*, **78**: 1203–32.

Gigon, A., 1983, Typology and principles of ecological stability and instability, *Mountain Research and Development*, **3** (2): 95–102.

Gilbert, R., 1975, Sedimentation in Lillooet Lake, British Columbia, *Canadian Journal of Earth Sciences*, **12**: 1697–1711.

Gilbert, R. and Shaw, J., 1981, Sedimentation in proglacial Sunwapta Lake, Alberta, *Canadian Journal of Earth Sciences*, **18**: 81–93.

Glen, J.W., 1953, Experiments on the deformation of ice, *Journal of Glaciology*, **2**: 111–14.

Glen, J.W., 1955, The creep of polycrystalline ice, *Proceedings of the Royal Society of London, A*, **228**: 519–38.

Gorbunov, A.P., 1978, Permafrost investigations in high-mountain regions, *Arctic and Alpine Research*, **10**: 282–94.

Gorshkov, G.S., 1959, Gigantic eruption of the volcano Bezymianny, *Bulletin Volcanologique*, **20**: 77–112.

Goudie, A.S. 1984, Salt efflorescences and salt weathering in the Hunza Valley, Karakoram mountains, Pakistan. In Miller, K.J. (ed.), *The International Karakoram Project*, vol. II, Cambridge University Press, Cambridge, 607–15.

Graf, W.L. 1970, The geomorphology of the glacial valley cross-section, *Arctic and Alpine Research*, **2**: 303–12.

Graf, W.L., 1971, Quantitative analysis of Pinedale Landforms, Beartooth Mountains Montana and Wyoming, *Arctic and Alpine Research*, **3** (3): 253–61.

Grant, P.J., 1981, Major periods of erosion and sedimentation in the North Island, New Zealand since 13th century. In *Erosion and Sediment Transport in Pacific Rim Steepland*, International Association of Hydrological Sciences Publication **132**: 288–304.

Gray, J.T., 1972, Debris accretion on talus slopes in the central Yukon Territory. In Slaymaker, O. and McPherson, H.J. (eds), *Mountain Geomorphology: Geomorphological Processes in the Canadian Cordillerra*, Tantalus Press, Vancouver, 75–92.

Gray, J.T., 1973, Geomorphic effects of avalanches and rockfalls on steep mountain slopes in the Central Yukon Territory. In Fahey, B.D. and Thompson, R.D. (eds) *Proceedings 3rd Guelph Symposium*, Geo Abstracts, Norwich, 107–17.

Green, J. and Short, N., 1971, *Volcanic Landforms and Surface Features*, Springer-Verlag, New York.

Griffiths, G.A., 1979, High sediment yields from major rivers of the western Southern Alps, New Zealand, *Nature*, **282**: 61–3.

Griffiths, G.A., 1981. Some suspended sediment yields from South Island catchments, New Zealand, *Water Resources Bulletin*, **17**(4): 662–71.

Groom, G.E., 1959, Niche glaciers in Bunsow-land, Vestspitsbergen, *Journal of Glaciology*, **3**: 369–76.

Grosval'd, M.G. and Kotlyakov, V.M., 1969, Present-day glaciers in the USSR and some data on their mass balance, *Journal of Glaciology*, **22**: 186–8.

Grove, J.M., 1988, *The Little Ice Age*, Methuen, London.

Gubler, H., 1980, Simultaneous measurements of stability indices and characteristic parameters describing the snow cover and weather in fracture zones of avalanches, *Journal of Glaciology*, **26**: 65–74.

Guest, J.E. and Murry, J.B., 1979, An analysis of hazard from Mt. Etna, Volcano, *Journal of the Geological Society of London*, **136**: 347–54.

Gurnell, A.M., 1982, The dynamics of suspended sediment concentrations in a pro-glacial stream. In Glen, J.W. (ed.), *Hydrological Aspects of Alpine and High Mountain Areas*, International Association of Hydrological Sciences Publication, **138**: 319–30.

Gurnell, A.M., 1983, Downstream channel adjustments in response to water abstraction for hydro-electric power generation from alpine glacial melt-water streams, *Geographical Journal*, **149**: 342–54.

Gurnell, A.M., 1987, Suspended sediment. In Gurnell, A.M. and Clark, M.J. (eds), *Glacio-fluvial Sediment Transfer*, Wiley, Chichester, 305–54.

Gutenberg, B. and Richter, C.F., 1954, *Seismicity of the Earth and Associated Phenomena*, Princeton University Press, Princeton, NJ.

Guymon, G.L., 1974, Regional sediment yield analysis of Alaska streams *American Society of Civil Engineers, Journal of the Hydraulics Division*, **100**: 41–51.

Haantjens, H.A. and Bleeker, P., 1970, Tropical weathering in the Tertiary of Papua and New Guinea, *Australian Journal of Soil Research*, **8**: 157–77.

Hadley, J.B., 1964, Landslides and related phenomena accompanying the Hebgen Lake earthquake of August 17th, 1959, *United States Geological Survey Professional Paper*, **435**: 107–38.

Haeberli, W., 1979, Holocene push-moraines in alpine permafrost, *Geografiska*

Annaler, **61A**: 43–8.

Haefeli, R., 1970, Changes in the behaviour of the Unteraargletscher in the last 125 years, *Journal of Glaciology*, **9**: 195–212.

Hagen, J.O., Wold, B., Wiestol, O., Ostrem, G. and Sollid, J.L., 1983, Subglacial processes at Banhusbreen, Norway; preliminary results, *Annals of Glaciology*, **4**: 91–8.

Haigh, M.J., 1984, Landslide prediction and highway maintenance in the Lesser Himalaya, India, *Zeitschrift für Geomorphologie*, Supplementband, **51**: 17–37.

Hall, K., 1975, Nivation processes at a late-lying, north-facing snowpatch site in Austre Okstinbredalen, Okstinden, northern Norway. Unpublished MSc thesis, Reading University.

Hall, K., 1980, Freeze-thaw activity at a nivation site in northern Norway, *Arctic and Alpine Research*, **12** (2): 183–94.

Hallet, B., 1979, A theoretical model of glacial abrasion, *Journal of Glaciology*, **23**: 39–50.

Hallet, B., 1981, Glacial abrasion and sliding: their dependence on the debris concentration in basal ice, *Annals of Glaciology*, **2**: 23–8.

Hamilton, L.S. and Pearce, A.J., 1988, Soil and water impacts of deforestation. In Ives, J. and Pitt, D.C. (ed.), *Deforestation: Social Dynamics in Watersheds and Mountain Ecosystems*, Routledge, London, 75–98.

Hammer, K.M. and Smith, N.D., 1983, Sediment production and transport in a proglacial stream: Hilda Glacier, Alberta, Canada, *Boreas*, **12**: 91–106.

Hammond, E.H., 1954, Small-scale continental landform maps, *Annals of the Association of American Geographers*, **44**: 33–42.

Hammond, E.H., 1964, Classes of land-surface form in the forty eight states, USA, *Annals of the Association of American Geographers*, **54** (1): Map Suppl. 4.

Hansen, A., 1984, Landslide hazard analysis. In Brunsden, D. and Prior, D.B. (ed.), *Slope Instability*, Wiley, Chichester, 523–602.

Harland, W.B., 1957, Exfoliation joints and ice action, *Journal of Glaciology*, **3**: 8–10.

Harris, C., 1972, Processes of soil movement in turf-banked solifluction lobes, Okstinden, northern Norway. In Price, R.J. and Sugden, D.E. (eds), *Polar Geomorphology*, Institute of British Geographers Special Publication **4**: 155–74.

Harris, C., 1981, *Periglacial Mass-wasting: A Review of Research*, British Geomorphological Research Group, Research Monograph Series 4, Geo Abstracts, Norwich.

Harris, S.A., 1979, Ice caves and permafrost zones in south-west Alberta, *Erdkunde*, **33**: 61–70.

Harris, S.A., 1981a, Distribution of active glaciers and rock glaciers compared to permafrost landforms, based on freezing and thawing indices, *Canadian Journal of Earth Sciences*, **18**: 376–81.

Harris, S.A., 1981b, Climatic relationship of permafrost zones in areas of low winter snow cover, *Arctic*, **34**: 64–70.

Harris, S.A., 1985, Distribution and zonation of permafrost along the eastern ranges of the Cordillera of North America, *Biuletyn Peryglacjalny*, **30**: 107–18.

Harris, S.A., 1986, *The Permafrost Environment*, Croom Helm, Beckenham.

Harris, S.A. and Brown, R.J.E., 1978, Plateau Mountain — a case study of alpine permafrost in the Canadian Rocky mountains, *Proc. Canadian Conference on Permafrost*, National Research Council of Canada, Ottawa, **1**: 385–91.

Harris, S.A. and Brown, R.J.E., 1982, Permafrost distribution along the Rocky

Mountains in Alberta, *Procs. 4th Canadian Conference on Permafrost*, National Research Council of Canada, Ottawa, 59–67.

Harrison, A.E., 1964, Ice surges on the Muldrow glacier, *Journal of Glaciology*, **5**: 265–368.

Harrison, J.V. and Falcon, N.L., 1937, An ancient landslip at Saidmarreh, Iran, *Journal of Geology*, **46**: 296–309.

Hartline, B.K., 1979, Snow physics and avalanche prediction, *Science*, **203**: 346–8.

Hast, N., 1967, The states of stresses in the upper part of the Earth's crust, *Engineering Geology*, **2**: 5–17.

Hastenrath, S., 1966, Certain aspects of the three-dimensional distribution of climate and vegetation belts in the mountains of Central America and southern Mexico. In Troll, C. (ed.), *Geo-ecology of the Mountainous Regions of the Tropical Americas*, Ferd. Dimmlers Verlag, Bonn, 122-30.

Hastenrath, S., 1971, On the Pleistocene snow-line depression in the arid regions of the South American Andes, *Journal of Glaciology*, **10**: 255–67.

Hastenrath, S., 1978, Heat budget measurements on the Quelccaya Ice Cap, Peruvian Andes, *Journal of Glaciology*, **20**: 85–97.

Hastenrath, S., 1981, *The Glaciation of the Ecuadorian Andes*, Balkema, Rotterdam.

Hastenrath, S., 1984, *The Glaciers of Equatorial East Africa*, Reidel, Dordrecht.

Haupt, H.F., 1967, Infiltration, overland flow and soil movement on frozen and snow covered plots, *Water Resources Research*, **3**: 145–61.

Hay, J.E., 1971, Computational model for radiative fluxes, *Journal of Hydrology* (NZ), **10**: 36–48.

Hay, R.L., 1960, Rate of clay formation and mineral alteration in a 4000-year-old volcanic ash soil on St. Vincent, British West Indies, *American Journal of Science*, **258**, 354–68.

Haynes, V.M., 1968, The influence of glacial erosion and rock structure on corries in Scotland, *Geografiska Annaler*, **50A**: 221–34.

Hayward, J.A., 1979, Mountain stream sediments. In Murray, D.L. and Ackroyd, P. (eds), *Physical Hydrology*, New Zealand Hydrological Society, 193–212.

Hayward, J.A., 1980, *Hydrology and Stream Sediments from Torlesse Stream Catchment*, Tussock Grasslands and Mountain Lands Institute, Lincoln College, New Zealand, Special Publication 17.

Heede, B.H., 1972a, *Flow and Channel Characteristics of Two High Mountain Streams*, United States Department of Agriculture Forest Service Research Paper, RM 92.

Heede, B.H., 1972b, Influences of a forest on the hydraulic geometry of two mountain streams, *Water Resources bulletin*, **8**: 3.

Heim, A., 1919, *Geologie der Schweiz*, Tauchnitz, Leipzig.

Heim, A., 1932, *Bergsturz und Menschenleben*, Fretz and Wasmuth, Zurich.

Heine, K., 1977, Beobachtungen und Überlegungen zur Eiszeitlichen Depression von Schneegrenze und Strukturbodengrenze in den Tropen und Subtropen, *Erdkunde*, **31**: 161–77.

Henrick, R.L., Filgate, B.D. and Adams, W.M., 1971, Application of environmental analysis to watershed snowmelt, *Journal of Applied Meteorology*, **10**: 418–29.

Heuberger, H., Masch, L., Preuss, E. and Schrocker, A., 1984, Quaternary landslides and rock fusion in central Nepal and in the Tyrolean Alps, *Mountain Research and Development*, **4** (4): 345–62.

Hewitt, K., 1967, Studies in the geomorphology of the mountain regions of the Upper Indus Basin. Unpublished PhD thesis, University of London.

Hewitt, K., 1968, The freeze-thaw environment of the Karakoram Himalaya, *Canadian Geographer*, **12**: 85–98.

Hewitt, K., 1972, The mountain environment and geomorphic processes. In Slaymaker, O. and McPherson, H.J. (eds), *Mountain Geomorphology: Geomorphological Processes in the Canadian Cordillerra*, Tantalus Press, Vancouver, 17–34.

Hewitt, K., 1982, Natural dams and outburst floods of the Karakoram Himalaya. In Glen, J.W. (ed.), *Hydrological Aspects of Alpine and High Mountain Areas*, International Association of Hydrological Sciences Publication **138**: 259–69.

Higuchi,K., Ageta, Y., Yasunai, T. and Inoue, J., 1982, Characteristics of precipitation during the monsoon in high mountain areas of the Nepal Himalaya. In Glen, J.W. (ed.), *Hydrological Aspects of Alpine and High Mountain Areas*, International Association of Hydrological Sciences Publication **138**: 21–30.

Hirano, M. and Aniya, M., 1988, A rational explanation of cross-profile morphology for glacial valleys and of glacial valley development, *Earth Surface Processes and Landforms*, **13**: 707–16.

Hoblitt, R.P. and Miller, C.D., 1984, Comment, *Geology*, **12**: 692–3.

Hoblitt, R.P., Miller, C.D. and Vallance, J.W., 1981, Origin and stratigraphy of the deposit produced by the May 18 directed blast. In Lipman, P.W. and Mullineaux, D.R. (eds), *The 1980 Eruptions of Mount St Helens, Washington*, United States Geological Survey Professional Paper, **1250**: 401–9.

Hollermann, P., 1973a, Some reflections on the nature of high mountains with special reference to the western United States, *Arctic and Alpine Research*, **5** (3): 149–60.

Hollermann, P., 1973b, Some aspects of the geoecology of the basin and range province (California Section), *Arctic and Alpine Research*, **5** (5): A85–98.

Hollingshead, A.B., 1971, Sediment transport measurements in a gravel river, *American Society of Civil Engineers Journal of Hydrology Division*, **97** (11): 1817–34.

Hooke, R.L., Wold, B. and Hagen, J.O., 1985, Subglacial hydrology and sediment transport at Bondhusbreen, southwest Norway, *Bulletin Geological Society of America*, **96**: 388–97.

Hoppe, G. and Ekman, S.R., 1964, A note on the alluvial fans of Ladtjovarre, Swedish Lapland, *Geografiska Annaler*, **46**: 338–42.

Hoshai, M. and Kobayashi, K., 1957, A theoretical discussion on the so-called 'snow line' with reference to the temperature reduction during the last glacial age in Japan, *Japan Journal of Geology and Geography*, **28**: 61–75.

Houghton, B.F., Latter, J.H. and Hackett, W.R., 1987, Volcanic hazard assessment for Ruapehu composite volcano, Tanpo Volcanic Zone, New Zealand, *Bulletin Volcanologique*, **49**: 737–51.

Howarth, P.J. and Bones, J.G., 1972, Relationships between process and geometric form on high arctic debris slopes, S.W. Devon Island, Canada. In Price, R.J. and Sugden, D.E. (eds), *Polar Geomorphology*, Institute of British Geographers Special Publication, **4**: 139–55.

Howe, E., 1909, *Landslides in the San Juan Mountains, Colorado, Including a Consideration of their Causes and their Classification*, United States Geological Survey Professional Paper, **67**: 58 pp.

Howe, J., 1971, Temperature test readings in test boreholes, *Mt. Washington Observatory News Bulletin*, **12** (2): 37–40.

Hsu, K.H., 1975, Catastrophic debris streams (sturzstroms) generated by rockfalls, *Bulletin Geological Society of America*, **86**: 129–40.

Huang, T.K., 1945, On major tectonic forms in China, *Memoir Geological Survey of China, Series* A, no. 20.

Hughes, O.L., 1972, Surficial geology and land classification. Mackenzie Valley transportation corridor, *Procs. Canadian Northern Pipeline Research Conference*, Ottawa, 1972, National Research Council of Canada, Technical memorandum, **104**: 17–24.

Hunt, C.B., 1966, *Plant ecology of Death Valley, California*, United States Geological Survey Professional paper, **494-A**, 162 pp.

Hurni, H., 1982a, Soil erosion in Huai Thung Choa, Northern Thailand: concerns and constraints, *Mountain Research and Development*, **2** (2): 141–56.

Hurni, H., 1982b, *Simen Mountains — Ethiopia: Climate and the Dynamics of Altitudinal Belts from the Last Cold Period to the Present Day*, Geographia Bernesia G13, Institute of Geography, Berne University.

Hutter, K. and Olunloyo, V.O.S., 1981, Basal stress concentrations due to abrupt changes in boundary conditions: a cause for high till concentration at the bottom of a glacier, *Annals of Glaciology*, **2**: 29–33.

Hyers, A.D., 1980, Mesoscale relationships of talus and insolation, San Juan Mountains, Colorado. Unpublished PhD thesis, Arizona State University.

Hyers, A.D., 1981, Mesoscale relationships between theoretical insolation and talus fragment dimension, San Juan Mountains, Colorado. In Brazel, A.J. (ed.), *Research Papers in Climatology*, Geographical Publication no. 1, Department of Geography, Arizona State University, 149–67.

Iken, A., Rothlisberger, H., Flotron, A. and Haeberli, W., (1983), The uplift of Unteraargletscher at the beginning of the melt season — a consequence of water storage at the bed?, *Journal of Glaciology*, **29**: 28–47.

Isakov, Yu.A., Zimina, R.P., Nokolaeva, L.P. and Panfilov, D.V., 1972, Structure and biocoenotic role of the animal population in the main mountain landscapes of the Caucasus. In Troll, C. (ed.), *Geoecology of the High-Mountain Regions of Eurasia*, Franz Steiner Verlag, Wiesbaden, 177–81.

Isard, S.A., 1983, Estimating potential direct insolation to alpine terrain, *Arctic and Alpine Research*, **15**: 77–89.

Ives, J.D., 1966, Blockfields, associated weathering forms on mountain tops and the nunatak hypothesis, *Geografiska Annaler*, **48**: 220–3.

Ives, J.D., 1974, Permafrost. In Ives, J.D. and Barry, R.G., (eds), *Arctic and Alpine Environments*, Methuen, London, 159–94.

Ives, J.D., 1978, Remarks on the stability of timberline. In Troll, C. and Lauer, W. (eds.), *Geoecological Relations between the Southern Temperate Zone and the Tropical High Mountains*, Franz Steiner Verlag, Wiesbaden, 313–17.

Ives, J.D., 1980, *Geoecology of the Colorado Front Range: A Study of Alpine and Subalpine Environments*, Westview Press, Boulder, Co.

Ives, J.D., 1985, Mountain Environments, *Progress in Physical Geography*, **9**: 425–33.

Ives, J.D., 1987, The mountain lands. In Clark, M.J., Gregory, K.J. and Gurnell, A.M. (eds), *Horizons in Physical Geography*, Macmillan London, 232–49.

Ives, J.D. and Barry, R.G. (eds), 1974, *Arctic and Alpine Environments*, Methuen, London.

Ives, J.D. and Bovis, M.J., 1978, Natural hazard maps for land-use planning, San Juan mountains, Colorado, USA, *Arctic and Alpine Research*, **10**: 185–212.

Ives, J.D. and Fahey, B.D., 1971, Permafrost occurrence in the Front Range, Colorado Rocky Mountains, USA, *Journal of Glaciology*, **10**: 105–11.

Ives, J.D., Mears, A.I., Carrara, P.E. and Bovis, M.J., 1976, Natural hazards in mountain Colorado, *Annals of the Association of American Geographers*, **66**: 129–44.

Ives, J.D. and Messerli, B., 1981, Mountain hazards mapping in Nepal: introduction to an applied mountain research project, *Mountain Research and Development*, **1** (3–4): 223–30.

Ives, J.D. and Messerli, B., 1989, *The Himalayan Dilemma: Reconciling Development and Conservation*, Routledge, London.

Iwata, S., 1980, Types and intensity of the processes in the high mountain regions of Shirouma-dake, the Japan Alps, *Journal of Geography (Tokyo Geographical Society)*, **89**: 319–35.

Jahn, A., 1960, Some remarks on evolution of slopes on Spitsbergen, *Zeitschrift für Geomorphologie*, Supplementband, **1**: 49–58.

Jain, M.S., 1966, Geological reconnaissance of the Lish and the Ramthu Valleys, Kalimpong Subdivision Darjeeling District, *Bulletin of the Geological Survey of India Series B*, **15**: (1).

Jeanneret, F., 1975, Blockgletscher in den Südalpen Neuseelands, *Zeitschrift für Geomorphologie*, NF 19: 83–94.

Jenks, W.F., 1948, *Geology of the Arequipa Quadrangle of the Carta Nacional del Perú*, Geological Institute of Peru and Bolivia, **9**, 204 pp.

Jenks, W.F., 1956, *Handbook of South American Geology*, Memoir of the Geological Society of America, **65**, 378 pp.

Johnson, A.M., 1970, *Physical Processes in Geology: A Method for Interpretation of Natural Phenomena, Intrusions in Igneous Rocks, Fractures and Folds, Flow of Debris and Ice*, Freeman Cooper & Co., San Francisco.

Johnson, K., Olson, E.A. and Manandhar, S., 1982, Environmental knowledge and response to natural hazards in mountainous Nepal, *Mountain Research and Development*, **2** (2): 175–88.

Johnson, P.G., 1975, Mass movement processes in Metalline Creek, southwest Yukon Territory, *Arctic*, **28**: 130–9.

Johnson, P.G., 1978, Rock glacier types and their drainage systems, Grizzly Creek, Yukon Territory, *Canadian Journal of Earth Sciences*, **15**: 1496–1507.

Johnson, P.L. and Billings, W.D., 1962, The alpine vegetation of the Beartooth Plateau in relation to cryopedogenic processes and patterns, *Ecological Monographs*, **32**: 105–35.

Johnson, R.B., 1967, Rock streams and Mount Mestas, Sangre de Cristo Mountains, Southern Colorado, *United States Geological Survey Professional Paper*, **575D**: 217–20.

Joly, F., 1952, Le Haut Atlas Oriental, *XIX International Geological Congress, Regional Monograph*, **3**: 67–80.

Jones, D.K.C., Brunsden, D. and Goudie, A.S, 1983, A preliminary geomorphological assessment of the Karakoram Highway, *Quarterly Journal of Engineering Geology*, **16**: 331–55.

Jonsson, O., 1974, Landslides and mudflows, *Jokull*, **24**: 63–76.

Joshi, S.C., 1986, *Nepal Himalaya: Geo-ecological Perspectives*, Himalayan Research Group, Naini Tal, India.

Judson, A., 1983, On the potential use of index paths for avalanche assessment, *Journal of Glaciology*, **29**: 178–84.

Judson, A., Leaf, C.F. and Brink, G.E., 1980, A process-oriented model for stimulating avalanche danger, *Journal of Glaciology*, **26**: 53–63.

Jungerius, J.D., 1975, The properties of volcanic ash soils in dry parts of the Colombian Andes and their relation to soil erodibility, *Catena*, **2**: 69–80.

Kadomura, H., Imagawa, T. and Yamamoto, H., 1983, Eruption-induced rapid erosion and mass movements on Usu Volcano, Hokkaido, *Zeitschrift für Geomorphologie*, Supplementband, **46**: 123–42.

Kahn, M., 1966, Considerations préliminaires sur la repartition chronologique des avalanches de neige, *Association Internationale d'Hydrologie Scientifique*, **69**: 332–40.

Kanasewich, E.R., 1963, Gravity measurements on the Athabaska Glacier, Alberta, Canada, *Journal of Glaciology*, **4**: 617–31.

Kang Zicheng and Li Jing, 1987, Erosion processes and effects of debris flows. In *Erosion and Sedimentation in the Pacific Rim*, International Association for Scientific Hydrology Publication, **165**: 233–42.

Karrasch, H., 1973, Microclimatic studies in the Alps, *Arctic and Alpine Research*, **5** (3): A55–63.

Kasser, P., 1959, Der Einfluss von Gletscherrückgang und Gletschervorstoss auf den Wasserhaushalt, *Wasser und Energiewirtschaft*, **6**: 155–68.

Kasser, P., 1981, Rezente Gletscherveränderungen in den Schweizer Alpen, Gletscher und Klima, *Jahrbuch der Schweizerischen Naturforschenden Gesellschaft, Wissenchaftlicher Teil — Annuaire de la Société Helvétique des Sciences Naturelles, Partie Scientifique, 1978*, Birkhäuser Verlag, Basle, Boston and Stuttgart, 106–38.

Kear, D., 1957, Erosional stages of volcanic cones as indicators of age, *New Zealand Journal of Science and Technology*, **B38**: 671–82.

Keller, E.A. and Swanson, F.J., 1979, Effects of large organic material on channel form and fluvial process, *Earth Surface Processes*, **4**: 361–80.

Kellerhals, R., 1970, Runoff routing through steep natural channels, *American Society of Civil Engineers, Journal of Hydraulics Division*, **96**: 2201–17.

Kellerhals, R., 1972, Hydraulic performance of steep natural channels. In Slaymaker, O. and McPherson, H.J. (eds), *Mountain Geomorphology: Geomorphological Processes in the Canadian Cordillerra*, Tantalus Press, Vancouver, 131–9.

Kent, P., 1966, The transport mechanism of catastrophic rockfalls, *Journal of Geology*, **74**: 79–83.

Kesseli, J.E., 1941, Rock streams in the Sierra Nevada, California, *Geographical Review*, **31**: 203–27.

Khaire, V.A, 1975, SPR project in Nepal, *Journal of Indian Roads Congress*, **35** (4): 725–851.

Khomenyk, Y.V., 1978, Method of predicting snow avalanches by means of image recognition programs, *Soviet Hydrology*, **17**: 67–73.

Khosla, A.N., 1953, *Silting of Reservoirs*, Central Board of Irrigation and Power (India), Publication **51**, 203 pp.

Kieffer, S.W., 1984, Seismicity at Old Faithful Geyser: An isolated source of geothermal noise and possible analogue of volcanic seismicity, *Journal of Geophysical Research*, **93**: 8157–63.

Kienholz, H., 1977, Kombinierte Geomorphologische Gefahrenkarte 1:10 000 von Grindelwald, *Catena*, **3**: 265–94.

Kienholz, H., 1978, Maps of geomorphology and natural hazards of Grindelwald, Switzerland, scale 1:10 000, *Arctic and Alpine Research*, **10**: 169–84.

Kienholz, H., 1984, Natural hazards: A growing menace. In Brugger, E.A, Furre, G., Messerli, B. and Messerli P. (eds), *The Transformation of Swiss Mountain Regions*, Verlag Paul Haupt, Berne and Stuttgart, 385–405.

Kienholz, H., Hafner, H. and Schneider, G., 1984, Stability, instability and conditional instability. Mountain ecosystem concepts based on a field survey of the Kakani Area in the Middle Hills of Nepal, *Mountain Research and Development*, **4**(1): 55–62.

Kienholz, H., Hafner, H., Schneider, G. and Tamraker, R., 1983, Mountain hazards mapping in Nepal's Middle Mountains with maps of land use and geomorphic damage (Kathmandu–Kakani area), *Mountain Research and Development*, **3** (3): 195–220.

Kienholz, H., Schneider, G., Bichsel, M., Grunder, M. and Mool, P., 1984, Mapping of mountain hazards and slope stability, *Mountain Research and Development*, **4** (3): 247–66.

King, C.A.M. and Gage, M., 1961, Note on the extent of glaciation in part of west Kerry, *Irish Geographer*, **4**: 202–8.

King, C.A.M. and Lewis, W.V., 1961, A tentative theory of ogive formation, *Journal of Glaciology*, **3**: 913–39.

King, L.C., 1967, *The Morphology of the Earth*, 2nd edn, Oliver and Boyd, Edinburgh.

Kingdom-Ward, F., 1955, Aftermath of the great Assam earthquake of 1950, *Geographical Journal*, **121**: 290–303.

Kirkby, M.J. and Statham, I., 1975, Surface stone movement and scree formation, *Journal of Geology*, **83**: 349–62.

Kjarntansson, G., 1967, The Steinholtshlaup Central-South Iceland on January 15 1967, *Jokull*, **17**: 249–62.

Kling, G.W., Clark, M.A., Compton, H.R., Devine, J.D., Evans, W.C., Humphrey, A.M., Koenigsberg, E.J., Lockwood, J.P., Tuttle, M.L. and Wagner, G.N., 1987, The 1986 Lake Nyos gas disaster in Cameroon, West Africa, *Science*, **236**: 169–75.

Knighton, D., 1984, *Fluvial Forms and Processes*, Arnold, London.

Komarkova, V. and Webber, P.J., 1978, An alpine vegetation map of Niwot Ridge, Colorado, *Arctic and Alpine Research*, **1**: 1–29.

Koster, E.H., 1978, Transverse ribs: their characteristics, origin and paleohydraulic significance. In Miale, A.D. (ed.), *Fluvial Sedimentology*, Canadian Society of Petroleum Geologists, 161–86.

Kuusisto, E., 1984, *Snow Accumulation and Snowmelt in Finland*, Publications of the Water Research Institute, National Board of Waters, Finland, **55**: 149 pp.

Kvasov, D.D. and Verbitsky, M.Y., 1981, Causes of Antarctic glaciation in the Cenozoic, *Quaternary Research*, **15**: 1–17.

Laban, P., 1978, *Field Measurements on Erosion and Sedimentation in Nepal*, Department of Soil Conservation and Watershed Management, PAO/UNDP/IWM/SP/05.

Laban, P., 1979, *Landslide Occurrence in Nepal*, Phewa Tal Project Report no. 8P/13, Integrated Watershed Management Project, Kathmandu.

La Chapelle, E.R., 1977a, Snow avalanches: a review of current research and applications, *Journal of Glaciology*, **19**: 313–24.

La Chapelle, E.R., 1977b, Alternative methods for the artificial release of snow avalanches, *Journal of Glaciology*, **19**: 389–97.

Lall, J.S. and Moddie, A. (eds), 1981, *The Himalayas: Aspects of Change*, Oxford University Press, New Delhi.

LaMarche, V.C., 1968, Rates of slope degradation as determined from botanical evidence, White Mountains, California, *United States Geological Survey Professional Paper*, **352-I**: 341–77.

Lang, H., Schadler, B. and Davidson, G., 1977, Hydroglaciological investigations on the Ewigschneefeld-Grosser Aletschgletscher, *Zeitschrift für*

Gletscherkunde und Glazialgeologie, **12**: 109–24.

Lang, H., Leibundgut, C. and Festel, E., 1979, Results from tracer experiments on the water flow through the Aletschgletscher, *Zeitschrift für Gletscherkunde und Glazialgeologie*, **15**: 209–18.

Larsson, S., 1982, Geomorphological effects on the slopes of Longyear Valley, Spitsbergen after a heavy rainstorm in July 1972, *Geografiska Annaler*, **64A**: 105–25.

Latter, J.H., 1986, Volcanic risk and surveillance in New Zealand, *New Zealand Geological Survey Record*, **10**: 5–22.

Lauer, W., 1973, The altitudinal belts of the vegetation in the Central Mexican Highlands and their climatic conditions, *Arctic and Alpine Research* **5** (3): A99–A113.

Lauer, W., 1984a, Natural potential and land-use system of the Kallawaya in the Upper Charazani Valley (Bolivia). In Lauer, W. (ed.), *Natural Environments and Man in Tropical Mountain Ecosystems*, Franz Steiner Verlag, Wiesbaden, 173–96.

Lauer, W. (ed.), 1984, *Natural Environments and Man in Tropical Mountain Ecosystems*, Franz Steiner Verlag, Wiesbaden.

Lautridou, J.P., 1971, Conclusions générales des récherches de gelifraction experimentale du Centre de Géomorphologie, *Bulletin du Centre de Géomorphologie du CNRs*, **10**: 63–84.

Lautridou, J.P., 1988, Recent advances in cryogenic weathering. In Clark, M.J. (ed.), *Advances in Periglacial Geomorphology*, Wiley, Chichester, 33–47.

Lawson, A.C., 1915, Epigene profiles of the desert, *Bulletin California University Department of Geological Science*, **9**: 23–48.

Lawson, D.E., 1982, Mobilisation, movement and deposition of active subaerial sediment flows, Matanuska Glacier, Alaska, *Journal of Geology*, **90**: 279–300.

Lee, R., 1964, Potential insolation as a topoclimatic characteristic of drainage basins, *International Association of Scientific Hydrology*, **9**: 27–41.

Lehmann, O., 1933, Morphologische Theorie der Verwitterung von Steinschlagwanden, *Vierteljahrsschrift der Naturforschende Gesellschaft in Zürich*, **87**: 83–126.

Lehmann, O., 1934, Über die morphologischen Folgen der Wandwitterung, *Zeitschrift für Geomorphologie*, **8**: 93–9.

Lehre, A.K., 1981, Sediment budget of a small California Coast Range drainage basin near San Francisco. In Davies, T.R.H. and Pierce, A.J. (eds), *Erosion and Sediment Transport in Pacific Rim Steeplands*, International Association of Hydrological Sciences Publication **132**: 123–39.

Lehre, A.K., Collins, B. and Dunne, T., 1983, Post-eruption sediment budget for the North Fork Toutle River drainage, June 1980–June 1981, *Zeitschrift für Geomorphologie*, Supplementband, **46**: 143–63.

Lewis, W.M. and Grant, M.C., 1979, Relationships between stream discharge and yield of dissolved substances from a Colorado mountain watershed, *Soil Science*, **128**: 353–63.

Lewis, W.M. and Grant, M.C., 1980, Relationships between snow cover and winter losses of dissolved substances from a mountain watershed, *Arctic and Alpine Research*, **12**: 11–17.

Lewis, W.V., 1936, Nivation, river grading and shoreline development in southeast Iceland, *Geographical Journal*, **88**: 431–7.

Lewis, W.V., 1938, A melt-water hypothesis of cirque formation, *Geological Magazine*, **75**: 249–65.

Lewis, W.V., 1939, Snow-patch erosion in Iceland, *Geographical Journal*, **94**: 153–61.

Lewis, W.V., 1954, Pressure release and glacial erosion, *Journal of Glaciology*, **2** (16): 417–22.

Lewis, W.V. (ed.), 1960, Norwegian cirque glaciers, *Royal Geographical Society Research Series*, **4**, 104 pp.

Li, J., 1980, Recent advances of the study on existing glaciers in Qing-Zang Plateau, *Glaciology and Crypoedology*, **2** (2): 11–14.

Li, Y.H., 1976, Denudation of Taiwan Island since the Pliocene epoch, *Geology*, **4**: 105–7.

Libby, W.G., 1968, Rock glaciers in the north Cascade Range, Washington (abs.), *Geological Society of America Special Paper*, **101**: 318–19.

Lied, K. and Bakkehoi, S., 1980, Empirical calculations of snow avalanche runout distance based on topographic parameters, *Journal of Glaciology*, **29**: 165–78.

Liestol, O., 1955, Glacier dammed lakes in Norway, *Norsk Geografisk Tidsskrift*, **15**: 122–49.

Limbird, A., 1985, Genesis of soils affected by discrete volcanic ash inclusions, Alberta, Canada, *Catena*, Supplement 7: 120–30.

Linton, D.L., 1957, Radiating valleys in glaciated lands, *Tijdschrift van het Kononklijke Nederlandsche Aardrijkskundig Genootschap*, **74**: 297–312.

Linton, D.L., 1963, The forms of glacial erosion, *Transactions of the Institute of British Geographers*, **33**: 1–28.

Linton, D.L., 1964, Landscape evolution. In Priestley, R., Adie, R.J. and Robin G. de Q. (eds), *Antarctic Research*, Butterworth, London, 85–99.

Linton, D.L. and Moisley, H.A., 1960, The origin of Loch Lomond, *Scottish Geographical Magazine*, **76**: 27–37.

Lisle, T.E., 1986, Stabilisation of a gravel channel by large streamside obstruction and bedrock bends, Jacoby Creek, northwestern California, *Bulletin of the Geological Society of America*, **97**: 999–1011.

Lisle, T.E., 1987, Overview: channel morphology and sediment transport in steepland streams. In *Erosion and Sedimentation in the Pacific Rim*, International Association of Hydrological Sciences Publication, **165**: 287–97.

Litvan, G.G., 1980, Freeze-thaw durability of porous materials, Durability of building material components, *American Society Testing Materials Special Technical Publication*, **691**: 455–63.

Liu Shuzheng and Zhong Xianghao, 1987, The geomorphology of the Hengduan Mountains, China. In Gardiner, V. (ed.), *International Geomorphology 1986*, Pt II, Chichester, 229–37.

Lliboutry, L., Arnao, B.M., Pautre, A. and Schneider, B., 1977, Glaciological problems set by the control of dangerous lakes in Cordillerra Blanca, Peru: I. Historical failures of morainic dams, their causes and prevention, *Journal of Glaciology*, **18**: 239–54.

Lobeck, A.K., 1926, *Panorama of Physiographic Types*, Geographical Press, Columbia University, New York.

Lobeck, A.K., 1939, *Geomorphology*, McGraw-Hill, New York.

Lockwood, J.P., Costa, J.E., Tuttle, M.L., Nini, J. and Tebor, S.G., 1988, The potential for catastrophic dam failure at Lake Nyos maar, Cameroon, *Bulletin Volcanologique*, **50**: 340–9.

Loffler, E., 1972, Pleistocene glaciation of Papua and New Guinea, *Zeitschrift für Geomorphologie*, NF Supplementband, **13**: 32–58.

Loffler, E., 1977, *Geomorphology of Papua New Guinea*, Australian National University, Canberra.

Lomnitz, C., 1974, *Global Tectonics and Earthquake Risk*, Elsevier, Amsterdam.

Lorenzo, J., 1969, Minor periglacial phenomena among the high volcanoes of Mexico. In Pewe, T.L. (ed.), *The Periglacial Environment*, McGill-Queen's University Press, Montreal, 161–75.

Loughran, R.J., Campbell, B.L. and Elliott, G.L., 1981, Sediment erosion, storage and transport in a small steep drainage basin at Pokolbin, NSW, Australia. In Davies, T.R.H. and Pierce, A.J. (eds), *Erosion and Sediment Transport in Pacific Rim Steeplands*, International Association of Hydrological Sciences Publication, **132**: 252–68.

Loup, J. and Lovie, C., 1967, Sur la fréquence des avalanches en Haute Tarentaise, *Revue de Geographie Alpine*, **55** (4): 587–604.

Love, D., 1970, Subarctic and subalpine: where and what? *Arctic and Alpine Research*, **2**: 63–73.

Luckman, B.H., 1971, The role of snow avalanches in the evolution of alpine talus slopes. In Brunsden, D. (ed.), *Slopes: Form and Process*, Institute of British Geographers Special Publication, **3**: 93–110.

Luckman, B.H., 1972, Some observations on the erosion of talus slopes by snow avalanches in Surprise Valley, Jasper National Park, Alberta. In Slaymaker, O. and McPherson, H.J. (eds), *Mountain Geomorphology: Geomorphological Processes in the Canadian Cordillerra*, Tantalus Press, Vancouver, 85–92.

Luckman, B.H., 1976, Rockfalls and rockfall inventory data: some observations from Surprise Valley, Jasper National Park, Canada, *Earth Surface Processes*, **1**: 287–98.

Luckman, B.H., 1977, The geomorphic activity of snow avalanches, *Geografiska Annaler*, **59A**: 31–48.

Luckman, B.H., 1978a, Geomorphic work of snow avalanches in the Canadian Rocky Mountains, *Arctic and Alpine Research*, **10**: 261–76.

Luckman, B.H., 1978b, The measurement of debris movement on alpine talus slopes, *Zeitschrift für Geomorphologie*, Supplementband **29**: 117–29.

Luckman, B.H. and Crockett, J.K., 1978, Distribution and characteristics of rock glaciers in the southern part of Jasper National Park, Alberta, *Canadian Journal of Earth Sciences*, **15**: 540–50.

Luhr, J.F. and Carmichael, I.S.E., 1982, The Colima volcanic complex, Mexico, III. Ashfall and scoria-fall deposits from the upper slopes of Volcan Colima, *Contribution Mineralogy and Petrology*, **80**: 262–75.

Mabbutt, J.A., 1966, Landforms of the Wester Macdonnell Ranges. In Dury, G.H. (ed.), *Essays in Geomorphology*, Heinemann, London, 83–119.

Macdonald, G., 1972, *Volcanoes*, Prentice Hall, Englewood Cliffs, NJ.

Machida, H., 1967, The recent development of the Fuji volcano, Japan, *Geographical Report, Tokyo Metropolitan University*, **2**: 11–20.

Mackay, J.R. and Mathews, W.H., 1973, Geomorphology and Quaternary history of the Mackenzie River Valley near Fort Good Hope NWT, Canada, *Canadian Journal of Earth Sciences*, **10**: 26–41.

Madole, R.F., 1972, Neoglacial facies in the Colorado Front Range, *Arctic and Alpine Research*, **4**: 119–30.

Maizels, J.K., 1978, Débit des eaux de fonte, changes sédimentaires et taux d'érosion dans le massif du Mont Blanc, *Revue de Géographie Alpine*, **66**: 65–91.

Malaurie, J., 1960, Gélifraction, éboulis er ruissellemenet sur la côte N.O. de Groenland, *Zeitschrift für Geomorphologie*, NF Supplementband **1**: 59–68.

Male, D.H. and Granger, R.J., 1979, Energy mass fluxes at the snow surface in a prairie environment. In Colbeck, S.C. and Ray, M. (eds), *Proceedings*

Modelling Snow Cover Runoff, CRREL, Hanover, USA, 101–24.

Malgot, J. and Mahr, T., 1979, Engineering geological mapping of the West Carpathian landslide areas, *Bulletin of the International Association of Engineering Geologists,* **19**: 116–21.

Manley, G., 1959, The late-glacial climate of North-West England, *Liverpool and Manchester Geological Journal,* **2**: 188–215.

Marchand, D.E., 1970, Soil contamination in the White Mountains, eastern California, *Bulletin of the Geological Society of America,* **81** (8): 2497–505.

Marcus, M.G. and Ragle, R.H., 1970, Snow accumulation in the Krefield Ranges, St. Elias Mountains, Yukon, *Arctic and Alpine Research,* **4**: 277–92.

Marr, J.W., 1967, *Ecosystems of the East Slope of the Front Range in Colorado,* Series in Biology no. 8, University of Colorado Press, Boulder.

Marr, J.W., Johnson, A.W., Osburn, W.S. and Knorr, O.A., 1968, *Data on Mountain Environments, Front Range, Colorado; Four Climax Regions, part 2, 1953–1958,* Series in Biology no. 28, University of Colorado Press, Boulder.

Marr, J.W., Clark, J.M., Osburn, W.S. and Paddock, M.W., 1968, *Data on Mountain Environments, Front Range; Colorado; Four Climax Regions, Part 3, 1959–1964,* Series in Biology no. 29, University of Colorado Press, Boulder.

Martinelli, M., Lang, T.E. and Mears, A.I., 1980, Calculations of avalanche friction coefficients from field data, *Journal of Glaciology,* **26**: 109–19.

Mathews, W.H., 1955, Permafrost and its occurrence in the southern coast mountains of British Columbia, *Canadian Alpine Journal,* **38**: 94–8.

Mathews, W.H., 1956, Physical limnology and sedimentation in a glacial lake, *Bulletin Geological Society of America,* **67**: 537–52.

Mathews, W.H. and Mackay, J.R., 1963, Snowcreep studies, Mount Seymour, British Columbia: preliminary field investigations, *Geographical Bulletin,* **20**: 58–75.

Mathews, W.H. and Mackay, J.R., 1975, Snow creep: its engineering problems and some techniques and results of its investigation, *Canadian Geotechnical Journal,* **12**: 187–98.

Matthes, F.E., 1900, Glacial sculpture of the Bighorn Mountains, Wyoming, *United States Geological Survey 21st Annual Report 1899–1900,* 167–90.

Matthes, F.E., 1930, Geologic history of the Yosemite Valley, *United States Geological Survey Professional Paper,* **160**: 54–103.

Matthes, F.E., 1938, Avalanche sculpture in the Sierra Nevada of California, *Bulletin of the International Association of Scientific Hydrology,* **23**: 631–7.

Mauch, S.P., 1983, Key processes for stability and instability of mountain ecosystems. Is the bottleneck really a data problem? *Mountain Research and Development,* **3** (2): 113–19.

May, D.C.E., 1973, Models for predicting composition and production of alpine tundra vegetation from Niwot Ridge, Colorado. Unpublished MA thesis, University of Colorado, Boulder.

Mayewski, P.A. and Hassinger, J., 1980, Characteristics and significance of rockglaciers in southern Victoria Land, Antarctica, *Antarctic Journal of the United States,* **5**: 68–9.

Mayewski, P.A., Jeschke, P.A. and Ahmad, N., 1981, An active rockglacier, Wavbal Pass, Jammu and Kashmir Himalaya, India, *Journal of Glaciology,* **27**: 201–2.

Mayo, L.R., Meier, M.F. and Tangborn, W.V., 1972, A system to combine stratigraphic and annual mass-balance systems: a contribution to the International Hydrological Decade, *Journal of Glaciology,* **11**: 3–14.

McCabe, L.H., 1939, Nivation and corrie erosion in West Spitsbergen, *Geographical Journal*, **94**: 447–65.

McGown, A. and Derbyshire, E., 1977, Genetic influences on the properties of tills, *Quarterly Journal of Engineering Geology*, **10**: 389–410.

McGreevy, J.P., 1981, Perspectives on frost shattering, *Progress in Physical Geography*, **5**: 56–75.

McGreevy, J.P., 1982, Frost and salt weathering: further experimental results, *Earth Surface Processes*, **7**: 475–88.

McGreevy, J.P. and Whalley, W.B., 1982, The geomorphic significance of rock temperature variations in cold environments: a discussion, *Arctic and Alpine Research*, **14**: 157–62.

McPherson, H.J., 1971, Downstream changes in sediment character in a high energy mountain stream channel, *Arctic and Alpine Research*, **3**: 65–79.

McPherson, H.J. and Hirst, F., 1972, Sediment changes on two alluvial fans in the Canadian Rocky Mountains. In Slaymaker, O. and McPherson, H.J. (eds), *Mountain Geomorphology: Geomorphological processes in the Canadian Cordillerra*, Tantalus Press, Vancouver, 161–75.

McSaveney, E.R., 1972, The surficial fabric of rockfall talus. In Morisawa, M.E. (ed.), *Quantitative Geomorphology: Some Aspects and Applications*, Allen & Unwin, London, 181–97.

Meade, R.H., Emmett, W.W. and Myrick, R.M., 1981, Movement and storage of bed material during 1979 in East Fork River, Wyoming, USA, *International Association of Scientific Hydrology Publication*, **132**: 225–35.

Meier, M.F., 1960, *Mode of Flow of Saskatchewan Glacier*, United States Geological Survey Professional Paper, **351**.

Meier, M.F., 1961, Mass budget of South Cascade Glacier 1957–60, *United States Geological Survey Professional Paper*, **424B**: 206–11.

Meier, M.F. and Evans, W.E., 1975, Comparison of different methods of estimating snow cover in forested, mountainous basins using Landsat images. In *Operational Applications of Satellite Snowcover Observations*, Proceedings NASA Symposium 1975, NASA SP–391, 215–34.

Meier, M.F. and Johnson, A., 1962, The kinematic wave on Nisqually glacier, Washington, *Journal of Geophysical Research*, **67**: 886.

Meier, M.F. and Post, A.S., 1962, Recent variations in mass net budgets of glaciers in Western North America, *International Association of Scientific Hydrology Publication*, **58**: 63–77.

Meier, M.F. and Post, A.S., 1969, What are glacier surges?, *Canadian Journal of Earth Sciences*, **6** (4): 807–17.

Meier, M.F., Tangborn, W.V., Mayo, L.R. and Post, A.S., 1971, *Combined Ice and Water Balances of Gulkana and Wolverine Glaciers, Alaska and South Cascade Glacier, Washington, 1965 and 1966 Hydrologic Years*, United States Geological Survey Professional Paper, **715A**.

Mellors, R.A., Waitt, R.B. and Swanson, D.A., 1988, Generation of pyroclastic flows and surges by hot-rock avalanches from the dome of Mount St Helens volcano, USA, *Bulletin Volcanologique*, **50**: 14–25.

Menard, H.W., 1961, Some rates of regional erosion, *Journal of Geology*, **69**: 154–61.

Merill, R.K. and Pewe, T.L., 1977, *Late Cenozoic Geology of the White Mountains, Arizona*, Arizona Bureau of Geology and Mineral Technology, Special Paper 1, 65 pp.

Messerli, B., 1973, Problems of vertical and horizontal arrangement in the high mountains of the extreme arid zone (Central Sahara), *Arctic and Alpine*

Research, **5**(3): A139–47.

Messerli, B., 1983, Stability and instability of mountain ecosystems: Introduction to the workshop, *Mountain Research and Development*, **3**(2): 81–94.

Messerli, B. and Ives, J.D., 1984a, Gongga Shan (7556 m) and Yulongxue Shan (5596 m). Geoecological observations in the Hengduan Mountains of Southwestern China. In Lauer, W. (ed.), *Natural Environments and Man in Tropical Mountain Ecosystems*, Franz Steiner Verlag, Wiesbaden, 55–77.

Messerli, B. and Ives, J.D., 1984b, *Mountain Ecosystems: Stability and Instability*, International Mountain Society, Boulder, Co.

Meybeck, M., 1976, Total mineral dissolved transport by world major rivers, *Hydrological Sciences Bulletin*, **21**: 265–84.

Milhous, R.T. and Klingeman, P.C., 1973, Sediment transport system in a gravel-bottomed stream, *Proceedings of the 21st Annual Specialty Conference of the Hydraulics Division, American Society for the Testing of Materials*, 293–303.

Miller, J.P., 1958, *High Mountain Streams: Effects of Geology on Channel Characteristics and Bed Material*, State Bureau of Mines and Mineral Resources, New Mexico Institute of Mining and Technology, Socorro, NM, Memoir 4.

Miller, K.J. (ed.), 1984, *The International Karakoram Project*, Cambridge University Press, Cambridge, 2 vols.

Miller, M.M., 1973, Entropy and self-regulation of glaciers in the Arctic. In Fahey B.D. and Thompson, R.D. (eds), *Research in Polar and Alpine Geomorphology*, Third Guelph Symposium on Geomorphology, Geo Books, Norwich, 136–58.

Mills, R.H., 1979, Some implications of sediment studies for glacial erosion on Mt. Rainier, *Northwest Science*, **53**: 190–9.

Ming-ko Woo, 1972, A predictive model for snow storage in the temperate forest zone of the Coastal Mountains. In Slaymaker, O. and McPherson, H.J. (eds), *Mountain Geomorphology: Geomorphological Processes in the Canadian Cordillerra*, Tantalus Press, Vancouver, 207–14.

Muzuyama, T., 1984, Catastrophic change of a mountain river and its hydraulic explanation, *Transactions of the Japanese Geomorphologists Union*, **5**: 195–203.

Moberly, R., Jr, 1963, Rate of denudation in Hawaii, *Journal of Geology*, **71**: 371–5.

Molloy, B.P.J., 1964, Soil genesis and plant succession in the subalpine zone of Torlesse Range, Canterbury, New Zealand, *New Zealand Journal of Botany*, **2**: 143–76.

Monasterio, M. and Sarmiento, G., 1984, Ecological diversity and human settlements in the tropical northern Andes. In Lauer, W. (ed.), *Natural Environments and Man in Tropical Mountain Ecosystems*, Franz Sterner Verlag, Wiesbaden, 295–306.

Moon, B.P., 1984, Refinement of a technique for determining rock mass strength for geomorphological purposes, *Earth Surface Processes and Landforms*, **9**: 189–93.

Moon, B.P. and Selby, M.J., 1983, Rock mass strength and scarp forms in southern Africa, *Geografiska Annaler*, **65A**: 135–45.

Moore, J.G., 1967, Base surge in recent volcanic eruptions, *Bulletin Volcanologique*, **30**: 337–63.

Moore, J.G. and Albee, W.C., 1981, Topographic and structural changes, March–July 1980 — photogrammetric data. In Lipman, P.W. and Mullineaux,

D.R. (eds), *The 1980 Eruptions of Mount St. Helens, Washington*, United States Geological Survey Professional Paper, **1250**: 123–34.

Moore, J.G., Nakamura, K. and Alcaraz, A., 1966, The 1965 eruption of Taal volcano, *Science*, **151**: 955–60.

Moore, J.G. and Sisson, T.W., 1981, Deposits and effects of the May 18 pyroclastic surge. In Lipman, P.W. and Mullineaux, D.R. (eds), *The 1980 Eruptions of Mount St Helens, Washington*, United States Geological Survey Professional Paper, **1250**: 421–38.

Moore, R.D. and Owens, I.F., 1984, Controls on advective snowmelt in a maritime alpine basin, *Journal of Climate and Applied Meteorology*, **23**: 135–42.

Moran, S.R., Clayton, L., Hooke, R.L., Fenton, M.M. and Andriashek, L.D., 1980, Glacier-bed landforms of the prairie region of North America, *Journal of Glaciology*, **25**: 457–76.

Morisawa, M., 1985, *Rivers*, Longman, London and New York.

Moriya, I., 1970, History of Akagi volcano, *Bulletin Volcanological Society of Japan*, Series 2, **15**: 120–31.

Moriya, I., 1979, Geomorphological developments and classification of the Quaternary volcanoes in Japan, *Geographical Review of Japan*, **52**: 479–501.

Morris, S.E., 1981, Topoclimatic factors and the development of rock glacier facies, Sangre de Cristo Mountains, southern Colorado, *Arctic and Alpine Research*, **13**: 329–38.

Morton, D.M., 1971, Seismically triggered landslides in the area above the San Fernando Valley, *United States Geological Survey Professional Paper*, **733**: 99–104.

Moser, M., 1978, Proposals for geotechnical maps concerning slope stability potential in mountain watersheds, *International Association of Engineering Geologists Bulletin*, **17**: 100–8.

Mosley, M.P., 1978, Erosion in the south-eastern Ruahine range: its implications for downstream river control, *New Zealand Journal of Forestry*, **23**: 21–48.

Mugridge, S.J. and Harvey, R.Y., 1983, Disintegration of shale by cyclic wetting and drying and frost action, *Canadian Journal of Earth Sciences*, **20**: 568–76.

Muir-Wood, R., 1984, A science of mountains. In Miller, K.J. (ed.), *The International Karakoram Project*, vol. 1, Cambridge University Press, Cambridge, 53–6.

Mukhopadhyay, S.C., 1982, *The Tista Basin: A Study in Fluvial Geomorphology*, K.P. Bagchi & Co., Calcutta.

Muller, F. and Iken, A., 1973, Velocity fluctuations and water regime of arctic valley glaciers, *International Association of Scientific Hydrology Publication*, **95**: 165–82.

Murai, I., 1961, A study of the textural characteristics of pyroclastic flow deposits in Japan, *Bulletin of the Earthquake Research Institute, Tokyo University*, **39**: 133–248.

Myers, J.S., 1976, Erosion surfaces and ignimbrite eruption, measures of Andean uplift in northern Peru, *Journal of Geology*, **11**: 29–44.

Nakamura, F., Araya, T. and Higashi, S., 1987, Influence of river channel morphology and sediment production on residence time and transport distance. In *Erosion and Sedimentation in the Pacific Rim*, International Association of Hydrological Sciences Publication, **165**: 355–64.

Naranjo, J.A. and Francis, P., 1987, High velocity debris avalanche at Lastarria volcano in the north Chilean Andes, *Bulletin Volcanologique* **49**: 509–14.

Nichols, R.L., 1969, Geomorphology of Inglefield Land, North Greenland, *Meddelelser om Grønland*, **188**: 1–109.

Nossin, J.J., 1967, Comparative study of the Kalagarh landslip, southern Himalayas, *Zeitschrift für Geomorphologie*, **11**: 356–67.

Numata, M., 1972, Ecological interpretation of vegetational zonation of high mountains, particularly in Japan and Taiwan. In Troll, C. (ed.), *Geoecology of the High-mountain Regions of Eurasia*, Franz Steiner Verlag, Wiesbaden, 288–99.

Nye, J.F., 1952, The mechanics of glacier flow, *Journal of Glaciology*, **2**: 82–93.

Nye, J.F., 1959, The deformation of a glacier below an ice fall, *Journal of Glaciology*, **3**: 387–408.

Nye, J.F., 1960, The response of glaciers and ice sheets to seasonal and climatic changes, *Proceedings of the Royal Society A*, **256**: 559–84.

Nye, J.F., 1965a, The flow of a glacier in a channel of rectangular, elliptical or parabolic cross-section, *Journal of Glaciology*, **5**: 661–90.

Nye, J.F., 1965b, The frequency response of glaciers, *Journal of Glaciology*, **5**: 567–87.

Nye, J.F. and Martin, P.C.S., 1968, Glacial erosion, *International Association of Scientific Hydrology Publication*, **79**: 78–86.

Oberlander, T., 1965, *The Zagros Streams*, Syracuse Geography Series no. 1, Syracuse University, 168 pp.

Obled, C. and Good, W., 1980, Recent development of avalanche forecasting by discriminant analysis techniques: a methodological review and some applications to the Pansen area (Davos, Switzerland), *Journal of Glaciology*, **25**: 315–46.

Obled, C. and Harder, H., 1979, A review of snowmelt in the mountain environment. In Colbeck, S.C. and Ray, M. (eds), *Proceedings, Modelling of Snowcover Runoff*, US Army Cold Regions Research and Engineering Laboratory, Hanover, 179–204.

O'Connor, K.F., 1980, The use of mountains: a review of New Zealand experience. In Anderson, A.G. (ed.), *The Land Our Future: Essays On Land Use and Conservation in New Zealand*, Longman Paul, New Zealand Geographical Society Inc., 193–222.

O'Connor, K.F., 1984, Stability and instability of ecological systems in New Zealand mountains, *Mountain Research and Development*, **4** (1): 15–29.

O'Dell, A.C., 1956, *Railways and Geography*, Hutchinson, London.

Oerter, H. and Moser, H., 1982, Water storage and drainage within the firn of a temperate glacier (Vernagtferner, Oetztal Alps, Austria). In Glen, J.W. (ed.) *Hydrological Aspects of Alpine and High Mountain Areas*, International Association of Hydrological Sciences Publication **138**: 71–81.

O'Gilvie, R.T. and Baptie, B., 1967, A permafrost profile in the Rocky Mountains of Alberta, *Canadian Journal of Earth Sciences*, **4**: 744–5.

Ohmori, H., 1978, Relief structure of the Japanese mountains and their stages in geomorphic development, *Bulletin Department of Geography*, University of Tokyo, **10**: 31–85.

Ohmori, H., 1983, Characteristics of the erosion rate in the Japanese mountains from the viewpoint of climatic geomorphology, *Zeitschrift für Geomorphologie*, Supplementband **46**: 1–14.

Okunishi, K. and Okimura, T., 1987, Groundwater models for mountain slopes. In Anderson, M.G. and Richards, K.S. (eds), *Slope Stability*, Wiley, Chichester, 265–85.

Okunishi, K., Okuda, S. and Suwa, H., 1987, A large-scale debris avalanche as an episode in slope-channel processes. In *Erosion and Sedimentation in the Pacific Rim*, International Association for Hydrological Sciences Publication, **165**: 225–32.

Ollier, C.D., 1969, *Volcanoes*, MIT Press, London.

Ollier, C.D. and Brown, M.J.F., 1971, Erosion of a young volcano in New Guinea, *Zeitschrift für Geomorphologie*, **15**: 12–28.

Ollier, C.D. and Mackenzie, D.E., 1974, Subaerial erosion of volcanic cones in the tropics, *Journal of Tropical Geography*, **9**: 63–71.

O'Loughlin, C.L., 1969, Stream bed investigations in a small mountain catchment, *New Zealand Journal of Geology and Geophysics*, **12**: 684–706.

Olyphant, G.A., 1983, Analysis of the factors controlling cliff burial by talus within Blanca massif, Southern Colorado, U.S.A., *Arctic and Alpine Research*, **15**: 65–75.

Osborn, E.G., 1975, Advancing rockglacier in the Lake Louise area, Banff National Park, Alberta, *Canadian Journal of Earth Sciences*, **12**: 1060–2.

Ostrem, G., 1974, Present alpine ice cover. In Ives, J.D. and Barry, R.G. (eds), *Arctic and Alpine Environments*, Methuen, London, 226–50.

Ottersberg, R.J. and Nielsen, G.A., 1977, Recognition of volcanic ash influenced soils by soil scientists in western Montana and parts of Idaho, *Soil Survey Horizons*, **18**: 8–13.

Owens, I.F., 1972, Morphological characteristics of alpine mudflows in the Nigel Pass area. In Slaymaker, O. and McPherson, H.J. (eds), *Mountain Geomorphology: Geomorphological Processes in the Canadian Cordillerra*, Tantalus Press, Vancouver, 93–100.

Pain, C.F. and Bowler, J.M., 1973, Denudation following the November 1970 earthquake at Madang, Papua New Guinea, *Zeitschrift für Geomorphologie*, Supplementband **18**: 92–104.

Pal, S.K. and Bagchi, K., 1975, Recurrence of floods in Brahmaputra and Kosi Basins: a study in climatic geomorphology, *Geographical Review of India*, **37**: 242–8.

Parker, G. and Klingeman, P.C., 1982, On why gravel bed streams are paved, *Water Resources Research*, **18** (5): 1409–23.

Parker, G., Klingeman, P.C. and McLean, D.G., 1982, Bedload and size distribution in paved gravel-bed streams, *American Society of Civil Engineers, Journal of Hydraulics Division*, **108** (4): 544–71.

Parker, G.P. and Peterson, A.W., 1980, Bar resistance of gravel-bed streams, *American Society of Civil Engineers, Journal of Hydraulics Division*, **106** (10): 1559–75.

Parsons, R.G., 1978, Soil–geomorphology relations in mountains of Oregon, U.S.A., *Geoderma*, **21**: 25–39.

Paterson, T.T., 1951, Physiographic studies in Northwest Greenland, *Meddelelser om Grønland* **188** (1): 1–109.

Paterson, W.S.B., 1969, *The Physics of Glaciers*, Pergamon, Oxford.

Patton, H.B., 1910, Rock streams of Veta Park, Colorado, *Bulletin of the Geological Society of America*, **21**: 663–76.

Patzelt, G., 1986, Begleitworte zur Karte des Gurgler Ferners 1981, *Zeitschrift für Gletscherkunde und Glazialgeologie*, **22** (2): 163–70.

Pawson, I.G., Stanford, D.D., Adams, V.A. and Nurbu, M., 1984, Growth of tourism in Nepal's Everest Region. Impact on the physical environment and structure of human settlements, *Mountain Research and Development*, **4** (3): 237–46.

Peattie, R., 1936, *Mountain Geography*, Harvard University Press, Cambridge, MA.

Peev, C.D., 1966, Geomorphic activity of snow avalanches, *International Association of Scientific Hydrology Publication*, **69**: 357–68.

Penck, W., 1953, *Morphological Analysis of Landforms*, Hafner, London.

Perla, R.I., 1970, On contributory factors in avalanche hazard evaluation, *Canadian Geotechnical Journal*, **7**: 414–19.

Perla, R.I., Cheng, T.T. and McLung, D.M., 1980, A two parameter model of snow avalanche motion, *Journal of Glaciology*, **26**: 197–208.

Peters, T. and Mool, P.K., 1983, Geological and petrographic base studies for the Mountain Hazards Mapping Project in the Kathmandu–Kakani area, Nepal, *Mountain Research and Development*, **3** (3): 221–6.

Peterson, D.F. and Mohanty, P.K., 1960, Flume studies in steep rough channels, *American Society of Civil Engineers, Journal of Hydraulics Division*, **86**: 55–76.

Pewe, T.L., 1983, Alpine permafrost in the contiguous United States: A review, *Arctic and Alpine Research*, **15** (2): 145–56.

Pickup, G. Higgins, R.J. and Warner, R.F., 1981, Erosion and sediment yield in Fly River drainage basins, Papua New Guinea. In Davis, T.R.H. and Pierce, A.J. (eds), *Erosion and Sediment Transport in Pacific Rim Steeplands*, International Association of Hydrological Sciences, Publication **132**: 438–56.

Pierce, K.L., 1979, *History and Dynamics of Glaciation in the Northern Yellowstone National Park Area*, United States Geological Survey Professional paper, **729F**, 90 pp.

Pierce, W.G., 1961, Permafrost and thaw depressions in a peat deposit in the Beartooth Mountains, northwestern Wyoming, *United States Geological Survey Professional Paper*, **424-B**: 154–6.

Pierdie, R.W. and Noble, I.R., (eds), 1983, Mountain ecology in the Australia region, *Proceedings of Ecological Society of Australia*, **12**.

Pierson, T.C., 1985, Initiation and flow behaviour of the 1980 Pine Creek and Muddy Rivers lahars, Mount St Helens, Washington, *Bulletin of the Geological Society of America*, **96**: 1056–69.

Pierson, T.C. and Scott, K.M., 1985, Downstream dilution of a lahar — transition from debris flow to hyperconcentrated streamflow, *Water Resources Research*, **21**: 1511–24.

Pissart, A., 1964, Vitesse des mouvements du sol an Chambeyron (Basses Alpes), *Biuletyn Peryglacjalny*, **14**: 303–9.

Ponton, J.R., 1972, Hydraulic geometry in the Green and Birkenhead Basins, British Columbia. In Slaymaker, O. and McPherson, H.J. (eds), *Mountain Geomorphology: Geomorphological Processes in the Canadian Cordillerra*, Tantalus Press, Vancouver, 151–60.

Poser, H., 1936, Talstudien aus Westspitzbergen und Ostgrønland, *Zeitschrift für Gletscherkunde und Glazialgeologie*, **24**: 43–98.

Potter, N., Jr, 1972, Ice-cored rock glacier, Galena Creek, northern Absaroka Mountains, Wyoming, *Bulletin of the Geological Society of America*, **83**: 3025–58.

Powell, J.W., 1876, *Report on the Geology of the Eastern Part of the Uinta Mountains*, Washington, DC, 218 pp.

Powell, J.W., 1895, Physiographic features, *National Geographic Society Monograph*, **1**: 34–40.

Preston, James, 1935, *An Outline of Geography*, Ginn & Co., Boston.

Price, L.W., 1973, Rates of mass wasting in the Ruby Range, Yukon Territory. In *Permafrost*, National Academy of Science Publication **2115**: 235–45.

Price, L.W., 1981, *Mountains and Man*, University of California Press, Berkeley.

Priesnitz, K., 1988, Cryoplanation. In Clark, M.J. (ed.), *Advances in Periglacial Geomorphology*, Wiley, Chichester, 49–68.

Prucker, H., 1976, Unterlagen für die Beurteilung der Gefahrdung des Ortes Burs durch die Schesa, Diplomarbeit Bodenkulter, Wien.

Ranalli, G., 1975, Geotechnic relevance of rock-stress determinations, *Tectonophysics*, **29**: 49–58.

Ranganathan, S., 1979, *Agro-forestry: Employment for Millions*, Tata Press, Bombay.

Rango, A. and Martinec, J., 1979, Application of a snowmelt-runoff model using Landsat data, *Nordic Hydrology*, **10**: 225–38.

Rantz, S.E., 1964, *Snowmelt Hydrology of a Sierra Nevada Stream*, United States Geological Survey Water Supply Paper 1779R.

Rapp, A., 1959, Avalanche boulder tongues in Lappland: a description of little-known landforms of periglacial debris accumulation, *Geografiska Annaler*, **41**: 34–48.

Rapp, A., 1960a, Recent development of mountain slopes in Karkevagge and surroundings, northern Scandinavia, *Geografiska Annaler*, **41**: 65–200.

Rapp, A., 1960b, *Talus Slopes and Mountain Walls at Tempelfjorden, Spitsbergen*, Norsk Polarinstitutt 119, 96 pp.

Rapp, A., 1975, Studies of mass wasting in the arctic and in the tropics. In Yatsu, E., Ward, A.J. and Adams, F. (eds), *Mass Wasting*, Proceedings of the Fourth Guelph Symposium in Geomorphology, Geo Abstracts, Norwich, 79–103.

Rapp, A., 1985, Extreme rainfall and rapid snowmelt as causes of mass movements in high latitude mountains. In Church, M. and Slaymaker, O. (eds), *Field and Theory: Lectures in Geocryology*, University of British Columbia Press, Vancouver, 36–55.

Rapp, A. and Nyberg, R., 1981, Alpine debris flows in northern Scandinavia: Morphology and dating by lichenometry, *Geografiska Annaler*, **63A**: 183–96.

Raymond, C.F., 1971, Flow in a transverse section of Athabaska glacier, Alberta, Canada, *Journal of Glaciology*, **10**: 55–84.

Reger, R.D. and Pewe, T.L., 1976, Cryoplanation terraces: indicators of a permafrost environment, *Quaternary Research*, **6**: 99–109.

Reiner, E., 1960, The glaciation of Mount Wilhelm, Australian New Guinea, *Geographical Review*, **50**: 491–503.

Reneau, S. and Dietrich, W.E., 1987, Size and location of colluvial landslides in a steep forested landscape. In *Erosion and Sedimentation in the Pacific Rim*, International Association of Hydrological Sciences Publication, **165**: 39–48.

Retzer, J.L., 1965, Present soil-forming factors and processes in Arctic and alpine areas, *Soil Science*, **99**: 38–44.

Reynaud, L., 1973, Flow of a valley glacier with a solid friction law, *Journal of Glaciology*, **12**: 251–8.

Reynolds, R.C., 1971, Clay mineral formation in an alpine environment, *Clays and Clay Minerals*, **19**: 361–74.

Reynolds, R.C. and Johnson, N.M., 1972, Chemical weathering in the temperate glacial environment of the Northern Cascade Mountains, *Geochimica et Cosmochimica Acta*, **36**: 537–54.

Richards, A.F., 1959, Geology of the Islas Revillagigedo, Mexico, 1. Birth and development of Volcan Barcena, Isla San Benedicto (1), *Bulletin Volcanologique*, **22**: 73–123.

Richards, A.F., 1965, Geology of the Islas Revillagigedo, Mexico, 3. Effects of erosion on Isla San Benedicto 1952–1961 following the birth of Volcan Barcena, *Bulletin Volcanologique*, **28**: 381–403.

Richards, K.S. and Lorriman, N.R., 1987, Basal erosion and mass movement. In

Anderson, M.G. and Richards, K.S. (eds), *Slope Stability*, Wiley, Chichester, 331–57.

Richardson, D., 1968, Glacier outburst floods in the Pacific Northwest, *United States Geological Survey Professional Paper* **600D**, 79–86.

Richmond, G.M., 1962, *Quaternary Stratigraphy of the La Sal Mountains, Utah*, United States Geological Survey Professional Paper 324.

Richter, H., Haase, G. and Barthel, H., 1963, Die Goletzterrassen, *Petermanns Geographische Mitteilungen*, **107**: 183–92.

Rieger, H.C., 1981, Man versus mountain: the destruction of the Himalaya ecosystem. In Lall, J.S. (ed.), *The Himalaya: Aspects of Change*, Oxford University Press, New Delhi, 351–76.

Roberts, R.G. and Church, M., 1986, The sediment budget in severely disturbed watersheds, Queen Charlotte Ranges, British Columbia, *Canadian Journal of Forestry Research*, **16**: 1093–1106.

Robin, C. and Boudal, C., 1984, Une éruption remarquable par son volume: l'événement de type Saint-Helens du Popocatepetl (Mexique), *Comptes Rendus de l'Academie des Sciences, Paris*, **229**: 881–6.

Robin, G. de Q., 1976, Is the basal ice of a temperate glacier at the pressure melting point? *Journal of Glaciology*, **16**: 183–96.

Robin, G. de Q. and Barnes, P., 1969, Propagation of glacier surges, *Canadian Journal of Earth Sciences*, **6** (4): 969–77.

Roots, E.F. and Glen, J.W., 1982, Preface. In Glen, J.W. (ed.), *Hydrological Aspects of Alpine and High Mountain Areas*, International Association of Hydrological Sciences Publication **138**: v–vi.

Rose, W.I. and Stuiber, R.E., 1969, The 1966 eruption of Izako Volcano, El Salvador, *Journal of Geophysical Research*, **74**: 3119–30.

Rothacher, J., 1971, Regimes of streamflow and their modification by logging. In Krygier, J.T. and Hall, J.D. (eds), *Forest Land Uses and the Stream Environment*, Oregon State University, Corvallis, 40–54.

Rothlisberger, H., 1974, Möglichkeiten und Grenzen der Gletscherüberwachung, *Neue Zürcher Zeitung*, **196**: 2–15.

Rothlisberger, H. and Iken, A., 1981, Plucking as an effect of water pressure variations at the glacier bed, *Annals of Glaciology*, **2**: 57–62.

Rothlisberger, H. and Lang, H., 1987, Glacial hydrology. In Gurnell, A.M. and Clark, M.J. (eds), *Glacio-fluvial Sediment Transfer*, Wiley, Chichester, 207–84.

Rouse, W.C., 1984, Flowslides. In Brunsden, D. and Prior, D.B. (eds), *Slope Instability*, Wiley, Chichester, 491–522.

Rudberg, S., 1962, A report on some field observations concerning periglacial geomorphology and mass movement on slopes in Sweden, *Biuletyn Peryglacjalny*, **11**: 311–23.

Rudberg, S., 1964, Slow mass movement processes and slope development in the Norra Storfjäll area, southern Swedish Lapland, *Zeitschrift für Geomorphologie*, Supplementband 5: 192–203.

Rudolph, R., 1962, Abflussstudien an Gletscherbächen, *Veröffentlichung des Museum Ferdinandeum, Innsbruck*, no. 41: 118–266.

Russell, S.O., 1972, Behaviour of steep creeks in a large flood. In Slaymaker, O. and McPherson, H.J. (eds), *Mountain Geomorphology: Geomorphological Processes in the Canadian Cordillerra*, Tantalus Press, Vancouver, 223–7.

Rutkis, J., 1971, *Tables on Relative Relief in Middle and Western Europe*, Uppsala University Naturgeografisk Institutt, report no. 9.

Ruxton, B.P., 1968, Rates of weathering of Quaternary volcanic ash in northeast Papua, *Transactions 9th International Congress Soil Science, Adelaide*, **4**: 367–76.

Ruxton, B.P. and McDougall, I., 1967, Denudation rates in northeast Papua from potassium-argon dating of lavas, *American Journal of Science*, **265**: 545–61.

St. Lawrence, W.F. and Bradley, C.C., 1977, Spontaneous fracture initiation in mountain snow packs, *Journal of Glaciology*, **19**: 411–17.

St. Lawrence, W.F., Lang, T.E., Brown, R.L. and Bradley, C.C., 1973, Acoustic emission in snow at constant rates of deformation, *Journal of Glaciology*, **12**: 144–6.

Salisbury, R.D., 1907, *Physiography*, Henry Holt & Co., New York.

Salway, A.A., 1979, Time series modelling of avalanche activity from meteorological data, *Journal of Glaciology*, **22**: 513–28.

Sarkar, R.L. and Lama, M.P. (eds), 1983, *The Eastern Himalayas: Environment and Economy*, Atma Ram & Sons, New Delhi.

Sawada, T., Ashida, K. and Takahashi, T., 1983, Relationship between channel pattern and sediment transport in a steep gravel bed river, *Zeitschrift für Geomorphologie*, Supplementband **46**: 55–66.

Schaerer, P.A., 1972, Terrain and vegetation of snow avalanche sites at Roger's Pass, British Columbia. In Slaymaker, O. and McPherson, H.J. (eds), *Mountain Geomorphology: Geomorphological Processes in the Canadian Cordillerra*, Tantalus Press, Vancouver, 215–22.

Schafer, J.P. and Hartshorn, S.H., 1965, The Quaternary of New England. In Wright, H.E., Jr and Frey, D.G. (eds), *The Quaternary of the United States*, Princeton University Press, Princeton, 113–28.

Scheidegger, A.E., 1961a, Mathematical models of slope development, *Bulletin of the Geological Society of America*, **72**: 37–50.

Scheidegger, A.E., 1961b, *Theoretical Geomorphology*, Prentice Hall, Englewood Cliffs, NJ.

Scheidegger, A.E., 1963, On the tectonic stresses in the vicinity of a valley and a mountain range, *Proceedings of the Royal Society of Victoria*, **76**: 141–5.

Scheidegger, A.E., 1970, The large scale tectonic stress field in the earth. In Johnson, H. and Smith, B.L. (eds), *The Megatectonics of Continents and Oceans*, Rutgers University Press, New Brunswick, NJ, 223–40.

Schick, R., 1981, Source mechanisms of volcanic earthquakes, *Bulletin Volcanologique*, **44** (3): 491–7.

Schommer, P., 1976, Wasserspeigelmessungen im Firn des Ewigschneefeldes (Schweizer Alpen) 1976, *Zeitschrift für Gletscherkunde und Glazialgeologie*, **12**: 125–41.

Schommer, P., 1978, Rechnerische Nachbildung von Wasserspiegelganglinien im Firn und Vergleich mit Feldmessungen im Ewigschneefeld (Schweizer Alpen), *Zeitschrift für Gletscherkunde und Glazialgeologie*, **14**: 173–90.

Schumm, S.A., 1963, *A Tentative Classification of Alluvial River Channels*, United States Geological Survey Professional Paper, **352C**.

Schumm, S.A., 1965, Quaternary paleohydrology. In Wright, H.E. and Frey, D.G. (eds), *The Quaternary of the United States*, Princeton University Press, Princeton, NJ, 783–93.

Schweinfurth, U., 1957, Die horizontale und vertikale Verbreitung der Vegetation in Himalaya, *Bonner Geographische Abhandlung H20*.

Schweinfurth, U., 1981, The vegetation map of the Himalaya 1957 — a quarter of a century later, *Documents de Cartographie Ecologiques (Grenoble)*, **24**: 19–23.

Schweinfurth, U., 1984, The Himalaya: complexity of a mountain system manifested by its vegetation, *Mountain Research and Development*, 4 (4): 339–44.

Schytt, V., 1959, The glaciers of the Kebnekajse Massif, *Geografiska Annaler*, 41: 213–27.

Scott, W.E., 1984, Assessments of long-term volcanic hazards, *United States Geological Survey Open File Report*, 84–760: 447–98.

Scotter, G.W., 1975, Permafrost profiles in the Continental Divide region of Alberta and British Columbia, *Arctic and Alpine Research*, 7: 93–5.

Seddon, B., 1957, Late-glacial cwm glaciers in Wales, *Journal of Glaciology*, 3: 94–9.

Sekiya, S. and Kikuchi, Y., 1889, The eruption of Bandai-san, *Tokyo Imperial University College Science Journal*, 3 (2): 91–172.

Selby, M.J., 1966, Soil erosion on the pumice lands of the Central North Island, *New Zealand Geographer*, 22: 194–6.

Selby, M.J., 1980, A rock mass strength classification for geomorphic purposes: with tests from Antarctic and New Zealand, *Zeitschrift für Geomorphologie*, NF 24: 31–51.

Selby, M.J., 1982, Controls on the stability and inclinations of hillslopes formed on hard rock, *Earth Surface Processes and Landforms*, 7: 449–67.

Selby, M.J., 1985, *Earth's Changing Surface*, Oxford University Press, Oxford.

Selby, M.J., 1987, Rock slopes. In Anderson, M.G. and Richards, K.S. (eds), *Slope Stability*, Wiley, Chichester, 475–504.

Seshadri, T.N., 1960, *Report on Kosi River and its Behaviour*, Investigation and Research Directorate, Khosi Project, Birpur.

Sharp, R.P., 1942, Mudflow levees, *Journal of Geomorphology*, 5: 222–7.

Sharp, R.P., 1949, Studies of superglacial debris on valley glaciers, *American Journal of Science*, 289–315.

Sheridan, M.F. and Malin, M.C., 1983, Application of computer assisted mapping to volcanic hazard evaluation of surge eruptions, Vulcano, Lipari and Vesuvius, *Journal of Volcanology and Geothermal Research*, 17: 187–202.

Shields, A., 1936, *Anwendung der Ähnlichkeitsmechanik und Turbulenzforschung auf die Geshiebebewegung*, Report 26, Mitteilungen Preuss. Versuchsanst., Wasserbau und Schiffsbau, Berlin.

Shreve, R.L., 1966, Sherman landslide, Alaska, *Science*, 1639–43.

Shreve, R.L., 1968, Leakage and fluidisation in air-layer lubricated avalanches, *Bulletin of the Geological Society America*, 79: 653–8.

Shumskii, P.A., 1950, *The Energy of Glaciation and the Life of Glaciers*, SIPRE Translation 7, United States Army Corps of Engineers.

Shumskii, P.A., Krenke, A.N. and Zetikov, I.A., 1964, Ice and its changes, *Research in Geophysics*, 2: 425–60.

Siebert, L., 1984, Large volcanic debris avalanches: characteristics of source areas, deposits and associated eruptions, *Journal of Volcanology and Geothermal Research*, 22: 163–97.

Siebert, L., Glicken, H. and Ui, T., 1987, Volcanic hazards from Bezymianny- and Bandai-type eruptions, *Bulletin Volcanologique*, 49: 435–59.

Sigurdsson, H., 1982, In the volcano, *Natural History*, 91 (3): 61–7.

Sigurdsson, H., Carey, S.N. and Fisher, R.V., 1987, The 1982 eruptions of the El Chichón volcano, Mexico (3): Physical properties of pyroclastic surges, *Bulletin Vocanologique*, 49: 467–88.

Simonett, D.S., 1967, Landslide distribution and earthquakes in the Bewani Torricelli mountains, New Guinea: a statistical analysis. In Jennings, J.N. and

Mabbutt, J. (eds), *Landform Studies from Australia and New Guinea*, Cambridge University Press, Cambridge, 64–84.

Singh, T., and Kalra, Y.P., 1984, Predicting solute yields in the natural waters of a subalpine system in Alberta, Canada, *Arctic and Alpine Research*, **16** (2): 217–24.

Singh, T. and Kaur, J. (eds), 1985, *Integrated Mountain Research*, Himalayan Books, New Delhi.

Sinha, B.N., 1975, An engineering geological approach to landslides and slope failures, *Proceedings of Seminar on Landslides and Toe Erosion Problems with Special Reference to the Himalayan Region*, Gangtok, Sikkim, 54–65.

Slaymaker, O., 1974, Alpine hydrology. In Ives, J.D. and Barry, R.G. (eds), *Arctic and Alpine Environments*, Methuen, London, 134–55.

Small, R.J., 1982, Glaciers – do they really erode? *Geography*, **6**: 9–14.

Small, R.J., 1987, Englacial and supraglacial sediment: transport and deposition. In Gurnell, A.M. and Clark, M.J. (eds), *Glacio-fluvial Sediment Transfer*, Wiley, Chichester, 111–45.

Smith, N.D., Venol, M.A. and Kennedy, S.K., 1982, Comparison of sedimentation regimes in four glacier-fed lakes of western Alberta. In Davidson-Arnott, R., Nickling, W. and Fahey, B.D. (eds), *Glacial, Glaciofluvial and Glaciolacustrine Systems*, Proceedings of the sixth Guelph Symposium on Geomorphology, Geo Books, Norwich, 203–38.

Soil Survey Staff, 1975, *Soil Taxonomy — A Basic System of Soil Classification for Making and Interpreting Soil Surveys*, United States Department of Agriculture Soil Conservation Service, Agriculture handbook 436.

Solenko, V.P., 1972, Seismogenic destruction of mountain slopes, *Proceedings International Geological Congress* 24, section 13 (engineering geology), 284–90.

Soons, J.M. and Rayner, J.N., 1968, Micro-climate and erosion processes in the Southern Alps, New Zealand, *Geografiska Annaler*, **50A**: 1–15, 120–2.

Souchez, R.A. and Lemmens, M.M., 1987, Solutes. In Gurnell, A.M. and Clark, M.J. (eds), *Glacio-fluvial Sediment Transfer*, Wiley, Chichester, 285–303.

Souchez, R.A. and Lorrain, R.D., 1987, The subglacial sediment system. In Gurnell, A.M. and Clark, M.J. (eds), *Glacio-fluvial Sediment Transfer*, Wiley, Chichester, 147–64.

Sparks, R.S.J., 1976, Grain size variations in ignimbrites and implications for the transport of pyroclastic flows, *Sedimentology*, **23**: 147–88.

Sparks, R.S.J. and Walker, G.P.L., 1973, The ground surge deposit: a third type of pyroclastic rock, *Nature*, **241**: 62–4.

Spencer, A.C., 1900, A peculiar form of talus, *Science*, **11**: 188.

Stablein, G., 1977a, Rezente Morphodynamik und Vorzeitrelief-influenz bei der Hang- und Talentwicklung in Westgønland, *Zeitschrift für Geomorphologie*, Supplementband, **28**: 181–99.

Stablein, G., 1977b, Arktische Boden West-Grønlands, Pedovarianz in Abhängigkeit vom geoökologischen Milieu, *Polarforschung*, **47** (1/2): 11–25.

Stablein, G., 1979, Boden und Relief in Westgrønland, *Zeitschrift für Geomorphologie*, Supplementband, **33**: 232–45.

Stablein, G., 1982, Morphoklimatische Messungen im periglazialen Bereich Ost-Grønlands, *Zeitschrift für Geomorphologie*, Supplementband, **43**: 5–17.

Stablein, G., 1984, Geomorphic altitudinal zonation in the Arctic-alpine mountains of Greenland, *Mountain Research and Development*, **4** (4): 319–31.

Stangl, K.O., Roggensack, W.D. and Hayley, D.W., 1982, Engineering geology of surficial soils, eastern Melville Island, *Proceedings 4th Canadian Permafrost*

Conference, Calgary Alberta, 1981, National Research Council of Canada, Ottawa, 136–47.

Starkel, L., 1970, Cause and effects of a heavy rainfall in Darjeeling and in the Sikkim Himalayas, *Journal of the Bombay Natural History Society*, **67** (1): 45–50.

Starkel, L., 1972, The role of catastrophic rainfall in the shaping of the relief of the lower Himalaya (Darjeeling Hills), *Geographica Polonica*, **21**: 103–47.

Statham, I., 1973, Scree slope development under conditions of surface particle movement, *Transactions of the Institute of British Geographers*, **59**: 41–53.

Statham, I., 1976, A scree slope rockfall model, *Earth Surface Processes and Landforms*, **1**: 43–62.

Stearns, H.T., 1966, *Geology of the State of Hawaii*, Panin, Palo Alto, Ca.

Stiny, J., 1910, *Die Muren*, Wagnerschen Universitäts-Buchandlung, Innsbruck, 17 pp.

Strahler, A.N., 1946, Geomorphic terminology and classification of land masses, *Journal of Geology*, **54**: 32–42.

Strakhov, N.M., 1967, *Principles of Lithogenesis* (English translation by J.P. Fitzsimmons), vol. 1, Consultants Bureau, New York (first published 1962).

Sturgl, J.R. and Scheidegger, A.E., 1967, Tectonic stresses in the vicinity of a wall, *Rock Mechanics and Engineering Geology*, **5**: 137–49.

Sugden, D.E., 1969, The age and form of corries in the Cairngorms, *Scottish Geographical Magazine*, **85**: 34–46.

Sugden, D.E., 1974, Landscapes of glacial erosion in Greenland and their relationship to ice, topographic and bedrock conditions, *Institute of British Geographers Special Publication*, **7**: 177–95.

Sugden, D.E. and John, B.S., 1976, *Glaciers and Landscape*, Arnold, London.

Suggate, R.P., 1950, Franz Josef and other glaciers of the Southern Alps New Zealand, *Journal of Glaciology*, **1**: 422–9.

Supan, A., 1930, *Grundzüge der physischen Erdkunde Vol 2(1) 'Das Land' (Allgemeine Geomorphologie)*, de Gruyter, Berlin and Leipzig.

Sutton, C.W., 1933, Andean mud slide destroys lives and property, *Engineering News Record*, **110**: 562–3.

Suzuki, T., 1975, Geomorphological aspects of volcanoes, *Bulletin of the Volcanological Society of Japan*, Series 2, **20**: 241–6.

Svensson, H., 1959, Is the cross-section of a glacial valley a parabola? *Journal of Glaciology*, **3**: 362–3.

Swanson, F.J. and Dyrness, C.T., 1975, Impact of clearcutting and road construction on soil erosion by landslides in the western Cascade Range, Oregon, *Geology*, **3**: 393–6.

Swanson, F.J. and James, M.E., 1975, *Geology and Geomorphology of the H.J. Andrews Experimental Forest, Western Cascades, Oregon*, United States Agriculture and Forest Service Research Paper, PNW–188.

Swanson, F.J., Janda, R.J. and Dunne, T., 1982, Summary: sediment budget and routing studies. In Swanson, F.J., Janda, R.J., Dunne, T. and Swanson, D.N. (eds), *Sediment Budgets and Routing in Forested Drainage Basins*, Report PNW–141, United States Department Agriculture Forest Service, Pacific Northwest Forest and Range Experimental Station, 157–65.

Swanston, D.N. and Swanson, F.J., 1980, Timber harvesting, mass erosion and steepland forest geomorphology in the Pacific Northwest. In Coates, D.R. (ed.), *Geomorphology and Engineering*, Allen & Unwin, London, 199–221.

Swett, K., Hambrey, M. and Johnson, D., 1980, Rockglaciers in northern Spitsbergen, *Journal of Glaciology*, **88**: 475–82.

Tabler, R.D., 1975, Predicting profiles of snowdrifts in topograhic catchments, *Proceedings of the Western Snow Conference*, 43: 87–97.

Tada, F., 1934, Relation between the altitude and relief energy of the mountain, *Geographical Review of Japan*, 10: 939–67.

Tanaka, M., 1976, Rate of erosion in the Tanzawa Mountains, central Japan, *Geografiska Annaler*, Series A, 3: 155–63.

Tarar, R.N., 1982, Water resources investigation in Pakistan with the help of Landsat imagery — snow surveys 1975–1978. In Glen, J.W. (ed.), *Hydrological Aspects of Alpine and High Mountain Areas*, International Association of Scientific Hydrology Publication, 138: 177–90.

Tarr, R.S. and Martin, L., 1914, *College Physiography*, Macmillan, New York.

Taylor, G.A.M., 1958, The 1951 eruption of Mount Lamington, Papua, *Bulletin of the Australian Bureau of Minerals Resource*, 38.

Temple, P.H., 1965, Some aspects of cirque distribution in the west-central Lake District, northern England, *Geografiska Annaler*, 47: 185–93.

Temple, P.H., 1972, Mountains. In Brunsden, B. and Doornkamp, J. (eds), *The Unquiet Landscape*, David and Charles, Newton Abbot, 15–20.

Temple, P.H. and Rapp, A., 1972, Landslides in the Mgeta area, western Uluguru Mountains, Tanzania, *Geografiska Annaler*, 54A: 157–94.

Terzaghi, K., 1960, Mechanisms of landslides, *Bulletin of the Geological Society of America, Berkeley Volume*, 83–122.

Terzaghi, K., 1962a, Dam foundations on sheeted rock, *Géotechnique*, 12: 199–208.

Terzaghi, K., 1962b, Stability of steep slopes on hard unweathered rock, *Géotechnique*, 12: 251–70.

Thomas, D.M. and Naughton, J.J., 1979, He/CO_2 ratios as premonitors of volcanic activity, *Science*, 204: 1195–6.

Thompson, M. and Warburton, M., 1988, Uncertainty on a Himalayan scale. In Ives, J. and Pitt, D.C. (eds), *Deforestation: Social Dynamics in Watersheds and Mountain Ecosystems*, Routledge, London, 1–53.

Thompson, W.F., 1960, The shape of New England mountains, *Appalachia*, 145–59.

Thompson, W.F., 1961, The shape of New England mountains, *Appalachia*, 316--35, 458–78.

Thompson, W.F., 1962, Cascade Alp slopes and Gipfelfluren as clima-geomorphic phenomena, *Erdkunde*, 14(2): 81–94.

Thompson, W.F., 1964, How and why to distinguish between mountains and hills, *Professional Geographer*, 16 (6): 6–8.

Thorarinsson, S., 1967, The eruptions of Hekla in historical times. In *The Eruption of Hekla 1947–1948, vol. I*, Society of Science Islendica 1967, 1–170.

Thorn, C.E., 1975, The influence of late-lying snow on rock-weathering rinds, *Arctic and Alpine Research*, 7: 373–8.

Thorn, C.E., 1976, Quantitative evolution of nivation in the Colorado Front Range, *Bulletin of the Geological Society of America*, 87: 1169–78.

Thorn, C.E., 1978, The geomorphic role of snow, *Annals Association of American Geographers*, 68: 414–25.

Thorn, C.E., 1979a, Bedrock freeze-thaw weathering regime in an alpine environment, Colorado Front Range, *Earth Surface Processes*, 4: 211–28.

Thorn, C.E., 1979b, Ground temperatures and surficial transport in colluvium during snowpatch meltout; Colorado Front Range, *Arctic and Alpine Research*, 11: 41–52.

Thorn, C.E., 1980, Alpine bedrock temperatures: an empirical study, *Arctic and Alpine Research*, **12**: 773–86.

Thorn, C.E., 1982, Bedrock microclimatology and the freeze-thaw cycle; A brief illustration, *Annals of the Association of American Geographers*, **72**: 131–7.

Thorn, C.E., 1988, Nivation: a geomorphic chimera. In Clark, M.J. (ed.), *Advances in Periglacial Geomorphology*, Wiley, Chichester, 3–31.

Thorn, C.E. and Darmody, R.G., 1980, Contemporary eolian sediments in the alpine zone, Colorado Front Range, *Physical Geography*, **1**: 162–71.

Thorn, C.E. and Darmody, R.G., 1985a, Grain-size distribution of the insoluble component of contemporary eolian deposits in the alpine zone, Front Range, Colorado, *Arctic and Alpine Research*, **17**: 433–42.

Thorn, C.E. and Darmody, R.G., 1985b, Grain-size sampling and characterization of eolian lag surfaces within alpine tundra, Niwot Range, Front Range, Colorado, USA, *Arctic and Alpine Research*, **17**: 443–50.

Thorn, C.E. and Hall, K., 1980, Nivation: an arctic-alpine comparison and reappraisal, *Journal of Glaciology*, **25**: 109–24.

Thorn, C.E. and Loewenherz, D.S., 1987, Alpine mass wasting in the Indian Peaks Area, Front Range, Colorado, In Graf, W. (ed.), *Geomorphic Systems of North America*, Geological Society of America, Boulder, Co., 238–47.

Thornes, J.B., 1971, State, environment and attribute in scree-slope studies. In Brunsden, D. (ed.), *Slopes: Form and Process*, Institute of British Geographers Special Publication, **3**: 49–63.

Tomasson, H., Palsson, S. and Ingolfsson, P., 1980, Comparison of sediment load transport in the Skeidara jokulhlaups in 1972 and 1976, *Jokull*, **30**: 21–33.

Tonkin, P.J., Harrison, J.B.J., Whitehouse, I.E. and Campbell, A.S., 1981, Methods for assessing late Pleistocene and Holocene erosion history in glaciated mountain drainage basins. In Davies, T.R.H. and Pierce, A.J. (eds), *Erosion and Sediment Transport in Pacific Rim Steeplands*, International Association of Scientific Hydrology Publication, **132**: 527–40.

Tourenq, C., 1970, *La Gélivité des roches. Application aux granulats.* Laboratoires des Ponts et Chaussées, Research Report no. **6**, 60 pp.

Tricart, J., 1957, Une lave torrentielle dans les Alpes Autrichiennes, *Revue de Géomorphologie Dynamique*, **8**: 161–5.

Tricart, J., 1970, *Geomorphology of Cold Regions*, (English translation by E. Watson) Macmillan, England.

Tricart, J., Dollfus, O. and Michel, M., 1962, Notes sur quelques aspects géomorphologiques de 'La Forêt de Pierre' de Huaron, *Revue de Géomorphologie Dynamique*, **13**: 125–9.

Troll, C., 1970, Landschaftsökologie (Geoecology) und Biogeocoenology, *Revue Roumaine de Géologie, Géophysique et Géographie Serie de Geographie* (Hommage au Professeur Vintila Mihailescu pour son 80ᵉ anniversaire) **14**: 9–18 (English translation in *Geoforum*, **8**, 1971, 43–6).

Troll, C., 1972a, Geoecology and world-wide differentiation of high-mountain ecosystems. In Troll, C. (ed.), *Geoecology of the High Mountain Regions of Eurasia*, Franz Steiner Verlag, Wiesbaden, 1–16.

Troll, C., 1972b, The three-dimensional zonation of the Himalayan system. In Troll, C. (ed.), *Geoecology of the High Mountain Regions of Eurasia*, Franz Steiner Verlag, Wiesbaden, 264–75.

Troll, 1973a, The upper timberlines in different climatic zones, *Arctic and Alpine Research*, **5** (3): 3–18.

Troll, C., 1973b, High mountain belts between the polar caps and the equator: their definition and lower limit, *Arctic and Alpine Research*, **5** (3): 19–28.

Tsukamoto, Y. and Minematsu, H., 1987, Hydrogeomorphological characteristics of a zero-order basin. In *Erosion and Sedimentation in the Pacific Rim*, International Association of Scientific Hydrology Publication, **165**: 61–70.

Tsuya, H., 1940, Geological and petrological studies of Volcano Fuji: III. Geology of the southwestern foot of Volcano Fuji, *Bulletin of the Earthquake Research Institute, Tokyo Imperial University*, **18**: 419–45.

Tsuya, H., 1955, Geological and petrological studies of Volcano Fuji: V. On the 1707 eruption of Volcano Fuji, *Bulletin of the Earthquake Research Institute, University of Tokyo*, **33**: 341–83.

Tushinskiy, G.K., 1966, Avalanche classification and rhythms in snow cover and glaciation of the Northern Hemisphere in historical times, *Association Internationale d'Hydrologie Scientifique*, **69**: 382–93.

Twidale, C.R., 1972, The neglected third dimension, *Zeitschrift für Geomorphologie*, NF **6**: 283–300.

Twidale, C.R., 1973, On the origin of sheet jointing, *Rock Mechanics*, **3**: 163–87.

Twidale, C.R., 1982, *Granite Landforms*, Elsevier, Amsterdam.

Ui, T., 1983, Volcanic dry avalanche deposits — identification and comparison with non-volcanic debris stream deposits, *Journal of Volcanology and Geothermal Research*, **18**: 135–50.

United Nations Development Project, 1966, *Feasibility Report on the Chisapani High Dam Project, Kathmandu.*

United States Army Corps of Engineers, 1956, *Snow Hydrology*, Summary Report of Snow investigations, North Pacific Corps of Engineers, Portland, Or.

Unwin, D.J., 1973, The distribution and orientation of corries in northern Snowdonia, Wales, *Transactions of the Institute of British Geographers*, **58**: 85–97.

Verstappen, H. Th., 1955, Geomorphologische Notizen aus Indonesien, *Erdkunde*, **9**: 134–44.

Vietoris, L., 1972, Über die Blockgletscher des äusseren Hochebenkars, *Zeitschrift für Gletscherkunde und Glazialgeologie*, **8**: 169–88.

Vinogradov, Yu.B., 1969, Some aspects of the formation of mudflows and methods of computing them, *Soviet Hydrology*, **5**: 480–500.

Vivian, R., 1970, Hydrologie et érosion sous-glaciaire, *Revue de Géographie Alpine*, **58**: 241–65.

Vivian, R., 1974, Les Débâcles glaciaires dans les Alpes Occidentales. In Castiglioni, G.B. (ed.), *Le calamità Naturali Nelle Alpi*, Instituto di Geografia dell'Università di Padova.

Vivian, R., 1979, *Les Glaciers Sont Vivants*, Denoel, Paris.

Vivian, R. and Zumstein, J., 1973, Hydrologie sous-glaciaire au glacier d'Argentière (Mont Blanc, France), *International Association of Scientific Hydrology Publication*, **95**: 53–64.

Voellmy, A., 1955, Über die Zerstörungskraft von Lawien, *Schweiz. Bauzeitung*, **73**: 159–65, 212–17, 246–9, 280–5.

Voight, B. (ed.), 1978, *Rockslides and Avalanches — 1. Natural Phenomena*, Elsevier, Amsterdam.

Voight, B., Glicken, H., Janda, R.J. and Douglass, P.M., 1981, Catastrophic rockslide avalanche May 18. In Lipman, P.W. and Mullineaux, D.R. (eds), *The 1980 Eruptions of Mount St Helens, Washington*, United States Geological Survey Professional Paper, **1250**: 347–78.

Warhrhaftig, C. and Cox, A., 1959, Rock glaciers in the Alaska Range, *Bulletin of the Geological Society of America*, **70**: 383–436.

Waitt, R.B., Jr, 1981, Devastating pyroclastic density flow and attendant air fall of May 18 — stratigraphy and sedimentology of deposits. In Lipman, P.W. and Mullineaux, D.R. (eds), *The 1980 Eruptions of Mount St Helens, Washington*, United States Geological Survey Professional Paper, **1250**: 439–58.

Waitt, R.B., Jr, 1984, Comment, *Geology*, **12**: 693.

Waitt, R.B., Jr, Pierson, T.C., MacLeod, N.S., Janda, R.J., Voight, B. and Holcomb, R.T., 1983, Eruption triggered avalanche, flood and lahar at Mount St Helens — effects of winter snowpack, *Science*, **221**: 1394–7.

Waldbaur, H., 1952, Die Reliefenergie in der morphographischen Karte, *Petermanns Geographische Mitteilungen*, **96**: 156–67.

Walker, G.P.L. and McBroome, L.A., 1983, Mount St Helens 1980 and Mount Pelee 1902 — flow or surge, *Geology*, **11**: 571–4.

Walling, D.E., and Kleo, A.H.A., 1979, Sediment yield of rivers in areas of low precipitation: a global view. In *The Hydrology of Areas of Low Precipitation*, International Association of Hydrological Sciences Publication, **128**: 479–93.

Wardle, P., 1973, New Zealand timberlines, *Arctic and Alpine Research*, **5** (3): A127–35.

Washburn, A.L., 1965, Geomorphic and vegetational studies in the Mesters Vig district, Northeast Greenland, *Meddeslelser om Grønland*, **166** (1): 1–60.

Washburn, A.L., 1967, Instrumental observations of mass-wasting in the Mesters Vig district, Northeast Greenland, *Meddelelser om Grønland*, **166** (4): 1–318.

Washburn, A.L., 1969, Weathering, frost action and patterned ground in the Mesters Vig district, Northeast Greenland, *Meddelselser om Grønland*, **176** (4), 1–303.

Washburn, A.L., 1970, Instrumental observations of mass-wasting in an arctic climate, *Zeitschrift für Geomorphologie*, Supplementband **9**: 102–18.

Washburn, A.L. and Goldthwait, R.P., 1958, Slushflows, *Bulletin of the Geological Society of America*, **69**: 1657–8

Waters, A.C. and Fisher, R.V., 1971, Base surges and their deposits: Capelinhos and Taal volcanoes, *Journal of Geophysical Research*, **76**: 5596–5614.

Wegman, E., 1957, Tectonique vivante, dénudation et phénomènes connexes, *Revue de Géographie Physique et Géologie Dynamique*, **2** (1): 3–15.

Weiser, S., 1875, Permanent ice in a mine in the Rocky Mountains, *Philosophical Magazine*, **49**: 77–8.

Wentworth, C.K., 1943, Soil avalanches on Oahu, Hawaii, *Bulletin of the Geological Society of America*, **53**: 53–64.

Wertz, J.B., 1966, The flood cycle of ephemeral mountain streams in the southwestern United States, *Annals of the Association of American Geographers*, **56** (4): 598–633.

Westercamp, F., 1980, Une méthode d'évaluation et de zonation des risques volcaniques à la Soufrière de Gaudeloupe, *Bulletin Volcanologique*, **43**: 431–52.

Whalley, W.B., 1974, *The Mechanics of High Magnitude–Low Frequency Rock Failure and its Importance in Mountainous Areas*, Reading University Geography Paper **27**.

Whalley, W.B., McGreevy, J.P. and Ferguson, R.I., 1984, Rock temperature observations and chemical weathering in the Hunza region, Karakoram: preliminary data. In Miller, K.J. (ed.), *The International Karakoram Project*, vol. 2, Cambridge University Press, Cambridge, 616–33.

Wheeler, D.A., 1984, Using parabolas to describe the cross-sections of glaciated valleys, *Earth Surface Processes and Landforms*, **9** (4): 391–4.

White, S.E., 1976a, Rock glaciers and blockfields, review and new data, *Quaternary Research*, **6**: 77–97.

White, S.E., 1976b, Is frost action really only hydration shattering? A review, *Arctic and Alpine Research*, **8**: 1–6.

White, S.E., 1981, Alpine mass movement forms (noncatastrophic) classification, description and significance, *Arctic and Alpine Research*, **13**: 127–37.

White, W.R. and Day, T.J., 1982, Transport of graded gravel bed material. In Hey, R.D., Bathurst, J.C. and Thorne, C.R. (eds), *Gravel-bed Rivers*, Wiley, Chichester, 181–213.

Whitehouse, I.E. and McSaveney, M.J., 1983, Diachronous talus surfaces in the southern Alps, New Zealand and their implications to talus accumulation, *Arctic and Alpine Research*, **15** (1): 53–64.

Whiting, P.J. and Dietrich, W.E., 1985, The role of bedload sheets in the transport of heterogeneous sediment (abstract), *Eos*, **66** (46): 910.

Whittaker, J.G., 1987, Modelling bed-load transport in steep mountain streams. In *Erosion and Sedimentation in the Pacific Rim*, International Association of Scientific Hydrology Publication **165**: 319–32.

Whittaker, J.G. and Jaeggi, M.N.R., 1982, Origins of step-pool systems in mountain streams, *American Society of Civil Engineers, Journal of Hydraulics Division*, **108** (6): 758–73.

Wiberg, P.L. and Smith, J.D., 1987, Initial motion of coarse sediment in streams of high gradient. In *Erosion and Sedimentation in the Pacific Rim*, International Association of Hydrological Sciences Publication, **165**: 299–308.

Wickman, F.E., 1966a, Repose-period patterns of volcanoes I. Volcanic eruptions regarded as random phenomena, *Arkiv for Mineralogi och Geologi*, **4**: 291–302.

Wickman, F.E., 1966b, Repose-period patterns of volcanoes V. General discussion and a tentative stochastic model, *Arkiv for Mineralogi och Geologi*, **4**: 353–66.

Willard, D.J., 1904, The profile of maturity in alpine glacial erosion, *Journal of Geology*, **12**: 569–78.

Williams, H. and McBirney, A.R., 1979, *Volcanology*, Freeman Cooper & Co., San Francisco.

Williams, J.E., 1949, Chemical weathering at low temperatures, *Geographical Review*, **39**: 129–35.

Williams, L.D., Barry, R.G. and Andrews, J.T., 1972, Application of computed global radiation for areas of high relief, *Journal of Applied Meteorology*, **11**: 526–33.

Williams, R.B.G. and Robinson, D.A., 1981, Weathering of sandstone by the combined action of frost and salts, *Earth Surface Processes*, **6**: 1–9.

Williams, V.S., 1977, Neotectonic implications of the alluvial record in the Sapta Kosi drainage basin, Nepalese Himalayas. Unpublished PhD thesis, Department of Geology, University of Washington.

Winder, C.G., 1965, Alluvial cone construction by alpine mudflow in a humid temperate region, *Canadian Journal of Earth Sciences*, **2**: 270–7.

Winiger, M., 1983, Stability and instability of mountain ecosystems, United Nations University Workshop 1981, *Mountain Research and Development*, **3** (2): 103–11.

Wirthmann, A., 1976, Reliefgeneration im unvergletscherten Polargebiet, *Zeitschrift für Geomorphologie*, **20** (4): 391–404.

Wood, A., 1942, The development of hillside slopes, *Proceedings of the Geologists Association*, **53**: 128–40.

Woodcock, A.H., 1974, Permafrost and climatology of a Hawaii volcano crater, *Arctic and Alpine Research*, **6**: 49–62.

Yamasaki, N., 1902, A trace of glacier in Japan, *Journal Geological Society Japan*, **9**: 361–9.

Yoshikawa, T., 1974, Denudation and tectonic movement in contemporary Japan, *Bulletin of the Department of Geography, University of Tokyo*, **6**: 1–14.

Yoshikawa, T., Kaizuka, S. and Ota, Y., 1981, *The Landforms of Japan*, University of Tokyo Press, Tokyo.

Young, A., 1972, *Slopes*, Oliver and Boyd, Edinburgh.

Young, A., 1974, The rate of slope retreat. In Brown, E.H. and Waters, R.S. (eds), *Progress in Geomorphology*, Institute of British Geographers Special Publication, **7**: 65–78.

Young, G.J. and Stanley, A.D., 1977, *Canadian Glaciers in the International Hydrological Decade Program 1965–1974. No. 4 Peyto Glacier, Alberta — Summary of Measurements*, Ottawa Fisheries and Environment Canada, Inland Waters Directorate Water Resources Branch (Scientific series no. **71**).

Zaruba, Q. and Mencl, V., 1969, *Landslides and Their Control*, Elsevier, Amsterdam.

Zimina, R.P., Isakov, Yu.A. and Panfilov, D.V., 1973, Geography of altitudinal belt ecosystems in the Caucasus, *Arctic and Alpine Research*, **5** (3): A33–5.

INDEX